AF760310

Instinctive Computing

Yang Cai

# Instinctive Computing

Springer

Yang Cai
Carnegie Mellon University
Pittsburgh, PA, USA

ISBN 978-1-4471-7276-5 ISBN 978-1-4471-7278-9 (eBook)
DOI 10.1007/978-1-4471-7278-9

Library of Congress Control Number: 2016953189

© Springer-Verlag London 2016
This work is subject to copyright. All rights are reserved by the Publisher, whether the whole or part of the material is concerned, specifically the rights of translation, reprinting, reuse of illustrations, recitation, broadcasting, reproduction on microfilms or in any other physical way, and transmission or information storage and retrieval, electronic adaptation, computer software, or by similar or dissimilar methodology now known or hereafter developed.
The use of general descriptive names, registered names, trademarks, service marks, etc. in this publication does not imply, even in the absence of a specific statement, that such names are exempt from the relevant protective laws and regulations and therefore free for general use.
The publisher, the authors and the editors are safe to assume that the advice and information in this book are believed to be true and accurate at the date of publication. Neither the publisher nor the authors or the editors give a warranty, express or implied, with respect to the material contained herein or for any errors or omissions that may have been made.

Cover image: "The Hunter," rock carving, by Yang Cai

Printed on acid-free paper

This Springer imprint is published by Springer Nature
The registered company is Springer-Verlag London Ltd.

The Cover Image: “The Hunter”, rock carving, by Yang Cai

# Preface

The idea for this book evolved from my research on how to incorporate self-awareness, intuition, and creativity into machines, which led me to explore primitive behaviors in humans, animals, and machines. Algorithms in nature are simple and elegant yet ultimately sophisticated. All behaviors are, in fact, connected to the basic instincts we take for granted. This book attempts to connect artificial intelligence to primitive intelligence. It explores the idea that a genuinely intelligent computer will be able to interact naturally with humans. To form this bridge, computers need the ability to recognize, understand, and even *have* instincts similar to humans.

Professor Herbert A. Simon, a pioneer of artificial intelligence, had a strong influence on my exploration of human and animal cognition. Simon introduced me to Professor Harold Cohen who created the first robotic artist AARON. This led me to have extensive discussions with Harold Cohen and his collaborator Ray Kurzweil, a visionary and technologist of artificial intelligence, about machine creativity.

Dr. Sylvia J. Spengler, the program director from CISE-IIS of the National Science Foundation, sponsored the first workshop on instinctive computing at Carnegie Mellon University in 2009. At the workshop, I had in-depth discussions with brilliant researchers in artificial intelligence and cognitive science, including Professors Kevin Warwick, Michael Leyton, Yvonne Masakowski, Chris Urmson, Christian Lebiere, Howard Lipson, Xavier Alamán, Manuel García-Herranz, Pablo A. Haya, Adam Bryant, Suguru Ishizaki, David Kaufer, and Paul Whitmore.

Robotics professor Takeo Kanade and psychology professor Jeffrey Cohen provided guidance and constructive criticism about the idea of facial reconstruction from crowdsourcing. Professor Cohen pointed me to the key reference about the nonverbal judgment model in experimental psychology. It took me 4 years to conduct a series of lab experiments to test the model. Professor Kanade spent untold hours helping me clear up the assumptions, experimental design, and resulting analysis of these studies.

Professor Howard Lipson inspired me to look into the computer virus Stuxnet and the survivability of cyber-physical systems. Michael Strougen contributed his expertise in network security analysis. Brian Zeleznik shared his philosophical point of view on human instinct. Statistics professor William Eddy's visualization work inspired me to explore the patterns of motion and events. Management science professor Thomas Saaty shared his insight about decision science and models of creativity. Professors Christof Koch, Brenda Bloodgood, and Robert M. Wightman provided important references about neural dynamics.

Drs. Donald Steiner, Paul Conoval, Robert Loce, and Wencheng Wu have been sponsors and collaborators on the related subjects. Professors Jose Bravo, Julio Abascal, Mel Siegel, David Martinelli, Kevin Dowling, Daniel P. Siewiorek, Virgil Gligor, Pradeep Khosla, Jose Morales, Andrew P. Moore, and Jonathan S. Minden provided guidance and feedback on research projects.

My assistants and students contributed prototypes, figures, and comments: Michal Gacka, Sebastian Peryt, Joseph Doyle, Joshua Antonson, Binbin Xiong, Shashank Deshpande, Zhixin Feng, Pu Liang, Chin Yang Oh, David Yuan, Clement Loh, and Elizabeth Chang. In particular, the composer Jakub Polaczyk contributed his knowledge about classic and contemporary music structures.

In addition, Professor Qifu Cai contributed his expertise about the history of ancient Chinese characters, and Liang Cai shared her knowledge about the ancient music instrument guqin. Shangcong Zeng, Alan Cai, Yizhen Shi, and Ming Cai reviewed the manuscript and supported my effort throughout the project.

Special thanks go to the editorial director Beverley Ford and editorial assistant James Robinson of Springer for their time and efforts to ensure the quality of this book. Michael Fuhrman and Elisabeth Solomon also devoted significant time reviewing, editing, and revising the manuscript, including illustrations and scientific content.

I am grateful to those who granted the permission for using their images: Joseph Doyle and Jason D. Lohn from Carnegie Mellon University, Lauren Amundson of the Lowell Observatory Archives, and Matthew Wight from WowWee.

The book project is in part sponsored by grants from the National Science Foundation, Xerox PARC, Northrop Grumman Corporation, the Army Research Office, CERT, Air Force Research Lab, Bosch, PNC Bank, CyLab, and the CIT Dean's Office of Carnegie Mellon University.

The content of this book does not reflect the official opinion of the research sponsors or collaborators. Responsibility for the information and views expressed herein lies entirely with the author.

Pittsburgh, PA, USA Yang Cai

# Contents

# Part I
# Primitive Problem-Solving

CHAPTER 1

# Introduction

We watch a mobile robot making his way to clean the floor in a furnished room. He starts moving in an outward spiral. Then, he bumps into the wall or a piece of furniture, and he bounces back in a random direction. After a while, he has covered all of the areas that he can reach. He is blind, without a camera. However, he can touch a surface to feel whether it is a wall or the stairs. When he feels hungry, he moves over to the power charger for a boost of energy. So as not to anthropomorphize his purposes, I used a long-exposure filming app on my phone to trace his path in a single image. The path is a laborious sequence of irregular, angular segments – not quite a random walk, nor optimal movement but complex and hard to describe. Everything seems spontaneous, but it works out eventually and the floor is cleaned. I then compare the traced path of this cleaning robot to that of a foraging ant, a stream of Internet packets, and even a malicious computer virus. I found that all of their paths are surprisingly similar: they have a general sense of purpose, but they cannot foresee all of the obstacles in between. Therefore, they must adapt their course repeatedly with allowable changes in direction and take detours around any barrier.

Ludwig Wittgenstein asserted that "The aspects of things that are most important for us are hidden because of their simplicity and familiarity. One is unable to notice something – because it is always before one's eyes. The real foundations of his enquiry do not strike a man at all. Unless that fact has at some time struck him. And this means: we fail to be struck by what, once seen, is most striking and most powerful." This common, shared knowledge and skill across species, which often goes unnoticed, is called *instinct*.

Instincts are primitive impulses, such as the urges for eating, exploring, and mating. Instincts motivate creatures to fulfill their primal needs. Freud stated that these biologically-based energies are the fundamental driving forces of our

© Springer-Verlag London 2016

Y. Cai, *Instinctive Computing*, DOI 10.1007/978-1-4471-7278-9_1

lives.[1] They act to protect us from danger and keep us fit and healthy, yet we are often barely aware of them. This book proposes that we give computers the *ability to recognize and to have instinctive behaviors*. It contains the following premise: *instincts are interconnected programs*. They appear to be rudimentary, but when they act as a whole, they present utterly sophisticated behaviors. *Instinctive computing*, or *spontaneous computing,* not only mimics instinctive behaviors found in nature, but also generalizes them to become an *operating system* across borders of cyber species, from mobile robots and social media, to security and privacy agents.

The book begins with a series of archeological discoveries of instinctive algorithms that have survived for decades in today's computing systems in many different fields such as mobile robotics, network protocols, and security systems. For example, the spontaneous alternation behavior (SAB) has already existed in the cleaning robot Roomba™ and it is also a part of the Internet protocol TCP/IP, evolutionary algorithms, and autonomous driving alogrithms. Then, we move on to explore new concepts and algorithms such as tool making, motion pattern discovery, primitive learning, machine creativity, figurative thinking, euphoria, self-awareness, sensuality, visual privacy, and stealth.

This book attempts to connect artificial intelligence to primitive intelligence. We ask questions like: how do natural instincts function in default mode? What are the principles for designing survivability and autonomy within a complex autonomous system? Is a computer virus an artificial life? How can security systems deter malicious software from the inside out? What is the true nature of privacy? Can machines make tools themselves? Are machines able to feel happiness or predict moments of euphoria in social media? Can machines be creative? How can we design robots that are attractive to humans?

## Historical Models of Instincts

Good ideas have deep roots. Understanding such roots is key to being able to generate new and valuable ideas while moving forward. The main motivation of this book is to connect the dots and enable the exploration of the roots of instinct.

In his book "Origin of Species," Charles Darwin discussed instinct in a standalone chapter.[2] He said: "I will not attempt any definition of instinct. It would be easy to show that several distinct mental actions are commonly embraced by the term; but everyone understands what is meant when it is said that instinct impels the cuckoo to migrate and to lay her eggs in other bird's nests." British philosopher Mary Midgley classifies instinct into two parts: closed and open instincts. According to Midgley, "closed instincts are behavioral patterns fixed genetically in every detail,

---

[1]Freud S (1915) Instincts and their Vicissitudes. Psychoanalytic Electronic Publishing, http://www.p-e-p.org/pepcd.htm

[2]Darwin S (2014) The Origin of Species. Barns and Noble Classic Books.

such as the honey dance of a bumblebee, a birdsong, or the nest-building patterns of weaver birds. Such genetic programming takes the place of intelligence; learning is simply maturation. On the other hand, open instincts are programs with a gap. Parts of the behavioral patterns are intuitively determined, but others are left to be filled in by experience."[3] Biological and psychological evidence supports this dual theory. For example, innate immune systems behave like a closed instinct. Conversely, adaptive immune systems behave like an open instinct. Take birds for example. Birds usually sing their own songs, but they may learn to mimic other sounds like human voices, thanks to the Baldwin Effect, discovered in 1896 by James Mark Baldwin. The Baldwin Effect suggests that an organism's ability to learn new behaviors will affect its offspring through natural selection.

1950 was a great year for science when Alan M. Turing published his seminal paper on "Computing Machinery and Intelligence."[4] That same year, the pioneer of ethology (the science of animal behavior), Konrad Lorenz published his psycho-hydraulic diagram, which was perhaps the first quantitative and measurable model for instinctive behavior. In his model, Lorenz illustrated a water tank with an inlet and outlet. The water pipe to the inlet continuously pours water into the tank, accumulating in the reservoir of "action-specific energy." At the outlet is a valve to discharge the energy. In this setup, both the accumulation of energy potential and the releasing forces tend to open the valve. The jet of liquid coming out of the tank corresponds to the instinctive reaction itself. Lorenz's model is actually a system dynamics model that can be formulated in a linear differential equation and digitally simulated on a computer. The dynamics reveals the temporal behaviors of instinct as a whole. For example, if the flow into the inlet were much larger than the flow leaving the outlet, the tank would overflow. If the outlet valve were repeatedly opened, then the action-specific energy reservoir would be pumped out. The energy reservoir saves integral memories, playing a critical role in behavioral dynamics. The timing and frequency of the releasing mechanism also shapes the characteristics of the action. See Fig. 1.1 for a simplified version of the Lorenz model.

That same year, 1950, Niko Tinbergen, another pioneer of ethology, published his hierarchical model that mimics the central nervous system.[5] His seminal book, *The Study of Instinct,* which was published the following year,[6] was the first and foremost work about the physiological causation of instinctive behavior. Tinbergen's model emphasizes the reflux-like *innate release mechanism* (IRM) in multiple layers of neurons, and the cooperation between ethology and neurophysiology. A graphical portrayal of his model shows innate releasing mechanisms acting on nervous cells, where circles represent *centers*; the elongated rectangles represent inhibitory blocks; and arrows represent causal factors. According to Tinbergen, the instinctive centers

[3]Midgley M (1995) Beast and Man – The roots of human nature. The Harvester Press Limited

[4]Turing, AM (1950) Computing machinery and intelligence. Mind, 59, 433–460.

[5]Tinberger, N (1950) The hierarchical organization of nervous mechanism underlying instinctive behavior. Symposia of the Society of Experimental Biology, vol. 4. pp.305 – 312, 1950

[6]Tinberger, N (1951) The study of instinct. Oxford: Clarendon Press, 1951

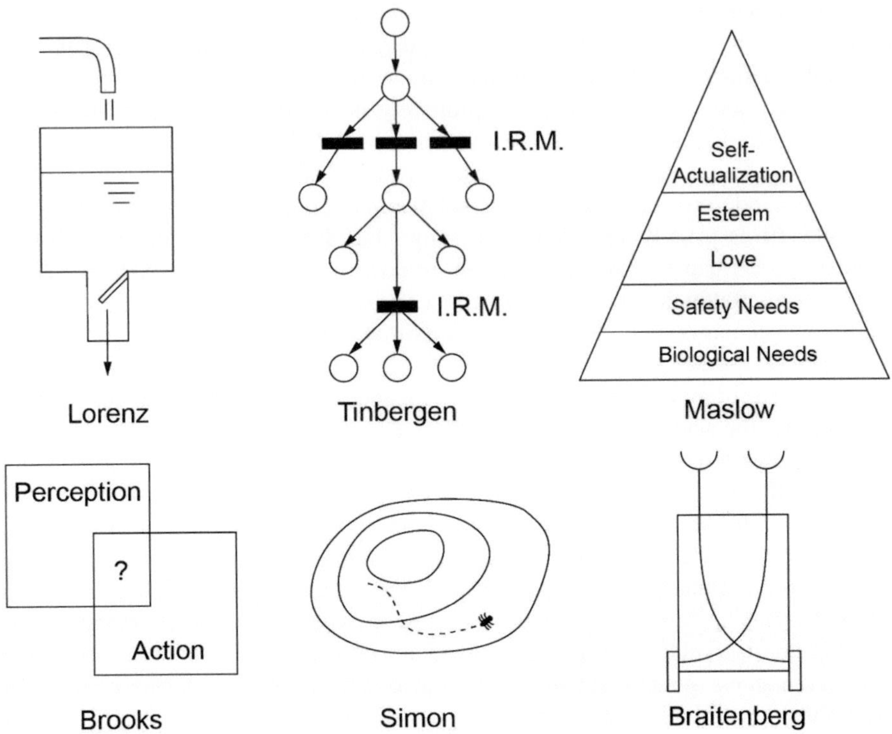


**Fig. 1.1** Historical models of instinct: Lorenz, Tinbergen, Maslow, Brooks, Simon and Braitenberg

are in a state of readiness. They are constantly being loaded from within, but a block prevents their discharge. IRM removes this block.

Both Lorenz and Tinbergen won the Nobel Prize for medicine and physiology in 1973. They established a new scientific field – ethology. For more information about their work, Richard W. Burkhardt Jr's book, *Pattern of Behavior* provides a thorough overview.[7]

In contrast to the models of Lorenz and Tinbergen, Maslow's model focuses on human motivation and instinctive needs. The earliest and most widespread version of his model includes a hierarchy of five needs within a pyramid, published in 1943[8] and 1954.[9] According to Maslow's theory, when someone's need is fulfilled, the person will seek to fulfill the next one, and so on. A person must fulfill a lower need before moving on to the upper needs. The original hierarchy of needs includes: (1)

[7]Burkhardt RW (2005) Patterns of behavior. University of Chicago Press, 2005

[8]Maslow AH (1943) A theory of human motivation. Psychological Review, vol. 50, no.4, pp.370–396.

[9]Maslow AH (1954) Motivation and personality. New York: Harper and Row.

biological and physiological needs: air, food, drink, shelter, warmth, sex, and sleep; (2) safety needs: security, stability, and freedom from fear; (3) love and belonging: friendship, affection, family and romantic relationships; (4) esteem needs: status and respect from others; and (5) self-actualization needs: achievement and fulfillment. This structure certainly distinguishes humans from animals (especially in the third through fifth levels), creating challenging tasks for today's computers and robots to understand those needs, let alone to possess those needs.

The cognitive scientist and father of artificial intelligence, Herbert A. Simon, pointed out that, as a behavioral system, humans and animals are intrinsically simple. The complexity of their behaviors results from the complexity of their external environments. Instead of carefully planning a foraging trip, an ant might simply explore the terrain or follow the pheromone trail, requiring very limited sensing, navigation, and communication resources.

Mathematical modeling of instinctive behavior began in 1950, when Norbert Wiener published the first model of instinctive behaviors shared by humans, animals, and machines alike. His models included the feedback control process, nervous oscillation movements, self-reproduction, and learning, which he called "Cybernetics".[10] Remarkably, Wiener used mathematics to abstract dynamic behaviors across species and disciplines, creating simple yet elegant representations. For example, Wiener used a single integral equation to represent the Gestalt phenomenon – that holistic perception is more than just a sum of many parts. Wiener's models show the power of mathematical generalization. The euphoria model in this book was inspired by Weiner's studies on the process of nervous oscillation.

In parallel, the pioneer of computer science John von Neumann published the cellular automata model to simulate self-reproduction.[11] The model is comprised of finite state cells interacting with one another in a neighborhood within a two-dimensional space. John Conway implemented cellular automata in his "Game of Life."[12] In his simulation game, an organism experiences its instinctual states of birth, movement, and eating, and finally death. Interesting patterns emerge from cell interactions such as blooming, oscillation, or extinction. Stephen Wolfram, the inventor of symbolic computing language Mathematica, further proved that many simple cellular interactions are able to produce rather complex patterns, such as chaotic behavior. He argues that interactive algorithms are more important than the underlying mathematical equations.[13] The spatial and temporal interaction among entities is key to understanding their complexity. Today, computational cellular

---

[10]Wiener N (1961) Cybernetics: or control and communication in the animal and the machine, The MIT Press, 1961

[11]von Neumann J (1966) Theory of self reproducing automata. Edited by A. Burks, Univ. of Illinois Press, Urbana. 1966

[12]Conway J (1970) Game of Life. Scientific American 223, Oct. 1970: 120–123

[13]Wolfram S (2000) The new kind of science, Wolfram Media

automata have become a powerful tool to reveal natural behavior, from microscopic cellular morphology[14] to mass panic movements in subway stations.[15]

In the 1980s, robotics researchers started to investigate the instinctive behaviors of mobile robots. Rodney Brooks from MIT's Artificial Intelligence Lab prototyped a variety of mobile robots that exhibit basic instincts such as walking and climbing over rough terrain without a world model, nor traditional planning or reasoning. Brooks referred to this as *Behavior-Based Robotics*. In his model, there is no stand-alone cognition per se. Instead, there are simply *fast, cheap, and out of control* perception and actuation components.[16] The cognition process only exists between the perception and actuation processes. See the diagram in Fig. 1.1. The same design philosophy has been implemented in the home cleaning robot, Roomba™, made by iRobot. Behavior-based robotics has been a growing trend in service, micro, and swarm robotics applications. Not only does it provide utility, but also fast and cheap solutions.[17]

Neuroscientist Valentino Braitenberg developed a series of synthetic psychological experiments using neuroscience knowledge to build a set of imaginary robotic vehicles that possess instinctive behaviors: mobility, fear, aggression, love, foresight, egotism, olfactory, evolution, memory, control, and vision.[18]

Figure 1.1 highlights key historical models of instinct. As we can see, these models represent common instinctive behaviors across species, from animals and humans, to robots. There is still plenty of room to connect the dots in different systems and to extract instinctive behaviors. For example, Lorenz, Tinbergen, and Maslow show the instinctual motivations and structures, but they did not elaborate on the interactions between the system and its environment. Brooks, Simon, and Braitenberg articulated the system's responses to its environment but they barely touched motivation and instinctual drives.

## Behavioral Tests

Ludwig Wittgenstein once said, "If a lion could talk, we would not understand him." Suppose one could build a robot that completely simulates instinctive behavior. Would it even be possible to measure its instinctual level? Would the robot be

[14]Kroc J (2006) Model of Mechanical Interaction of Mesenchyme and Epithelium in Living Tissues. 847–854, in Vassil N. Alexandrov, G. Dick van Albada, Peter M. A. Sloot, Jack Dongarra (Eds.): Computational Science – ICCS 2006, 6th International Conference, Reading, UK, May 28–31, 2006, Proceedings, Part IV. Lecture Notes of Computer Science. Vol. 3994 Springer 2006, ISBN 3-540-34385-7

[15]Regirer SA and Shapovalov DS (2003) Filling space in public transport by passengers, Automation and Remote Control, vol. 64, issue 8, August 2003

[16]Brooks RA and Flynn AM (1989) Fast, cheap and out of control: a robot invasion of the solar system. Journal of the British Interplanetary Society, Vol. 42, pp. 478–485, 1989

[17]Arkin RC (1998) Behavior-based robotics. The MIT Press, 1998

[18]Braitenberg V (1984) Vehicles – Experiments in synthetic psychology. The MIT Press, 1984.

aware of instinct and experience it as we humans do? Herbert A. Simon believed that computer science is an experimental science. Therefore, instinctive computing should be tested through behavioral experiments.

In 1950, Alan Turing proposed his famous *Turing Test* with the following steps: Put a computer in one room and a person in the other. Ask each of them questions without looking at each other. If the observer cannot tell by their answers which subject is which, then the program passes the Turing Test.[19] According to Turing, the test is passed successfully if a computer was mistaken for a human more than 30 % of the time during a series of 5-min keyboard conversations. Philosopher John Searle presented a thought experiment called the *Chinese Room Test* to challenge the Turing Test. Suppose one puts a native Chinese speaker in one room. In the other room, one puts someone who knows nothing about Chinese but is given books listing a complete set of rules for translating and producing Chinese characters and syntax. Next, one sends them questions in Chinese. The results would be indistinguishable. This actually proves Turing's point even further: if you cannot tell by outward appearance, how can you ever understand?[20]

The problem of the Turing Test and the Chinese Room Test is that each test assumes we are sitting in a sensory deprivation chamber without any real-world stimuli; and the context of the test is removed from the reality of an individual's survival. On the other hand, psychologist Wolfgang Köhler conducted a more realistic experiment in his famous Banana Test with chimpanzees in the 1910s. In one test, the food was suspended from the ceiling, out of reach of the chimpanzee below. The chimpanzee was supplied with a stick, however, it was too short to reach the food. The food could only be reached if two sticks, which could be fitted together end to end, were used. Köhler also tested the latch-opening capabilities of dogs and cats. Animals learned that even though the door may be unlatched, it must still be pushed. If a hand can do it, why not a paw?[21]

## The Instinctive Computing Model

The model of instinctive computing has a hierarchy of four layers of algorithms from bottom to top: Default Operations, Creative Operations, Social Behaviors and Self-Awareness. The Default Operations contain vital operations for survivability, such as perceiving motion and event patterns, spontaneous alternation, and the principle of least effort. Default Operations are often conducted during times of crisis. Social behaviors contain foraging, navigation, facial recognition, privacy, stealth,

[19]Turing AM (1950) Computing machinery and intelligence. Mind, 59, 433–460. http://loebner.net/Prizef/TuringArticle.html

[20]Searle J (1980) Minds, Brains and Programs, Behavioral and Brain Sciences, 3(3):417–457, 1980.

[21]Bekoff M, et al. (2002) The cognitive animal. The MIT Press, 2002.

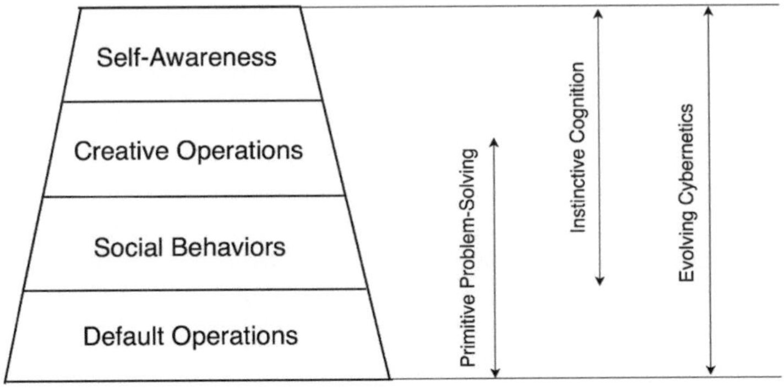


**Fig. 1.2** The model of the instinctive computing: default operations contain vital operations for survival. Social behaviors contain facial recognition, privacy, stealth, and autonomy. Creative operations include tool-making and primitive learning. Self-awareness algorithms include self-recognition, replication, and sensuality

and autonomy. Creative Operations contain more cognitive functions such as tool making, figurative thinking, primitive learning, and creativity. Self-Awareness algorithms include self-recognition for self or non-self identification, self-replication, sensuality, and euphoria. It is worth noting that the divisions of this hierarchy are not absolute. Many instinctive functions may spread over multiple layers. For example, learning, tool making, and creativity may also exist in social environments such as open source communities.

Instinctive computing aims to combine primitive problem-solving strategies with instinctive cognition to solve the evolving cybernetics problems in the digital age such as autonomous robots, data intelligence, security and privacy. Figure 1.2 illustrates the concept of the model.

In order to validate the concepts in this book, the author has conducted a series of empirical tests. They cover a broad spectrum, ranging from simple mental experiments, multi-year continuous surveys, instrumented lab experiments, all the way to developing computational models and the electronic circuit prototypes. These empirical tests include laboratory experiments, for example, common knowledge of default proportions (Chap. 2), airport passenger flow patterns (Chap. 3), Facebook personality tests (Chap. 4), a fun experiment with reverse goggles to navigate on campus and a computer simulated land mine detection game (Chap. 5), the mental equivalent of the banana test for robots (Chap. 6), the eye tracking and brainwave EEG tests of video viewers (Chap. 7), sensual voice and gesture tests (Chap. 8), and geometric facial feature recognition and reconstruction experiments that lasted for 3 years (Chap. 10). The longest experiment was the testing of words found with high frequency in ancient rock art, Chinese, and Google search engines lasting over 10 years (Chap. 11).

In addition, prototypes were developed and tested, for example, an electronic circuit of the euphoria model (Chap. 9), voice verification mobile phone app and network security data visualization (Chap. 13), visual privacy algorithms for autonomous transit and airport security systems (Chap. 15), and spontaneously navigating and covering algorithms for mobile service robots (Chap. 16).

## Book Overview

This book is organized in three parts. Part 1 provides an overview of primitive problem-solving methods in animals, humans, and machines. Chapters 2, 3, 4 and 5 cover default behaviors and commonsense as seen in our daily activities, such as spontaneous alternation, proportion-based reasoning, the principle of least effort, pheromones in nature and social media, and primitive learning strategies. This includes somebody walking with a pair of glasses that had turned his world, literally, upside-down! Chapter 6 argues that machines are capable of making tools themselves by scaffolding, self-assembly, and even self-replication. An alarming scenario is that autonomous tool-making may also be used for adversarial purposes such as developing computer viruses, and even biological viruses through gene editing – a plausible prediction of future cyber warfare.

Part 2 discusses behavioral models of instinctive cognition – how can a robot think like a caveman? Chapter 7 explores how to spot anomalous motion and events through animation and interactive games. Chapter 8 discusses how to give robots sensual voice, shape, and gait. Chapter 9 introduces the model of euphoria dynamics in the brain, in musical compositions, and in social media. Chapter 10 shows that computers are capable of co-creating new things with the aid of human experts, including painting, sculpting, and inventing, even though the computer does not do this for pleasure as humans do. Chapters 11 and 12 explore human figurative thinking in pictographic languages and face recognition and its applications in passcode design, text input, affective expression, facial forensics, and even programming in emoji!

Part 3 explores rapidly-evolving cybernetics in the digital age. Chapters 13, 14 and 15 cover instinctive computing for cybersecurity systems, including self-awareness models for behavioral-based identification and malicious behavior analysis, stealth methods, and visual privacy technologies for the Internet of Things (IoT). Chapters 16, 17 and 18 explore instinctive computing methods for designing autonomous systems, including the Hansel-and-Gretel algorithm for simultaneously navigating and covering, human-autonomous robot interaction, and last but not the least, survivability of cyber-physical systems.

CHAPTER 2

# Default Mode

We often take everyday common sense for granted, frequently acting without thinking. Many subconscious processes are instinctual, but they can also be learned and evolve. When a baby is born, its communication is limited to one mode – crying – which signals to the parents if the baby is feeling a lack of attention, hunger, or pain. When a creature has limited means of survival or is in risky situations that other means are not available, the creature typically reverts to a survival or backup state, so-called *default mode,* and performs survival actions. It uses simple algorithms to cope with complex and dynamic situations. In this chapter, we will explore how to use default mode algorithms and default knowledge to solve challenging problems in modern control, network, and vision systems such as collision handling in networks, exception-handling in autonomous vehicles, and object recognition.

## Default Operations

If a mouse is placed on a path connected to a T-intersection, the mouse must make the decision to go either left or right. This is a well-known psychology experiment called the "T-maze" problem (Fig. 2.1).[1] The mouse may also be placed before a Y-shaped intersection, the so-called "Y-maze" problem, which enables mice to move more smoothly. Even though mice have poor eyesight, they have other sensors to detect obstacles and move forward. As the movement of the mouse is a sequence of perception and action processes, it can be represented as a set of *production rules*, including conditions and actions. We can simply refer to them as the *control logic*.

In order to design the control logic for a mouse's movement, we have to painstakingly write down all possible conditions and related actions accordingly.

[1]Wikipedia (2016) T-Maze: https://en.wikipedia.org/wiki/T-maze

© Springer-Verlag London 2016
Y. Cai, *Instinctive Computing*, DOI 10.1007/978-1-4471-7278-9_2

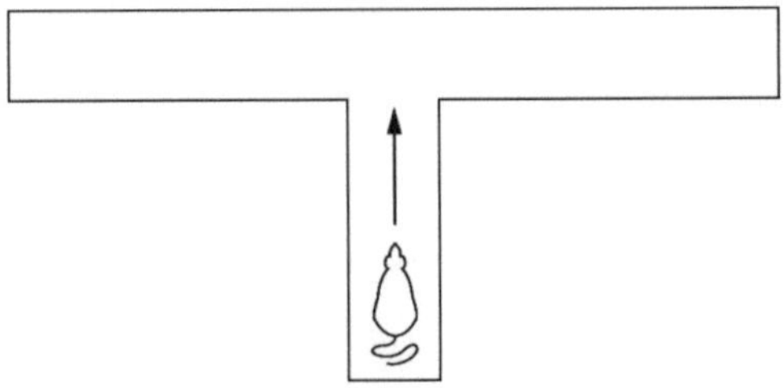

**Fig. 2.1** The T-Maze where a mouse faces an intersection problem

This is not always possible since we often have more unknown situations than known ones. To make matters even more complicated, sensors and communications may malfunction, which requires that the system is able to handle *exceptions*. In theory, given a control system, there are a lot of exceptions that we call "anything else." Therefore, we need a default action to handle the infinite number of exceptional cases.

Here, we have a set of rules for a mouse to move through the maze. If the mouse sees that the path is clear, then it moves forward; when it approaches a wall, it slows down with caution; when the mouse reaches the wall, it stops; when a new case emerges that does not match any of the conditions listed above, then the mouse uses its default mode to try to avoid collision by moving towards the right. The default movement is designed to keep moving when no input is received. This can save time and reduce the complexity of decision-making.

All contemporary programming languages contain a control logic expression, or a so-called *switch* statement. Let us look at the pseudo code of the default mode. The keyword "switch" is followed by a test parameter for equality with the following cases. When the expression in a "case" statement is satisfied, the related actions will be performed. If no case value matches the switch expression value, execution continues at the default clause. It is conventionally written after the last case because execution continues out of the switch.

```
Switch (road condition)
    Case 'clear road': keep moving;
    Case 'near wall':  slow down;
    Case 'at wall':    stop;
    Default:           move right;
End Switch
```

## *Anticipating Exceptions*

While coding a switch for known case-action rules is considered a science, defining default operations is a kind of art. Default design usually begins with assumptions about the environment that the system encounters. Almost all autonomous mobile robots have a default mode, including the popular floor-cleaning robot, Roomba, made by iRobot. This minimalist-designed robot begins with a spiral motion from the center outwards if placed inside of an empty space. The spiral-swiping motion enables the robot to systematically clear the floor without missing a spot. However, when the Roomba hits furniture or a wall, it moves along one side of the obstacle in the default mode. Default mode is usually used in the worst-case scenario of operation. Using default mode can reduce complexity in control logic design.

Perhaps the most important factor in default design is safety, which can be a baseline for default operation. Defaults contain static and dynamic conditions. Taking static defaults for example, the default position for a power switch should be the "off" position rather than the "on" position. The default position for a transmission gear in a vehicle should be the "parking" position instead of the "driving" position. Dynamic defaults should consider the worst-case scenarios. For instance, when a computer's accelerometer senses fast movement, it should shut down the power and activate the safety mode by default to protect the data. In the case of autonomous driving vehicles, any obstacle in front of the vehicle should be viewed as a hazard – a brick wall, a tree, a cow, or a shopping cart. Here, the default action should be to activate the brakes of the vehicle.

## *Graceful Degeneration*

In many cases, the goal of default design is based on the mission of the system. Default mode can be the fallback solution in order to keep a system running in unusual or dangerous situations. For example, the autonomous light rail transit system at the Pittsburgh International Airport contains two rail tracks with two trains running on parallel tracks in opposite directions. In the normal control mode, a computer synchronizes the positions of the two trains so that one train stops at the arrivals side while the other stops at the departures side. This ensures a smooth flow of traffic in both directions. When communication between the trains and the central control system is lost, each train will run in its default mode: move and stop independently without synchronizing with the other train. This accomplishes the goal of transporting passengers, but not in the most optimal way. For example, two trains arrive at the same station at the same time, while passengers at the other station have to wait for longer periods of time. See Fig. 2.2.

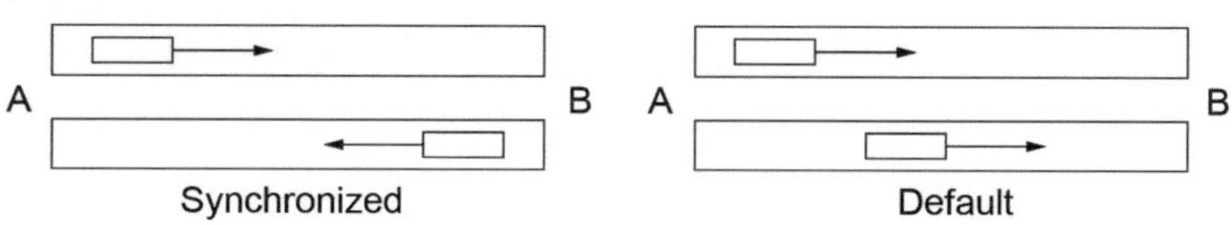


**Fig. 2.2** Graceful degeneration in the Default Mode

A default response is a reaction to anything else but the enumerated conditions. In many cases, the enumerated conditions are limited by the available sensors, design, and test time. The total number of conditions is normally large but unknown. A default mode might fail if the new or unexpected conditions occur outside of the estimated total condition set. There is always a temptation to design overly complicated default modes, for example, nested defaults, or *default of default*. This can be implemented with additional sensors in order to articulate the conditions for defaults. However, adding conditions to the default will diminish its overall purpose and ultimately confuse the system. Simple and unique default operations are most desirable.

Default operation is not foolproof. Sometimes, default operations may cause traffic jams or even deadlocks. For example, insects use pheromones to lay trails back to their nests. Incidentally, a group of ants may lose the pheromone trail and end up circling back on themselves. Other ants then follow, and soon the entire group is going around in endless circles. This is an example of the so-called "ant-mill" behavior pattern.

## Spontaneous Alternation Behavior

When an ant gets lost, it will randomly wander until it hits a trail marked with pheromones. A mouse does the same. Assume a mouse is placed at the start stem of a T-maze and given two trials with a brief inter-trial interval. First, the mouse would enter one arm and then the alternate arm with a significant frequency multiple times. In fact, this pattern occurs in tests with many different animals. It is called *spontaneous alternation behavior* (SAB).[2] SAB can be traced back to the microscopic mutation of genes. Spontaneous alternation of fraction of genes helps to create diversity in offspring and prevents genetic bottlenecking or deadlock.

From a macroscopic point of view, the dominant explanation of SAB is based on Hull's reactive inhibition theory, in which SAB is a response phenomenon – the mouse's first decision inhibits its second decision.[3] If it turns right first, then it would turn left in the second trial. Furthermore, experiments by Hughes et al. suggest that in addition to alternating responses, animals also appeared to be alternating between stimuli.[4] For example, when placed in a T-maze, mice usually process visual cues and only occasionally perceive secondary cues such as odor. When the salience of spatial cues is minimized, mice in the T-maze might alternate their attention to olfactory cues. In addition, animals may store and retrieve the cues necessary for

[2]Richman C and Dember WN (1986) Spontaneous alternation behavior in animals: a review. Current Psychological Research & Reviews, Winter 1986 – 87, vol. 5, no. 4, 358–391

[3]Hull CL (1943) Principles of behavior. New York: Appleton-Century

[4]Hughes RN (1967) Turn alternation in woodlice. Animal Behavior, 15, 282–286

"data mining" to estimate which direction would lead to more food. For example, a mouse could vary travelling down either arm of the T-maze based on the probability of food being present at the end of each arm.

## *Path Alternation*

Spontaneous alternation of paths for an autonomous robot, a search engine, or a problem-solving algorithm can help to explore new areas and avoid deadlock situations. Let us revisit the cleaning robot, Roomba. The Roomba's default mode enables the robot to move along one side of an obstacle. This may lead to a deadlock in the robot's path. It could potentially move in a repeating loop without exploring new areas. In order to avoid the possibility of a deadlock, whenever the robot collides into an obstacle, it could decide whether it is a frontal collision or a sideways collision, depending on the location of the collision sensor. If it were a frontal collision, the robot would turn at a random angle, which is a spontaneous alternation of its path, similar to SAB. If it were a sideways collision, then the robot would continue to move along one side of the obstacle until hitting a wall.

## *Collision Recovery*

Collisions can be found in many modern electronic systems in various fields, from autonomous driving vehicles to data communication protocols. There is a variation of the SAB strategy for collision recovery. When a collision occurs, the system spontaneously switches to different sensors or channels, or the system waits for random intervals and reconnects.

Let us take autonomous driving vehicles for example, in which several collision detection sensors are used for safety measures. When the sensory signals show obstacles within a dangerous range, the vehicle stops. At the third competition of the DARPA Grand Challenge, also known as the "Urban Challenge," took place at the site of the now defunct George Air Force Base in Victorville, California,[5] the course consisted of a 96 km urban area, which was to be completed in less than 6 hours. The autonomous vehicles were required to obey all traffic regulations while also negotiating with other traffic and obstacles. During the course of the race, one vehicle was ahead of another vehicle by a short distance as they were approaching an intersection. The second vehicle perceived two "obstacles": the static dust cloud behind the first vehicle, and a third vehicle waiting at the T-intersection. These two static obstacles created a deadlock for the second vehicle because it had no way to go forward, and had to stop. Luckily, the vehicle was programmed with a spontaneous

[5]DARPA Grand Challenge (2016) https://en.wikipedia.org/wiki/DARPA_Grand_Challenge

alternation behavior that enabled it to wait at random intervals and restart the system to move forward. After a few SAB trials, the second vehicle finally got back in the race after the dust cloud settled down. The third vehicle was able to cross the intersection after the road was clear.

We can also find spontaneous alternation behavior in the Ethernet – a fundamental component of the Internet. The Ethernet protocol is designed to share printers within a data network of an office. Sometimes, datagrams from multiple users sent to a printer will collide, causing a traffic jam. The protocol enables the computers to wait random intervals and then resend datagrams. Note that this operation does not require synchronization between computers. It only requires each computer to have a basic perception of datagram failure or data collision status. The "back down" and reconnect process is similar to SAB, which solves the problem of deadlock. The principle of the network protocol has become the backbone of today's Internet protocol TCP/IP that has survived for over 50 years.

## Least Effort

Studies of animal paths and human languages show that animals and humans follow the principle of least effort. The term least effort means least average probable work[6] or the behavior that involves the least expenditure of energy.[7] George K. Zipf argued that an individual's entire behavior is subject to the minimizing of effort; in other words, every individual's entire behavior is governed by the principle of least effort *all the time*. If two or more behavior sequences, each involving a different amount of work, have been equally well reinforced an equal number of times, the organism will gradually learn to choose the less laborious behavior sequence. Taking path selection as an example, animals and people often take short-cuts across a field to minimize the walking distance; when the short-cut is not available, people try to bypass the obstacle when possible. However, animals' or humans' choices are limited by the laws of physics and their problem-solving skills.[8]

The principle of least effort is not just limited to physical behaviors such as path finding. It also applies to cognitive behaviors such as word use and pattern recognition. R. S. Crutchfield defines this mental searching and matching criterion as a psychological distance, similar to the path choice in the physical world. For

[6]Zipf GK (1949) Human behavior and the principle of least effort. Addison-Wesley Press, 1949 and Martino Publishing, 2012.

[7]Tsai LS (1932) The laws of minimum effort and maximum satisfaction in animal behavior. Monograph of the National Institute of Psychology, Beijin, China, No. 1, 1932. Abstracted by Psychological Abstracts, vol. 6, 1932. No. 4329.

[8]Crutchfield RS (1939) Psychological distance as a function of psychological need. Journal of Comparative Psychology, vol. 29, pp.447 – 469, 1939.

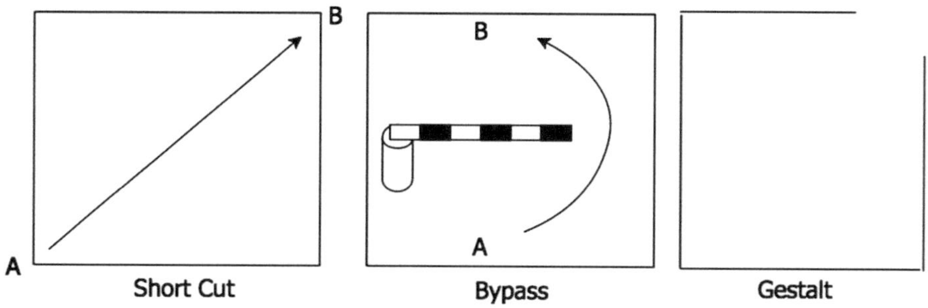


**Fig. 2.3** Examples of the principle of least effort: short cut, obstacle bypass, and image completion (Gestalt effect)

example, our vision system automatically completes the rectangular shape despite a part of the outline being missing. This phenomenon is often called *Gestalt*, or image completion. It turns out that our brain tends to use the least effort to store, recall and represent shapes. In this case, recognizing a rectangular shape is "easier" than recognizing the irregular shape that needs extra measurements. See Fig. 2.3 for the illustration of the cases above.

Zipf also studied the frequencies of word usage in difference languages and statistically proved that they follow the principle of least effort. Throughout this book, we will frequently cite Zipf's work for many applications. It is important for computers or robots to understand humans' least effort behavior and to apply the principle to their regular tasks.

## Alignment

Imitation or following is another economical behavior found in nature. *It simplifies the control logic for interactions between multiple entities.* During social interaction, it is very common for an act performed by entity A to be followed by a similar act by entity B. This is called *response matching*. In many situations, if A is instructed to make long utterances, B will also make long utterances. *Movement mirroring* is a spatial and temporal alignment of motions, which consists of the copying of another's movements.

The alignment can be in master-slave mode (one-way communication) or interactive mode (two-way communication). Figure 2.4 shows an illustration of these two modes. In the master-slave mode, entity B simply mimics A's most recent action without any feedback. This model assumes that each social act is determined by the last act of the other. This leads to the study of simple empirical sequences of social acts. In the interactive mode, on the other hand, the slave is able to communicate with the master to update about its current status.

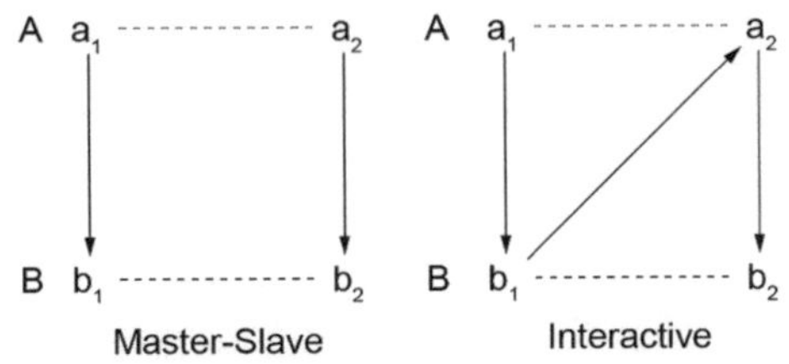


**Fig. 2.4** Illustration of "movement mirroring" in the master-slave mode versus interactive mode: *A* is a master with states $a_1$ and $a_2$; *B* is a slave with states $b_1$ and $b_2$. The *arrows* are communication paths

## *Following Behavior*

In autonomous train systems, there are safety doors on each vehicle and wayside. All of the doors have three states: open, close, and recycle. Recycle is a safety reaction in which passengers or obstacles hold the door and the door will open and close in several cycles. The synchronization between the two sets of doors is important for keeping the traffic of passengers flowing without jamming. Here, the vehicle door acts as a master and the wayside door acts as a slave. The slave follows the master's every move: open, close, or recycle. These instructions are sent through the wireless vehicle-to-wayside communication channel. Imagine if the two sets of doors operated independently. The control logic and communication design would be much more complex. For example, the communication would have to be two-way instead of one-way. Note that this is just a design problem for a train, which is much less complicated than the design for autonomous driving vehicles. Truck manufacturers, such as Volvo, have adopted the master-slave concept for autonomous fleet trucks. When two or more trucks travel together, the front-most truck would be the "master" and the following trucks would be the "slaves," which maintain the same speed as the master while keeping a safe distance.

Ethology is the study of animal behavior. It attempts to explain the causation, development, survival, and evolution of the behavior patterns of animals. Perhaps the most famous ethologists are Niko Tinbergen, Konrad Lorenz, and Karl von Frisch. Their hierarchical behavior models have often been studied by the Artificial Intelligence community.[9,10] Their theories have inspired an early movement of behavioral mobile robotics that directly connects perception to actuation.[11]

Delphine de Gerardin once said, "Instinct is the nose of the mind." Dogs are possibly the most instinctual helper to humans. Dogs are vigilant, empathic, and loyal. For thousands of years, dogs have acted as domestic assistants for the elderly and disabled, particularly for visually impaired people. In addition to their outstanding olfactory capacity, the motion intelligence of dogs is also

[9]Tinbergen N (1951) The study of instinct, Oxford University Press, Oxford, UK

[10]McFarland D (1985) Animal behavior, Benjamin/Cummings, Menlo Park, California

[11]Brooks R (1999) Cambrian Intelligence: The Early History of the New AI, The MIT Press

remarkable. Dogs can sense complex human activities, ranging from gestures and facial expressions to micro movement, such as shivering. Dogs use a rule-based system that is similar to the Internal Release Mechanism (IRM) in animals and insects that an action is devoted to a particular stimulus. Those rules can have been encoded in the DNA of a dog as so-called instincts.

A robotic dog model appears to be desirable here because of its canine senses, mobility, scalability, and multimodal interaction with humans. Instead of building a large sensor network for a smart environment, we intend to explore the potential of a mobile and interactive sensing platform. Here, we aim to use a dog as the metaphor for interactive and mobile sensing for smart buildings.

A dog often gives its owner undivided attention. A dog has the instincts to recognize and follow its owner. Ethologists call this phenomenon "imprint."[12] The artificial dog starts with a simplified version of the human following algorithm. Using just one infrared sensor and two servomotors, the robot can search for the target and follow the human within a predefined distance. The pseudo code of the control algorithm is as follows.

```
Turn on left and right infrared (IR) sensors on robot
Rotate the vehicle to search for human
Approach human within a safe distance
Switch based on IR signal strength
   Case: left IR = right IR, move forward
   Case: left IR > right IR, turn left
   Case: left IR < right IR, turn right
   Default: no IR, rotate robot to search for human
End Switch
Repeat Approach loop
```

## *Empathic Conversation*

Human interactions also follow the last actions of others. Therefore, it is possible to program a computer to simulate social interaction and cognitive alignment. The main difficulty is on the personal perception side, because a device is needed to decide how to categorize a particular action in terms of categories used. In the early artificial intelligence artifact ELIZA, the software was designed to simulate the conversation between a psychiatrist and a patient.[13] The patient types at a console and the computer's reply is automatically generated. Below is an example dialogue:

---

[12]Tinbergen N (1951) The study of instinct, Oxford University Press, Oxford, UK

[13]Eliza online chatbot (2016) http://nlp-addiction.com/chatbot/eliza/

```
P: Hello.
C: How do you do? Please tell me your problem.
P: I need more time to complete my book.
C: Is it important to you that you have your book?
P: Yes, I have spent a lot of time on it.
C: Are you sure?
P: I am sure.
C: You seem quite positive.
P: Of course I am.
C: I see.
```

As seen here, the computer has been programmed along the lines of non-directive therapy, and uses the technique of reflecting the emotions expressed by the patient. This then encourages the patient to talk more. The computer is able to respond to the patient's emotions and participate in an empathic conversation. The interaction alignment uses topics – if A remarks on a particular topic, then B will probably reply by speaking on the same topic. If he or she does not, and fails to explain why or goes on to change the subject, the conversation will discontinue.[14]

## Default Knowledge

When we encounter a new situation, we often select a structure from memory. This structure is called a *frame*,[15] which is a remembered framework to be adapted to fit reality by changing details as needed. Frame theory was created by Marvin Minsky in 1968. Frames represent knowledge as structured objects composed of labeled attributes or slots with their values, similar to an object-oriented approach. However, frames can contain broader information – not only static declarative, but procedural information as well. A frame can be represented as a network of nodes and relations. The top levels of a frame represent things that are commonly true about the supposed situation. The lower levels are slots or sub-frames that are filled with specific instances or data. Each attribute or slot can specify conditions its assignments must meet. Complex conditions can specify relations among the things assigned to several attributes (slots) or sub-frames. Collections of related frames are linked together into frame-systems. According to Minsky, the effects of common activities such as learning and reasoning are transformations between the frames of a system. The source of attribute values comes from initialization, user input, external programs, other frames, and inheritance. Figure 2.5 illustrates an example of the frame system in a living room. The abstract name "room" sits at the top of the diagram. Then, it points to its components: floor, door, and furniture. The sub-frame

[14] Argyle M (1969) Social interaction. Tavistock Publishing. 1969

[15] Minsky M (1975) "A Framework for Representing Knowledge", in P. Winston (Ed.) The Psychology of Computer Vision, (New York: McGraw-Hill), 211–277

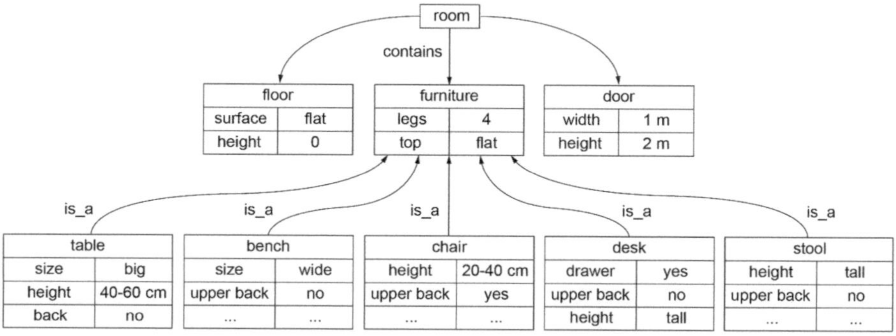


**Fig. 2.5** A frame network of a furnished room

of "furniture" contains table, chair, bench, desk, and stool. Each of them is specified with attributes that help to distinguish one from the others.

By far, framing has been the most versatile structural representation of knowledge. It covers visual and non-visual descriptions of things. For visual scene analysis, the different frames of a system describe the scene from different viewpoints. The transformations from one frame to another represent the effects of moving from place to place, or from zooming-out to zooming-in. For example, a room frame might contain slots for the walls, windows, ceiling, and floors, which describe the overall room from a visitor's perspective. For non-visual frames, the differences between the frames of a system can represent category relations, attributes, actions, or changes from a conceptual viewpoint. For example, a room frame can be a part of the root frame "home" containing slots to describe things such as temperature, size, and number of rooms. Perhaps much of the power of the frame lies in the inclusion of default knowledge. A frame's attribute slots are normally already filled with default assignments. The default assignments can be easily replaced by updated items with more specific information. This enables people to communicate and understand with partial information, which is common in verbal communication and visual thinking. For example, family members or close friends know each other so well that they can often complete each other's sentences. Our vision relies on so much default knowledge that we are literarily blind for most of time, because our eyes take snapshots from the continuous moving scene at a rate of less than 30 frames per second, and with an acute visual angle of only 2° or 3°. Therefore, most visual experience is based on our own imagination by filling in the missing data in frame slots, or even by replacing missing frames. Before we enter a library, for example, we usually know what to expect inside a library: book shelves, checkout desk, desks, and chairs. We often fill a default frame of a typical library in advance and then we update the frame with more details and more connections.

## *Inheritance*

People normally have enough common sense to cope with everyday situations. We do not have to be told everything explicitly because we can call upon default knowledge. We use default knowledge to handle situations where knowledge is incomplete. Default and inheritance are ways of achieving common sense.[16] Inheritance is a way of reasoning by default – when information is missing, the system falls back to its default settings.

Attribute slot names in frames often correspond to relation names. For example, the "is-a", "a-part-of", "a-kind-of", and "contains" frame links facilitate hierarchical representations and the inheritance of information, which are efficient for minimizing redundant information. A common approach is to create generic frames to represent the abstraction of specific entities, and then to create instance frames representing specific instances of these generic types. The instance frames are linked to the abstract frames via the instance relation. This approach allows attribute slots and their default values to be inherited by the instance frame from upper-level frames.

In Fig. 2.5, we have a parent frame called "furniture" followed by its child frames – "table," "chair," "bench," "desk," and "stool." By default, if we do not have any information about the child frames, we can always obtain their inherited attributes from their parent frames, for example, four legs and flat top surface, etc. We always fill the missing attributes with defaults. For example, we expect a typical chair to have four legs, a level seat, and an upper-back support. We also expect certain spatial relationships between those parts. For example, the legs must be below the seat and the back support must be above the seat. The legs must be supported by the floor. The top surface must be horizontal; the back must be nearly vertical, and so on. Now, suppose that this description does not match, e.g., the vision system finds four legs, a level plane, but no back. This suggests that the object is not a chair, but rather a table or bench. This structure is called a "similarity network." For example, a furniture network has a chair, couch, table, stool, and desk, including their similarities and differences. A table is too high to serve as a chair, while a box has no room for knees, etc.

## *Analogy*

Analogy is to use information from one domain to help solve a problem in another domain. Analogical thinking can be implemented based on similarities of attributes in different frames, and even different domains. Analogy can be applied to many applications when we do not have complete knowledge of the world. Taking Fig. 2.5

[16]Minsky M (1975) The Psychology of Computer Vision, P. Winston (Ed.), McGraw-Hill

for example, the entities table, chair, bench, desk, and stool are very similar. They all have four legs and a flat top surface. However, they also have subtle differences (e.g. size, height, and whether or not there is an upper-back support). If we search for a replacement for a chair with a tolerance of one or two errors in attributes, we might accept a bench, stool, table, or even a desk as alternatives. Indeed, people occasionally sit on tables or desks as chairs.

Children often use analogy. When they identify the similar key attributes of objects, they ignore any remaining attributes. For example, children often line up chairs in a row to form a train, as the line-up is a key attribute of a train.

Analogy enables thinking with imperfect data and images. It can be a progressively refined by better, but still usually imperfect ideas. Traditional logic cannot deal very well with realistic and complicated problems because it avoids approximations to solutions. Frames enable computers to jump to conclusions based on *partial* similarity matching. Partial matching is the key to information retrieval, as seen in error-tolerance string search algorithms for instance.

## *Stereotypes*

A stereotype is a representative model, position, appearance, or sequence of actions of a particular type of person or thing. A stereotype can be the result of accumulated experience in a particular area, or typical phenomena in society or culture. A stereotype of a chair has four legs and a flat top surface. However, there are variations such as three-legged chairs and indented top surfaces. Stereotypes can be used as defaults in frame-systems. They can fill in massive amounts of missing information about a scene or a thing, thanks to coherence in human activities. There are traceable data patterns in repeated routines and recorded images. Stereotype patterns can be summarized by simply averaging accumulated data. One could use, for example, the average pixel colors of multiple photographs to make a superimposed image, similar to using multiple exposures for film.

Figure 2.6 shows a few examples of a series of 100 superimposed images of a sunset, Taj Mahal, classical portrait paintings, newscasts, Jesus Christ, and street scenes. As we can see in the sunset images, the sun typically sets on the horizon at the lower part of the image. Taj Mahal is a structure with symmetrical architecture. Most people take photographs in a symmetrical way, so these superimposed images show a symmetrical view of the architecture in the center. Newscasters, classical portraits, and images of Jesus Christ are also stereotypically consistent. The figure typically appears in the center of the picture plane with a uniform background. However, the photographs of streets are not so consistent because the positions of the photographers were varied. Some individuals preferred to shoot from the middle of the street while others chose to shoot on one side of the street. Also, the skylines of the buildings are different in every photograph.

The pixel color average can only describe the emergence of a blurry pattern. For some important objects such as human faces, more articulated representations

**Fig. 2.6** Superimposed 100 images of sunset, Taj Mahal, classical portrait paintings, Jesus Christ, television newscasters, and street scenes (from *left* to *right* and *top* to *bottom*)

are needed to clearly define the structural layout of the facial elements, especially their positions in relation to each other. For example, the eyes are normally above the nose, the nose is normally above the mouth, and two ears are on both sides of the head. All of the faces we store in our memory are condensed to an average face, or stereotypical face. In order to represent the relationships among the facial feature elements, we can use a graph that contains nodes and links, assuming we can extract the feature points into nodes. The nodes and links form a mass-spring system wherein each node has its own weight and the links act like springs. They have attractive forces if they are pulled away, and resistant forces if they are pushed too close together. When a face structure is distorted, it creates mental stress because our mind is trying to align the stimulus with the stereotype. This mental stress may incite hilarious feelings. For example, a designer rearranged celebrity faces using the Fibonacci Sequence in his series "Dr. Fibonazi's Plastic Surgery Clinic." The series uses the Golden Ratio to warp the faces of celebrities like Jack Nicholson, Nicolas Cage, and Bryan Cranston.[17] The resulting images are humorous.

Psychological experiments also show that we often "rotate" our mental stereotypes in order to align them to the structure of stimuli. For example, we sometimes turn our head sideways in order to recognize a face in a horizontal orientation. Our mind is simply trying to minimize the distance between the stereotypical model and the appearance model. These phenomena also exist in broader human-machine interface design. When we move a computer mouse from left to right, we expect the cursor to also move from left to right. This visual-motor routine is called "spatial

[17] Wallace S (2016) Celebrity faces becomes hilarious distortions when made to fit the Golden Ratio. Smash. http://www.smash.com/celebrity-faces-become-hilarious-distortions-made-fit-fibonaccis-golden-ratio/

compatibility." Studies show that when the spatial compatibility is high, or the mental stereotype aligns well with the external stimuli, the user's response time is shorter and the number of errors is lower. Otherwise, the user's response time is longer and the number of errors is higher.

## *Proportions*

Our knowledge of proportions is an excellent example of everyday common sense. Intrinsic proportion measurements have been used in architecture and art for thousands of years. Roman architect Vitruvius stated that the proportions of a building should correspond to those of a person, and laid down what he considered to be the relative measurements of an ideal human. Similarly, in art, the proportions of the human body have a direct effect on the portrayal of the human figure in a statue or painting.

Proportions of objects and their relationships are hidden common sense to us. The author's lab conducted a survey of 50 subjects about proportion knowledge. Viewing a dog from the side, people from different countries had very consistent knowledge about the proportion of its height versus its length (mean 0.61 and standard deviation 0.15). On the other hand, when viewing an adult male standing in front of the camera, people had much more diverse estimations of the proportion of height versus width (mean 3.87 and standard deviation 1.45). Inanimate objects such as elevators have a worldwide standard proportion of height versus width (mean 1.96 and standard deviation 0.53). Our perceptions of the proportions between different objects are also amazingly consistent. For example, the proportions of human height versus the height of a 2-door car are similar (mean 1.53 and standard deviation 0.46). The results are shown in Fig. 2.7.

Artists often use analogous measurements to portray the world, measuring objects and their relationships based an intrinsic landmark. As a result, the measurement is invariant to any coordinate system. Artists frequently use the size of the human head in the picture to measure the height and width of a human body, and use the size of the eye to measure the height and width of a face.

```
Height of an adult human body:          6-8 heads
Width of an adult human body:           2-3 heads
Location of chest area:                 2-3 heads
Length of an arm:                         4 heads
```

The illustration on the left of Fig. 2.8 shows a sample of vertical proportion in a typical art book and the actual distribution of head-to-body proportion calculated from the CAESAR dataset.[18] The results show that the average human is six to eight

[18] Anthropometry Resource for CAESAR (2002) Final Report, Volume I: Summary, AFRL-HE-WP-TR-2002-0169, United States Air Force Research Laboratory, Human Effectiveness Directorate, Crew System Interface Division and SAE International.

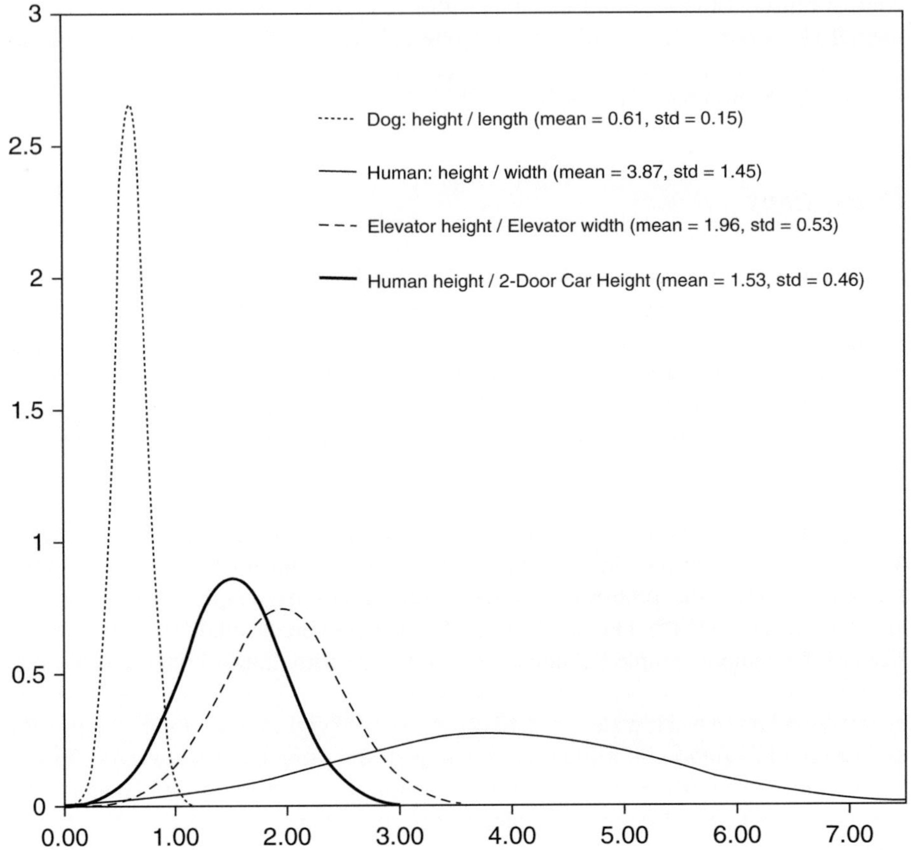


**Fig. 2.7** Normal distribution of proportions from a survey

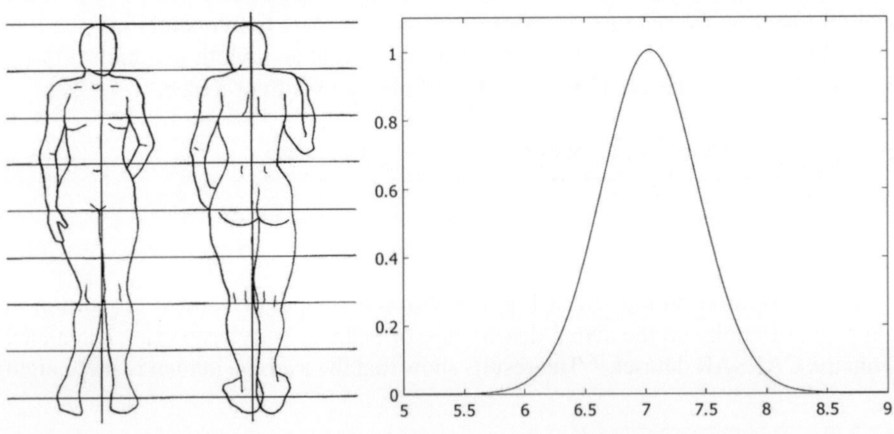


**Fig. 2.8** Body height measured by head (*left*) and normal distribution of head per body (*right*)

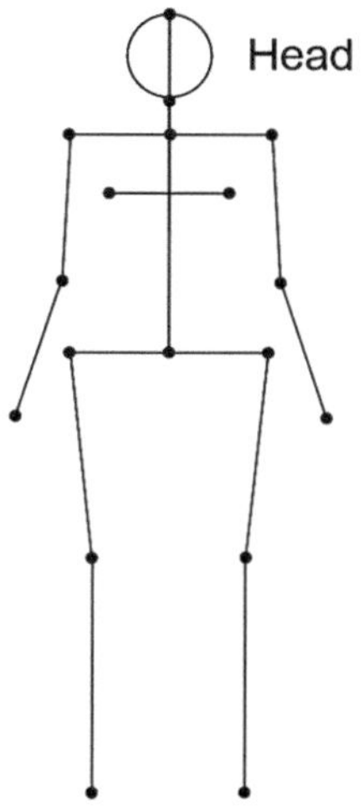


**Fig. 2.9** Analogia graph of a human body

heads tall. Based on our observations from one hundred 3D scan data sets of adults ages 16–65 including subjects from North America, Europe, and Asia, we found that the length of one and a half head units from the bottom of the head is enough to cover the chest area. In addition, the chest width is approximately three heads wide. The plot on the right of Fig. 2.8 shows an output from the intrinsic proportion calculation based on the sample set from the CAESAR database.

An analogia (Greek: $\alpha\nu\alpha\lambda o\gamma\acute{\iota}\alpha$, means 'proportion') graph is an abstraction of the proportion-preserving mapping of a shape. The idea is to use one edge of a graph to measure the rest of the edges. The analogia graph is commonly used in the arts. The Russian realist painter Ilya Yefimovich Repin once said that the secret of painting is "comparison, comparison, and comparison." To realistically represent objects in a picture, a painter must constantly measure and adjust the relationships among objects. "You should use the compass in your eyes, but also in your hands," according to Repin. Instead of using absolute measurement of distances and sizes, artists often use intrinsic landmarks within the scene to estimate relationships. Figure 2.9 is an analogia graph of a human body.

Why is the analogia graph so important to our visual experience? Most of us are not aware that our eyes make rapid movements, so called *saccades*, at a rate of four times per second, which helps to cover the full visual field. When people view an object, the saccades follow the shape of the object similar to the way blind people observe an object by touching it. The trajectory of those saccadic movements forms a graph. According to the philosopher George Berkeley, seeing is like touching at a distance. This idea was resurrected by neural scientist Rodney Cotterill.[19] Driven by instinct, vision enables us to react quickly to dangerous situations at a distance.

[19] Kringelbach ML (2009) The Pleasure Center, Oxford University Press, New York

The analogia graph is one of many visual reasoning techniques based on instinctual knowledge and heuristics about objects, such as the human figure.

From a computer vision standpoint, detecting features from 3D body scan data is nontrivial because human bodies are flexible and diverse. Function fitting has been used for extracting special landmarks such as ankle joints from 3D body scan data, similar to the method for extracting special points on a terrain. Curvature calculations use methods such as chain code and bending angles. However, sensitivity to data noise is a major problem in both the function fitting and curvature calculation methods because typical 3D scan data contain large amounts of noise. Many machine-learning algorithms for determining features are coordinate-dependent and limited by the training data space.

Using this artistic approach, we can create a graph where nodes represent regions and are connected to each other by edges, where weight is defined as the distance between the nodes in proportion to the height of the head. We can stretch the graph such that it overlays the entire body. We can then create a link between each node and its respective counterpart. We can link the head, shoulders, arms, elbows, hands, neck, breasts, waist, legs, knees, and feet to their respective regions. There is some tweaking required to assure that the waist region does indeed cover that area. Here, we run a quick top-down search through the plane slices until there are at least two disjointed areas, which we consider to be the middle of the waist. This change also makes modifications to where the arms, knees, and chest are, as well as the sizes of those regions.

We found that the intrinsic proportion method can reduce the search space for anthropometric landmarks by an order of magnitude. Our algorithm is 60 times faster than the record in the study.[20] In addition, this method reduces the risk of finding local optima while searching the whole body.

## Primitive Physics

Primitive physics is a type of common sense that we frequently take for granted. It exists in every level of creature, from cell and animal behavior, to human cognition. Here, we shall focus on three basic models: chemotaxis, multi-object interaction, and sense of gravity.

### *Chemotaxis*

*Chemotaxis* is the movement of an organism in response to a chemical stimulus. The word chemotaxis combines two parts: 'chemo' meaning chemical and 'taxis'

[20]Suikerbuik RH (2004) Tangelder, H. Daanen, A. Oudenhuijzen, Automatic feature detection in 3D human body scans, Proceedings of SAE Digital Human Modeling Conference, 2004, 04- DHM-52

meaning movement. Microscopic organisms such as somatic cells and bacteria direct their movements according to certain chemicals in their environment. Chemotaxis is critical to the movement of sperm cells towards the egg during fertilization, as well as the migration of neurons and lymphocytes. This behavior is also observed in bacteria when searching for food by swimming toward the highest concentration of food, or when fleeing from dangerous poisons.[21] Positive chemotaxis occurs if the movement is toward a higher concentration of the chemical; negative chemotaxis occurs if the movement is in the opposite direction.

Chemotaxis can be generalized as a diffusion and reaction process. The stimuli can be light, potential, or WiFi signal strength, in addition to chemical concentration. Perhaps the most striking instances of chemotaxis are hidden in our own cells. Biological studies suggest that mammalian cells indeed possess more intelligence than we can imagine.[22] Cell movement is not random. It is capable of immensely complex migration patterns that are responses to unforeseeable encounters. Cells can "see" (i.e., they can map the directions of near-infrared light sources in their environment), and direct their movements towards or away from various stimuli. No such 'vision' is possible without a very sophisticated signal processing system.[23]

Alan Turing not only invented the one-dimensional mathematical model of computing, but also worked on the two-dimensional model of the growing embryo – a fundamental biological process called *morphogenesis*.[24] Morphogenesis comes from the Greek word, morphê, meaning shape and genesis creation – literally, the "beginning of the shape." Turing used a reaction-diffusion model to simulate the formation of patterns or structures in a ring of cells. Remarkably, he discovered chemical "waves" on the surface of a sphere of a ring of cells. According to his theory, most of an organism, most of the time, is developing from one pattern into another, rather than from homogeneity into a pattern. In certain conditions, oscillatory patterns occur. Around the same time, Boris Belousov discovered a class of reactions that behave in a form of non-linear chemical oscillation, called a Belousov-Zhabotinsky reaction.[25] Computational diffusion and reaction models can be used to mimic certain properties of neurons, self-replicating cells, as well as electrical, magnetic, and chemical wave patterns, which are essential to the creation of a "wet computer."[26]

---

[21] WikiPedia: Chemotaxi: https://en.wikipedia.org/wiki/Chemotaxis

[22] Albrecht-Buehler G (1985) Is Cytoplasm intelligent too? In: Muscle and Cell Motility VI (ed. J. Shay) p. 1–21, 1985

[23] Albrecht-Buehler G (2007) http://www.basic.northwestern.edu/g-buehler/cellint0.htm, updated 21 Jan. 2007

[24] Turing AM (1952) The chemical basis of morphogenesis. Philosophical transactions of the Royal Society of London. Series B, Biological Sciences, Vol. 237, No. 641. August 14, 1952, pp.37–72

[25] WikiPedia: Belousov–Zhabotinsky_reaction.
https://en.wikipedia.org/wiki/Belousov--Zhabotinsky_reaction

[26] Palmer J (2010). Chemical computer that mimics neurons to be created. BBC Science News. http://news.bbc.co.uk/2/hi/science/nature/8452196.stm

## *Sense of Gravity*

Sense of gravity is an essential part of primitive physics. It is a part of the default cognitive model seen in animal behavior. For example, according to Beck's study,[27] herring gulls typically gain access to the edible interior of a shell only after using several collisions with a hard surface in order to break the shell. Large shells require being dropped many times over a considerable period of time. A sea wall normally contains hard-packed sand bars and stones. Most gulls drop food such as moon snails against the sea wall, which occupies only a small area, and perform very few drops over vast areas of soft sand. This example clearly demonstrates that the gulls are selective, generally dropping over the hardest available substrate, while avoiding softer, less suitable ones. The gull's cognitive model fits Newtonian physics well: shells dropped from greater heights are more likely to break, but only if they hit a hard surface. For example, gulls dropped shells from significantly greater heights (e.g. 6 m on average) in paved parking lots than they did at the sea wall (e.g. 4 m in average). This is because paved parking lots are large and thus easier to aim over, allowing the gulls to drop their shells from higher altitudes. The sea walls are much narrower, so the gulls have to fly at lower altitudes in order to successfully drop shells over the wall surface. The gulls appeared to increase dropping height, and thus the probability of breakage, in paved lots where hits were assured. In many cases, sense of gravity is essential to designing mobile robots. For instance, one could estimate a safe height for a drone to drop a package to the ground based on the material properties of the ground below – glass, concrete, soft dirt, or wood – as well as the area size needed for a given level of accuracy.

We often take it for granted when we see an object whether it is supported or if it is going to fall without enough support underneath it. This is indeed very primitive physics. Everything requires some sort of support from below, or it will fall due to gravity. For example, if we want to print a 3D ball, one must consider how the structure will be supported during the printing process. We normally have to print an object from its base, which is just a point or a tiny area. Then, a slightly larger slice or disk of material will be layered on top of the smallest one until it reaches the halfway point of the height of the ball. This certainly is not a safe design because large disks are resting on top of smaller ones. If left unsupported, gravity will take over and ruin the print. Some 3D printers have already incorporated the gravity factor into specific Computer-Aided Design (CAD) algorithms. For example, the MakerBot 3D printer checks the support structure and automatically adds scaffolding support around the lower part of the ball. After the printing process is complete, users can easily remove the support structure. Figure 2.10 illustrates an example of the scaffolding support needed for 3D printing a sphere.

Sense of gravity has a broader impact on human cognitive models. We often call this *psychophysics*. Let us use fall detection as an example, in which we often

[27] Beck, B. B. Animal tool behavior. Garland STPM Press. 1980

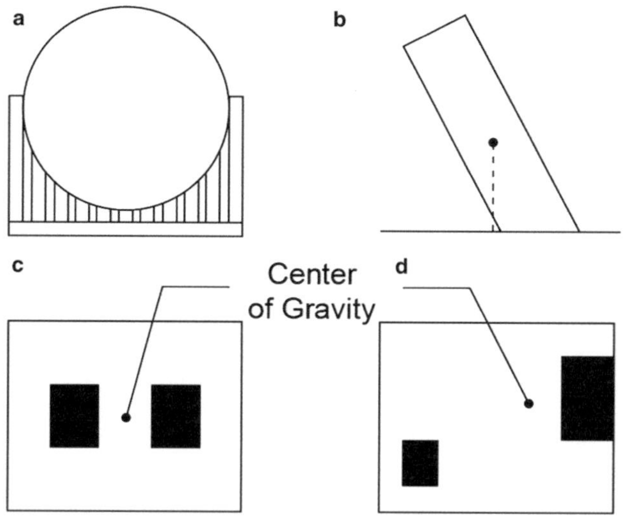


**Fig. 2.10** Sense of gravity: (**a**) scaffolding support for the 3D printed ball, (**b**) fall detection, (**c**) a balanced image, and (**d**) an unbalanced image

monitor the vertical lines of a pole, building, or wall to check whether it is safe to walk nearby. If the vertical line from the mass' center of gravity is outside of its supported area, then we anticipate the possibility that this object will fall.

Sense of gravity is also embedded into our design and aesthetic perception of paintings and photos. Even though they are in the two-dimensional world, we still can sense the invisible hand, which is gravity in the image. Balance is an essential psychophysical measurement of a visual composition. In a picture, each object carries a certain "weight", depending on its color, intensity, and role. The overall pictorial "balance" can be measured by calculating the central gravity of the image. The sense of gravity balance stimulates the viewer's emotions, inciting feelings such as calm, stress, or fear. Figure 2.10 shows examples of fall detection and balance compositions. Now, imagine you design an autonomous mobile robot that builds a bridge over a river. This robot should be able detect whether the structure is upright and balanced using machine vision. After a day of work, such a robot might stop by a web sites to learn from architectural masterpieces, and to appreciate classical and modern artwork, from the Leaning Tower of Pisa, to Frank Lloyd Wright's Fallingwater.

## Summary

The majority of our daily activities are based on default behaviors and common sense. A default mode can be designed as a fallback strategy whenever the conditions of reality do not meet our expectations. A simple default behavior can reduce the complexity of decision-making. However, it may also create a deadlock. Spontaneous alternation behavior (SAB) drastically changes the course of paths and/or retrial intervals to avoid deadlock whenever a collision occurs. These two strategies are simple to implement without requiring reinforced learning or memory.

The principle of least effort is a common behavior for animals and humans at all times! Such a behavior includes finding short-cut, bypassing obstacles, and completing images. It is important for a computer or robot to understand humans' least effort behavior, and even to have such a behavior.

Alignment by imitation, or following, is another example of default behavior that has a broader impact on multiple system coordination such as master-slave mode and a dog following a human. More advanced alignment strategies include empathic conversation in which a computer and a user mirror each other's phrases interactively.

Default behaviors emerge from default knowledge, or so-called common sense. Common sense can be represented as a network of frames of visual and non-visual content, such as semantic networks. For instance, there are three ways for people to gain knowledge: accumulation, inheritance, and crossover. Accumulation is a long process. For thousands of years, mankind has accumulated substantial knowledge in the form of stereotypical and proportional knowledge. Relatively speaking, inheritance and crossover are more efficient: taking known facts and scripts from parent frames and filling in attributes from outside frames that share similarities.

The most profound aspect of common sense perhaps is found in primitive physics. This includes chemotaxis, multiple object interaction, and sense of gravity. In order to enable a robot to work on organic cells, on the road, to build a house, or even to view a painting, we must embed primitive physics into its digital code. Ironically, people have fought for decades to move all computing from analog to digital. Now, we are struggling to make digital computers behave more like analog circuits.

# CHAPTER 3

# Pheromone Trails

Pheromones in nature are chemical messages that act within a species. It was Karlson and Lüscher who coined the term "pheromone" to cover the wide range of substances secreted to the outside world by individual organisms and received by other individuals of the same species. Pheromones are used widely by insects. The analogy between this mode of communication within an insect society and communication within the body by means of hormones, led to these substances also being referred to as "social hormones."[1] These chemical messages have diverse biological effects and differ widely in their modes of action. In practice, the term "pheromone" proves useful in describing behaviors such as trail formation, defensive secretions, and social coherence.

Pheromones are *external memories* that are physically projected onto the ground or into the air, and are shared within a group. The dynamics of depositing and vaporizing pheromones are in fact very sophisticated processes. It has long been known that social insects such as ants use pheromones to leave information on their trails for foraging food, leaving instructions for efficient routes, for searching, and for making recommendations (e.g. suggesting similar products based on the items in the online shopping cart). Many insects produce alerting pheromones as defense secretions for escaping from or repelling potential predators by means of aggregation and dispersal. Fire ants and honeybees use such pheromones during complex behavioral acts, from mass foraging, trail following, colony emigration, alarming others, to grooming and clustering. Pheromones also play a large role in social coherence within a species. They serve the purpose of attracting others from a distance.

In this chapter, we are not going to cover how to use chemical pheromones in nature. Rather, we will consider pheromones as biomorphic models to explore

[1] Wigglesworth VB (1970) Insect Hormones. pp. 134–pp. 141. W.H. Freeman and Company.

© Springer-Verlag London 2016

Y. Cai, *Instinctive Computing*, DOI 10.1007/978-1-4471-7278-9_3

trail following, alerting, and social attraction in digital worlds, where we leave "pheromones" in surveillance videos, social media, browsers, remote data clouds and connected devices.

## Pheromone Deposit and Decay

Simple pheromone-based movement can produce sophisticated dynamic patterns. Conventional Ant Colony Optimization (ACO) models assume that insects walk along paths that connect various nodes. Pheromones can be overlaid in multiple layers. Furthermore, pheromones decay at a certain rate. If they did not decay, ants would risk repeating the same route and not respond to a rapidly changing environment. Here, we generalize pheromone deposit and decay in a two-dimensional space, rather than on a path network. The amount of pheromones at a pixel position at time $t$ is:

$$p(t) = (1 - d) \cdot p(t - 1) + u(t) \tag{3.1}$$

where $d$ is a coefficient of decay, $0 \leq d < 1$. The first part of the expression represents the decay of existing pheromone; the second part of the expression $u(t)$ is the new pheromone deposit. When we have $m$ ants, the total amount of pheromone is:

$$u(t) = u_1(t) + u_2(t) + \cdots + u_m(t) \tag{3.2}$$

## Pheromone Map

The generalized pheromone update model can help us discover motion patterns in videos, which transform invisible patterns of moving objects into visible trails that accumulate or decay over time, much like a scent. Here, we will explore several methods to map motion patterns in a two-dimensional space with artificial pheromone updates.

### *Light Painting*

Light painting, which originated from the art of photography, records the movement of light using a long-exposure time when capturing an image. For example, Pablo Picasso used to draw pictures in the air with a light bulb and photographed the movement on a long-exposed film. In order to make a clear light painting, the

**Fig. 3.1** Light-painting of the traffic in a nursing home

background must be relatively dark so that the light trail can stand out from the rest of the image. Of course, one can set up the background to be completely black in order to capture a binary (black and white) image. However, one would lose the spatial reference coordinates for the moving light.

When capturing a digital image, for each pixel we can sum up the values of all the frames for each channel and then normalize the result to make sure that the values do not exceed the limit, e.g. less than 255 for a color or intensity channel. Figure 3.1 is a light painting of the traffic in the hallway of a nursing home. Luckily, many nurses and elderly were wearing light-colored clothing when this photograph was made. In the image, the white-colored area indicates areas of heavy traffic near the elevator. This suggests that it is not a good idea to store things such as wheelchairs in that area.

## *Traffic Heat Map*

A heat map is a map of accumulated pheromones left behind by moving objects. It shows the repeated motion events using either intensities or pseudo colors. A motion pheromone heat map must track the changes in pixel values. This can be done using the *frame subtraction* method in order to identify pixel values that change from frame to frame. Below is the pseudo code for live heat map visualization:

**Fig. 3.2** A traffic heat map of the Pittsburgh International Airport, where the lightest areas indicate the heaviest traffic. The heat map is superimposed on top of a frame from the original video to create reference points

```
Read in the first frame as the background frame
Initialize heat map
For each frame
   Get current frame
   Convert current frame to grayscale
   Calculate frame difference = |current frame - background frame|
   If difference > threshold, then update heat map
   Save current frame to background frame
   Superimpose the heat map on the current frame
End
```

Many streets, shops, schools and airports are equipped with closed circuit television systems (CCTV). The heat map method can be used for extracting behavioral patterns from recorded or live video from CCTV systems. Figure 3.2 shows an example of a traffic pattern at the Pittsburgh International Airport. The heat map shows the center lane in the check-in lobby has the heaviest traffic. The US Air check-in booth was also busy, while the other three booths had light traffic. The map also shows an area of light traffic upstairs where a trail was left by the airport staff. In this example, the frame subtraction-based motion detection method is noisy due to the non-uniform lighting in the image. The motion measurement values at the four brightest spots on the upstairs level appear brighter than the rest of the spots on the trail. This can be misleading because there was only one person who walked upstairs along the wall from the right to the left of the image.

## *Motion Energy Image – All Pheromones*

A Motion Energy Image (MEI) is an envelope of motion pixels. It is a special case of pheromone updating, assuming no pheromone decay and the output is a binary image – either 0 or 1. Given a sequence of video frames, MEI superimposes all of the moving pixels onto one image, creating an outline of the motion energy area. The pseudo code for plotting the MEI from a video is below:

```
Read in the first frame as the background frame
Initialize the motion energy image
For each frame
   Get current frame
   Convert current frame to grayscale
   Calculate frame difference = |current frame - background frame|
   If difference > threshold, then set the pixel in MEI to 1
   Save current frame to background frame
   Superimpose the MEI to the current frame
End
```

Let us consider road detection for example. Most highway cameras have pan, tilt and zoom functions. The orientation of a camera can be altered at any time. This poses a challenge to feature extraction algorithms because we must have a feature representation that is independent of any particular viewpoint. How do we determine a lane based on a set of dynamic vehicular trace points? Here, we can view *vehicles as insects* that leave pheromones along the road. The "vehicle brushes" create a rough lane shape that becomes refined over short periods of time. When a vehicle passes over the road, it will leave a trace of motion energy on our road map. Intuitively, the lanes are in the swept binary areas. Hence, we can use the motion energy model to get a summary of the vehicles' motion energy. After we extract the motion energy map, we can use a filter to find the outermost points that form the outlines of the lane. Figure 3.3 shows the progressive lane forming from a 30 s video clip.[2]

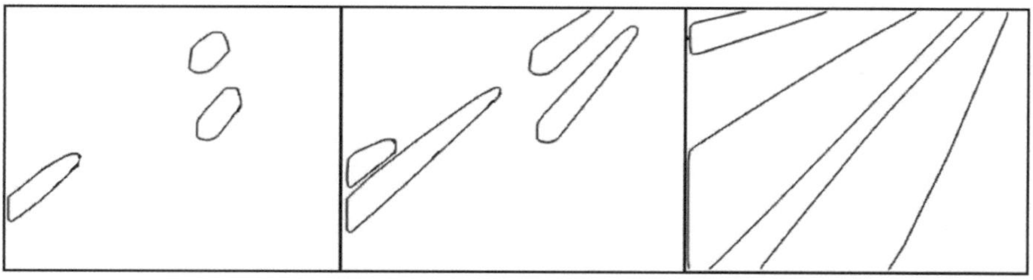

**Fig. 3.3** Adaptive highway lane detection with Motion Energy Image of vehicle "pheromones." The blobs outline the motion energy pheromones left behind by the vehicles

[2]Cai Y et al. (2013) Adaptive feature annotation for large video sensor networks. Journal of Electronic Imaging. 22(4), Oct–Dec 2013

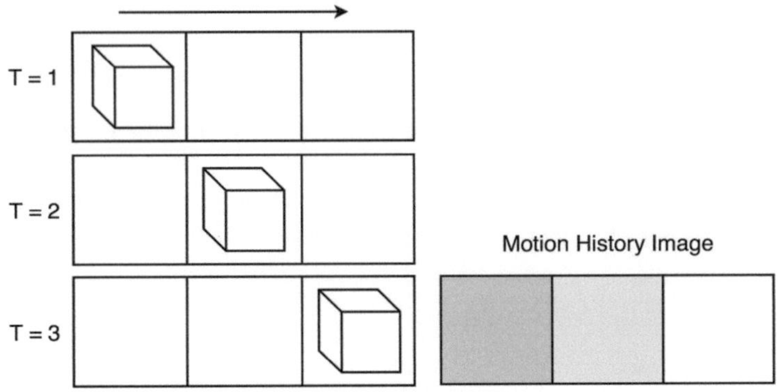


**Fig. 3.4** The Motion History Image shows an object moves from *left* to *right* like a comet tail

## *Motion History Image – Pheromone Decay*

The Motion Energy Image only summarizes moving pixels without considering their motion direction, speed or history. To analyze object motion directions, we need to employ the Motion History Image method (MHI), which deals with the decay of motion pheromones over time. Assume we have a sequence of images and we are able to detect the motion at the pixel level. At the current frame $t$, the pixel value (motion pheromone) will be set to the highest value if motion is detected. Then, if there is no motion detected at the pixel, the pixel value of the motion pheromone will start to decay at a rate of, say, 1, per time interval. Figure 3.4 shows an illustration of this process.

In general, the maximum value of a pixel is chosen as the constant 255, allowing the MHI to be easily represented as a gray scale image with a depth of one byte. Thus, an MHI pixel can have a wide range of values whereas an MEI pixel is limited to a binary version, which can easily be computed from an MHI by setting any non-zero value pixels to be 1. Below is the pseudo code for computing a motion history image.

```
Read in the first frame as the background frame
Initialize motion history image
For each frame
   Get current frame
   Convert current frame to grayscale
   Calculate frame difference = |current frame - background frame|
   If difference > threshold, then the pixel is moving
   If the pixel is moving, then set MHI pixel to highest value
   If the pixel is not moving and MHI pixel value > 0, then decay d unit
   If the pixel is not moving nor motion history, then keep MHI value 0
   Save current frame to background frame
   Superimpose the MHI to the current frame
End
```

## Deposit Sequences and Frequencies

In the digital world, our actions are constantly recorded somewhere in cyberspace. Internet browsing history becomes a pheromone trail. Social media platforms such as Facebook leave behind many pheromone trails for data miners.

Flocking is a swarm motion behavior seen in large groups of birds, insects or animals when they are foraging or in-motion. There are parallels with the schooling behavior of fish used for avoiding predators. Online flocking occurs when the collective motion of a large number of digital entities creates a swarm not governed by any central control unit. Large groups of singular events in cyberspace, e.g. tweeting, browsing, shopping, searching, moving and exiting, often exhibit flock-like patterns. Studies have shown that groups of Internet users exhibit behavioral patterns like that of a flock, where if a small number of the flock changes direction, the others follow suit,[3] similar to what we have discussed in the Default Mode chapter.

Computational simulations and mathematical models have been developed to mimic flocking behavior. We use a dynamic graph to visualize the layout of a "boid" – a bird-like object that can project approximate feature vector distances onto a 2D or 3D space. Basic flocking behaviors are controlled by a set of simple rules, for example, separation (avoiding crowding neighbors), alignment (steering towards the average direction of motion of neighbors) and cohesion (steering towards the average position of neighbors). The algorithm is in part a simulation of mass-spring physics, where each graph node is connected by a spring. The system consists of repulsive forces between all graph nodes and attractive forces between adjacent nodes. The force model corresponds to Hooke's law (of repulsive forces) and Coulomb's law (of attractive forces). The algorithm tries to reduce stress between graph nodes and dynamically plot the graph until equilibrium is reached. We define the distance between "boids" based on their time interval between two events, or the similarity or co-occurrence of attributes, which can then be formulated using the Jaccard Similarity Coefficients of the co-occurrence feature vectors. The closer the distance between two "boids", the higher the correlation. The pseudo code for the algorithm is:

```
Initialize the node positions.
Do iteratively:
   Calculate repulsive forces
   Calculate attractive forces
   Rearrange positions to fit the canvas
Stop when the node distance criterion is met
```

The force-directed graph is simple, intuitive and aesthetically pleasing. However, it is computationally expensive. It would be very slow if it were used to visualize large network situations involving millions of nodes. In addition, the algorithm may

---

[3]Nauert R (2015) http://psychcentral.com/news/2008/02/15/herd-mentality-explained/1922.html

run into local optima. To avoid this problem, we can randomize the initial starting positions and run the algorithm a few times to get better results.

In order to visualize the event transition history or motion history of the "boids," we use the analogy of "pheromone paths." To indicate the dynamic motion history between multiple "boids," thicker and darker traces are used to represent high correlation or heavy traffic. The pheromone paths may be customized to decay or remain persistent based on MHI or MEI.

To visualize attributes of "boids" in more than three dimensions, we can use a glyph shape with polar coordinates, where each corner represents a variable, and its radius represents the normalized value of the attribute. This solves the problem of mapping higher dimensional attributes onto a 2D or 3D visible space.

Figure 3.5 shows the visualization of UNIX computer users' command history, containing 9 sets of sanitized data drawn from 8 users over a period of 2 years.[4] The visualization shows that the common Unix commands "vi," "cd," "ls" and "rm" are clustered together. The thickest, darkest lines to EOF – that is, the most common last commands to be run – are from "who," "elm," and "ftp." "who" lets users see who else is currently logged in. This corresponds to a behavior pattern of seeing who

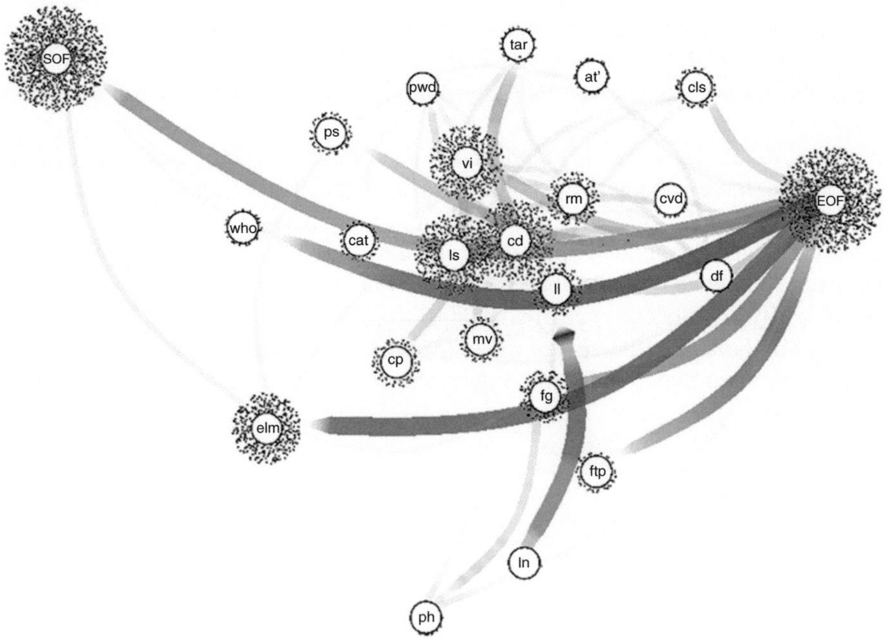


**Fig. 3.5** A snapshot of the flocking pattern in the UNIX computer user's command line history log file

[4]UNIX User Data (1998) http://kdd.ics.uci.edu/databases/UNIX_user_data/UNIX_user_data.html

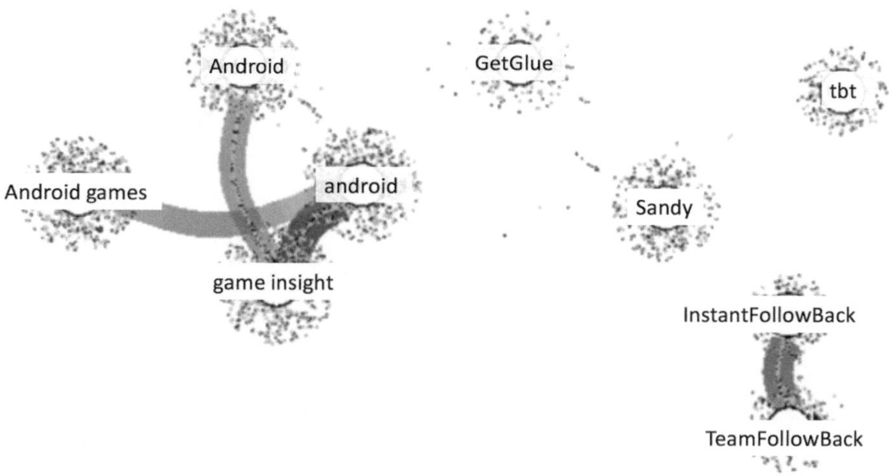


**Fig. 3.6** A snapshot of the flocking pattern in the Twitter dataset of November 2012

else is around before leaving. "elm" is an old email client – checking mail was the last thing these users did before logging out. "ftp" lets a user send files to a remote server, so the last action during these users' sessions was putting their work onto a different server. End-of-session behaviors show the importance of visualization in producing useful data insights. Even the developer stated that he would not have considered examining end-of-session behavior until he saw the visualization.

Figure 3.6 is an example of flocking pattern discovery in a Twitter dataset[5] – a collection of tweets from the month of November 2012 containing both hashtags and URLs. The dataset includes 610,000 tweets from 410,000 users over 3 days. The first pattern is a cluster of Android game-related hashtags. "Game Insight," an Android game development shop, had just recently opened a San Francisco location and was about to open offices in Russia. The next pattern is generated from the hashtag "Sandy," which was isolated from the rest of the "boids" – Superstorm Sandy was just beginning to dissipate at that point.

## Pheromones in Social Media

Social media collects digital pheromones in the form of posts, messages, photos, videos, events, and "likes" from users. These digital pheromones often form their own networks that are used for spreading posts, recruiting new "friends", creating user profiles, and streaming personalized commercial advertisements.

[5]Web Science 2114 Data Challenge: http://cnets.indiana.edu/groups/nan/webtraffic/websci14-data/

**Fig. 3.7** Superimposed images of Facebook posts from six users. The *top row*: an artist who often shared TryLife text posts (*left*), an artistic photographer who often shared black and white photos (*middle*), and a modeling page that shared images of human figures (*right*); The *bottom row*: a high school soccer player page (*left*), a college student's page of selfies and group photos (*middle*), and a college student's page with images of flowers (*right*)

## *Visual Pheromones*

Social media provides platforms for users to post visual content in chronological order. People are attracted to visual media because a picture is worth a thousand words and it only takes a fraction of a second to view a picture or click a share button to repost it. Some pictures have gone viral online because many viewers share similar values and tastes. A sequence of visual posts may reveal a user's style, theme, and mood. Here let us collect 20–40 images of six Facebook users over a period of 4 weeks. If we use the "light painting" method to superimpose those images, they would be very blurry, but one could tell the differences between users' styles and tastes, and even content.

Figure 3.7 shows visual pheromones of superimposed images of Facebook posts from six users. Subject A is an artist who liked to share posts from TryLife, which normally consist of text with a black background and a logo always located at the right bottom of the image. Subject B is a photographer who preferred classic black and white photos. Her superimposed image shows dark gray tones. Subject C is a video producer who often posts figurative studio scenes from videos. The superimposed image contains a lot of pinkish skin tones. Subject D is a high school soccer player who took many outdoor photos. One can see green tones all over the superimposed image and some hints of players in the center. Subject E is a college student who took many selfies and group photos. You can see the facial skin tones in the areas near the center. Finally, subject F is a college student who took photos of flowers and outdoor sceneries. The superimposed image shows flower colors.

Humans are visual animals. We make and follow visual pheromones in the real world and the digital world. Social media allows us to perceive visual pheromones across social networks. Studies show that the amalgamation of all of the images from the Internet superimposed on top of one another is orange! This indicates that people prefer to post more images of humans and warmly lit images than anything else. The "Light Painting" method allows us to have a rough impression of the "average" color distribution of visual posts. In order to have a better articulation of visual pheromones, we need to apply computer vision algorithms to recognize words, objects, and moods. For example, Optical Character Recognition (OCR) can be used for recognizing text in the post. Dynamic Time Warping (DTW) can be used for recognizing human gesture and expressions, and so on. Such "metadata" in the image can be extracted automatically, manually, and derived from the caption of the image. Social media like Facebook has already worked on developing these approaches, particularly by using the "tagging" function to encourage users to manually enter the names of users, location, and any other relevant information.

## *Psycholinguistic Pheromones*

In addition to visual content, text in social media also reveals characteristics of online users. The Linguistic Inquiry and Word Count (LIWC) program is a text analysis tool that measures word usage based on functional and emotional dimensions as well as ninety other linguistic dimensions. Traditional LIWC sorts out words based on a list of categories of social, psychological, and contextual significance. For example, "party" is a *Social Word*. "Courageous" and "calm" are *Positive Emotions*. "Hell" and "damn" are *Negative Emotions*. Furthermore, "horny" and "porn" are *Sexual Words*. The LIWC program records the percentage of words recognized in each document and displays the percentage of words that match the word stems of particular themes. The following example is a Facebook post from a woman who talked about her life.

> *Wow I'm so upset right now . . . it feels like my life is ruined . . . just plain over . . . I want to give up so badly. People already hate me and think the worst of me hell my own family doesn't even talk to me. They won't even beep if they see me walking down the street. I didn't mess up my life and abuse drugs to hurt or harm anyone I truly messed up. But does anyone ever think of how hurt I am. I'm not even ashamed to say publically on fb that I can't stop crying every night when my head hits the pillow because of who I am today. I am so alone its crazy and I have no one. But do I even deserve someone to care for me anymore?? I feel pathetic and I can't believe how hated I've become. I truly wish my family wasn't so distant from me . . . I miss them, man . . . Everyday I think of my little brother and sister my mom and my [grandma] nunie and it immediately brings tears to my eyes. Christmas used to be my favorite holiday now I'm scared its almost here. :(*

The LIWC analyzes the text above and tells us the word counts in 90 categories. First, the LIWC produces summary category scores:[6] Analytical Thinking (9.80), Clout (11.75), Authentic (97.85), and Tone (1). The analytical thinking dimension captures the degree to which people use words that suggest formal, logical, and hierarchical thinking patterns. It is a variable derived from eight functional word dimensions. In this case, the analytical thinking level is rather low.[7] Clout refers to confidence, leadership or social status that people display through their wording. The algorithm is based on the results from a series of studies of human interactions.[8] In our case, the words in the text indicate that the writer is pretty humble and powerless. Authenticity reveals the level of honesty in the writing. When people reveal themselves in an authentic way, they are more personal, humble, and vulnerable. The algorithm was derived from a series of studies of deception behaviors.[9,10] Here, we can see that the text is very authentic and honest. Emotional Tone summarizes the positive and negative emotional dimensions in one summary variable. The algorithm is built so that the higher the number, the more positive the tone. According to the LIWC, numbers below 50 suggest a more negative emotional tone. In this case, the emotional tone score 1 suggests that the emotional tone is extremely negative. The LIWC also displays the details of other variables such as number of pronouns (23); instances of "I" (16), "we" (0), "you" (0), "she"/"he" (0) and "they" (1); adverbs (10), affection (13.37), anger (1.98), sad (3.96), social (8.91), family (2.8), friends (0), cognitive process (13.37), focus on present (14.36), focus on future (0), and so on, for up to 90 dimensions.

The LIWC sorts words into multiple categories. For example, the word "ruined" in the post is put into the categories of affection, negative emotion, and sadness. The word "because" is classified as a function word, conjunction, cognitive process, and cause. The emoticon, :( is recognized as an affective word, negative emotion, informal expression, and network speaking style.

Linguistic dimensions also have psychological meanings. For example, personal pronouns[11] are relevant to personality, depression, even deception.[12] When someone

---

[6]LIWC, Interpreting LIWC Output. http://liwc.wpengine.com/interpreting-liwc-output/

[7]Pennebaker JW, Chung CK, Frazee J, Lavergne GM, and Beaver DI (2014) When small words foretell academic success: The case of college admissions essays. PLoS ONE 9(12): e115844. doi: 10.1371/journal.pone.0115844.

[8]Kacewicz W, Pennebaker JW, Davis M, Jeon M and Graesser AC (2013) Pronoun use reflects standings in social hierarchies. Journal of Language and Social Psychology. online version 19 September 2013, DOI: 10.1177/0261927X1350265.

[9]Newman ML, Pennebaker JW, Berry DS and Richards JM (2003) Lying words: Predicting deception from linguistic styles. Personality and Social Psychology Bulletin, 29, 665–675.

[10]Pennebaker JW (2011) The Secret Life of Pronouns: What Our Words Say About Us (NY: Bloomsbury).

[11]Tausczik YR, Pennebaker JW (2010) The psychological meaning of words: LIWC and computerized text analysis methods. Journal of Language and Social Psychology, 29, 24–54.

[12]Pennebaker JW (2011) Your Use of Pronouns Reveals Your Personality. https://hbr.org/2011/12/your-use-of-pronouns-reveals-your-personality

says, "I don't think I buy it," it reveals something about themselves in that statement. Pronouns tell us where people focus their attention. If someone uses the pronoun "I," it is a sign of self-focus. The "I think" may appear insignificant, but it is rather meaningful subconsciously. It indicates that the person is more focused on him or herself. In addition, depressed people often use the word "I" more often than emotionally stable people. People who are lower in status use "I" more frequently. Studies also show that a person who is lying tends to use "we" more, or to use sentences without a first person pronoun at all.

*Can words reveal a writer's personality*? Personality testing originated in the nineteenth century. Sir Francis Galton estimated the number of adjectives that described personality in the English dictionary. Thurstone refined Galton's list to 60 words that were commonly used for describing personality at the time.[13] Through factor analysis of responses from 1300 participants, Thurstone was able to eventually reduce these 60 adjectives into five common factors.

The Five Factor Model (FFM) or "Big Five" personality traits include *Openness*, *Conscientiousness*, *Extraversion*, *Agreeableness*, and *Neuroticism*. The acronym commonly used to refer to the five traits collectively is OCEAN. *Openness* reflects the nature of intellectual curiosity, creativity, novelty, or open-mindedness for new experiences versus closed-mindedness or pragmatism. *Conscientiousness* reflects the degree of organization, self-discipline, and stubbornness versus an easy-going nature, carelessness, and unreliability. *Extraversion* reflects sociability, energy, and an outgoing nature versus a reserved and solitary nature. *Agreeableness* reflects ability to trust, friendliness and compassion versus analytical and detached tendencies. *Neuroticism* reflects sensitivity, nervousness, and moodiness versus security, calmness, and confidence.

Statistical studies show that there are strong correlations between the LIWC categories and a writer's personality. Traditional personality tests are based mainly on self-reporting questionnaires. Rapidly-growing social media facilitates many more data sources for mining the correlation between the LIWC and personality. Let us consider an example of a large-scale analysis of personality and word use among 694 bloggers in 100,000 words.[14] The Spearman Rank coefficients indicate significant correlations ($p < 0.001$) among words and the Big Five personality traits. For example, the total number of pronouns in the LIWC category is negatively correlated to Openness (−0.21). The word "we" is positively correlated to Agreeableness (0.18). The word "I" is negatively correlated to Openness (−0.19). The word "you" is positively correlated to Extraversion (0.16).

Emotional words often raise flags for various personality traits. For example, positive emotion words correlate to Agreeableness (0.18) but negatively correlate

[13]Goldberg LR (1993) The structure of phenotypic personality traits. American psychology, 48(1), 26–34.

[14]Yarkoni T (2010) Personality in 100,000 words: A large-scale analysis of personality and word use among bloggers. Journal of Research in Personality, 44(3): 363–373, 2010

to Openness (−0.15). Optimism correlates to Agreeableness (0.15) and Conscientiousness (0.16). Negative emotions, on the other hand, correlate to Neuroticism (0.16), and negatively correlate to Agreeableness (−0.15) and Conscientiousness (−0.18). Anxiety correlates to Neuroticism (0.17), and anger negatively correlates to Agreeableness (−0.23) and Conscientiousness (−0.19).

Social and physical context in words also correlate to personality. For example, sexual expression in social media has a correlation to Extraversion (0.17). Social process correlates to Extraversion (0.15) but negatively correlates to Openness (−0.14). Friends correlate to Extraversion (0.15). Family correlates to Agreeableness (0.19) but negatively correlates to Openness (−0.17). Home correlates to Agreeableness (0.19). Time and motion have negative correlations to Openness (−0.22), but motion has a positive correlation to Agreeableness (0.14). Surprisingly, money and finance do not have significant correlations to personality, except some negative correlation to Agreeableness (−0.11)!

The largest psycholinguistic study about personality contains 700 million words, phrases, and topic instances collected from the Facebook messages of 75,000 volunteers, who also took standard personality tests.[15] This study attempts to find connections that are not captured within traditional closed-vocabulary word-category analyses such as LIWC. It adds new words to categories automatically after data-driven work-category clustering based on the Latent Dirichlet Allocation (LDA)[16] method. It uses the Least Squares linear regression method to find the correlation between each language feature and each psychometric outcome.

Facebook adopted the Big Five Model for profiling the personality of users based on collected posts and messages. Facebook also allows users to use the Five Labs Personality App inside Facebook to profile themselves. Figure 3.8 shows my

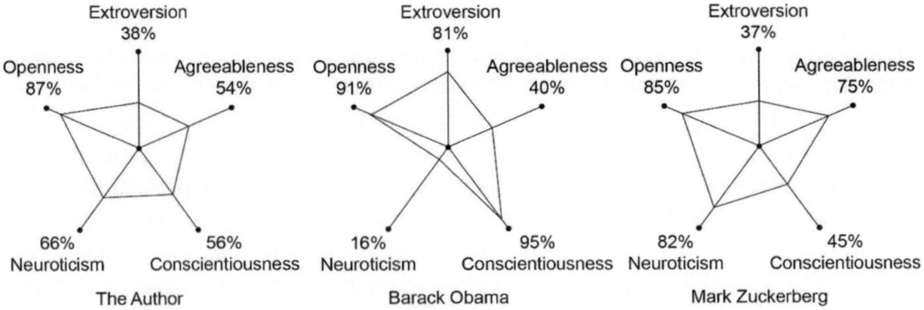


**Fig. 3.8** Personality test based on Facebook posts

[15]Schwartz HA, Eichstaedt JC, Kern ML, Dziurzynski L, Ramones SM, Agrawal M et al. (2013) Personality, Gender, and Age in the Language of Social Media: The Open-Vocabulary Approach. PLoS ONE 8(9): e73791. doi:10.1371/journal.pone.0073791

[16]Blei DM, Ng AY, Jordan MI (2003) Latent Dirichlet allocation. J. of Machine Learning Research 3:993–1022

personality profiling results based on my Facebook posts since 2013. The output uses a 5-point star shape to illustrate the score in each dimension. The figure also displays the profiles of some celebrities.

The Facebook personality test is one example of psycholinguistic pheromone modeling. The results are impressive, yet alarming. We learn that "Big Data" can help social media profile any user who has a substantial number of posts online. Social media can target commercial advertisements based on the profiles of potential customers, including personality, age, and gender. However, a large sample size does not imply a full understanding of textual phrases. The algorithm must also understand context in order to categorize words properly. Take this Facebook post for example:

> *I am so bad at living in this world. my damn wallet got stolen the other day with my last $22 in it. [I] lost my food stamps and damn photo ID. [I] can't take greyhound or apply for my passport (which was going to be my b-day present to myself this year.) now i've got to start all over again. Damn, damn, damn. I hate everything sometimes. FTW (Fuck the World).*

According to the LIWC 2015, the phrase "food stamps" is treated as two individual words, "food" and "stamps," which have no connection to anxiety, affection, well-being, or financial status. Instead, the LIWC classifies "food" simply as bio and ingest. The phrase "my last $22" is an outcry expression that is significant to anxiety, emotional tone, stress, and drives. Unfortunately, the LIWC 2015 does not pick up any connections to these categories. The word "lost" in this text is simply a verb but here the LIWC treats it as an emotional word: affective, negative emotion, sad, drives, achieve and focus-on-past. The phrase "photo ID" is also a critical object. Without it, the woman who posted could not take a Greyhound bus or apply for her passport. Again, the LIWC shows indifference toward the contextual objects, losing important meaning in the process.

Functional phrases play a more important role than individual words. For example, neurotic people disproportionally use the phrase "sick of". In many cases such two- or three-word phrases have a holistic meaning rather than simply a sum of the individual words. Linguist David Kaufer believes that phrases constitute meanings and writing styles. For decades, he has compiled a dictionary of 100 million of phrases in 100 categories. The system, which includes a graphical interface, is called DocuScope.[17]

---

[17]Kaufer D et al. (2005) Textural genre analysis and identification. In Cai Y (ed) Ambient Intelligence for Scientific Discovery, LNAI 3345, 2005

## Alarm Pheromones

Creatures such as fish and bees release alert pheromones to warn others of danger.[18,19,20] Although the existence of human alarm pheromones is still debatable, there is no doubt that our instinct often makes us aware of dangerous situations. Twitter, for example, was inspired by dispatch calls to emergency services. It acts much like an alert pheromone system. Normally, Twitter posts are short but concise. Tweets broadcast what, where, and when an event occurs. In order to be brief, Twitter sets the count limit to 140 words per post. As a result, Twitter can broadcast messages in real-time and at long-range.

### *Depression Alert*

Aside from Twitter, other social media such as Facebook can also send alert messages to broader readers. However, detecting risky situations in social media is a non-trivial task. In 2011, a 25-year-old mother named LaShanda Armstrong drowned herself and three of her four children in a minivan by driving into the freezing Hudson River. Before she loaded her children into the vehicle, she left a chilling message on Facebook:

> *I'm sorry everyone. Forgive me please for what I'm gonna do . . . This is it!!!*[21]

Photos posted on Armstrong's pages on social media appeared to portray a happy family. However, behind the façade, she was devastatingly depressed. Is it possible to perceive any signs of depression or to broadcast alerts through social media? Recent data mining studies show that there are correlations between the LIWC categories of words and depression, which is a facet of neuroticism. The top ten correlations are negative emotion (15), anger (15), negation (13), discrepancy (13), tentative (13), anxiety or fear (12), cognitive process (12), causation (12), feeling (12), and swear words (11). Normally, depression does not occur overnight. Rather, it evolves over time and leaves traces in private diaries, posts to friends in social media, but rarely in public blogs or writings.

---

[18]Smith R (1992) Alarm signals in fishes. Rev Fish Biol Fish 2:33–63, 1992

[19]McClintock MK (1984) Estrous synchrony: modulation of ovarian cycle length by female pheromones. Physiological Behavior 32, 701–705

[20]Wyatt TD (2003) Pheromones and Animal Behavior: Communication by Smell and Taste. Cambridge: Cambridge University Press. ISBN 0521485266.

[21]Bates D, Duell M (2011) 'Please forgive me for what I'm gonna do': Mother's Facebook message before drowning three of her four children by driving them into Hudson. http://www.dailymail.co.uk/news/article-1376427/LaShanda-Armstrong-drives-3-children-Hudson-River.html

Let us refer back to the Facebook message in the last section. The girl lost her wallet and ran into a chain of unfortunate events. She was definitely in a bad mood – her emotional tone is very low (1). She had anger (9.59) and negative emotion (12.33). She was stressed in cognitive processing (8.22). She used swear words five times, contributing to the LIWC categories affective words, negative emotion, anger, informal, and swearing. The statistics suggest signs of depression, even though the text has only 67 words. To make a more accurate diagnosis, we need to look into more data and more channels, including the rhythm of the posting intervals and content of images and videos. So far, the available algorithms can only analyze low-level language features. There is still a long way to go before such an algorithm can understand the sentence Miss Armstrong posted on Facebook. If a Facebook friend or a machine had picked up the alert and taken a prompt action, perhaps this tragedy could have been avoided. However, there is the polemic and ethical issue of privacy involved. If a machine finds out that a person is in a deep depression or is suicidal, should the machine inform the person's family, friends, or dial 911?

## *Ad-Hoc Safety Check*

Ad-hoc alert services are automatic message sharing systems for emergency alerts. The message receivers and message contents can be dynamically configured. According to the CDC, an estimated 19.3 % of women in the US have been raped during their lifetimes.[22] Kitestring,[23] Nirbhaya, and bSafe[24] are services that aim to make sure people get from point A to point B safely, notifying their emergency contacts. bSafe requires the user to touch anything on the smartphone as a response; Nirbhaya needs the user to shake their smartphone as a response. Kitestring, on the other hand, does not require an action to alert people. Instead, the user can tell Kitestring that the user is in a dangerous place or situation and give it a time frame of when to check in again. If the user does not reply back when it checks status, it will alert the user's emergency contacts with a preset message. For example,

> *Hey, this is Samantha. I walked from the concert tonight, but if you get this, it means I might not have made it back safely. Give me a call?*

The automated service is necessary here because when someone is in danger, he or she would not have the time or reflexes to enter their password on the phone, type a text message, or even hit a special button to alert someone.

[22]La Jeunesse M (2015) Tiny device aims to help protect women from sexual assault. http://mashable.com/2015/11/03/athena-indiegogo/?utm_cid=mash-com-fb-life-link#O864D.TZTEq4

[23]Matthews N (2014) This New App Could've Prevented My Friend's Rape. http://www.elle.com/culture/tech/a10/kitestring-app-safety-apps-women/

[24]bSafe web site (2016) http://getbsafe.com/

## *Earthquake Early Warning Systems (EEWS)*

Earthquake Early Warning Alarm (EEWS) is a sensory and communication system that sends early warning signals to nearby areas before earthquake shockwaves arrive. It is based on the principle of physics that an electrical signal travels faster than an earthquake wave. The early warning message can be distributed via computers, smartphones, televisions, and special receivers. The time difference of 10 s or more allows people to escape disaster. The EEWS can also provide EEW services to high-speed railway networks, subways, chemical plants, nuclear plants, et al., to save lives and reduce secondary disasters.

Early Warning Time (EWT) and Early Warning Radius (EWR) can measure the efficacy of an alert pheromone system. EWT is the interval between the alert triggering and the disastrous moment. EWT should be long enough for the person to escape a dangerous situation. Simulation models show that if the early warning time were 3 s, then it would save 14 % of casualties and injuries; if the early warning time were 10 s, then it would save 39 % of casualties and injuries.

The world's largest EEWS was built by the Institute of Care Life (ICL) in Western China in 2012. This EEWS consists of 5000 seismic sensors deployed in twenty-three provinces across China, covering an area of 400,000 $km^2$. This is comparable to the EEWS in Japan that covers an area of 377,000 $km^2$. The early warning can be triggered when an earthquake is stronger than 3.0 on the Richter Magnitude Scale (RMS). However, it has its limitations: once an earthquake occurs, it takes 7 s for the ICL's EEWS to publish the full earthquake warning data. The radius of the blind zone is the most critical feature of an EEWS. The radius of the blind zone of the ICL's EEWS is about 25 km from an actual epicenter. If the area is beyond 100 km, then the damage is too small to worry about. Thus, the effective region of the EEWS is a radius of 25–100 km from an epicenter. To prevent disastrous consequences from occurring within that range, it is expected that electricity and natural gas can be cut off automatically. This is vital to the safety of high-speed trains, oil and gas pipelines, chemical plants, and nuclear reactors.

Since 2012, over 1200 earthquakes have occurred in Western China. Some of them were destructive quakes that triggered the ICL's EEWS with no false alerts, including an earthquake with a magnitude 4.9 in Yunnan province on February 19, 2013. It gave 73 s of EEW time for Chengdu, a large city nearby. A 6.5 RMS earthquake at Lu Tian County, Yunnan Province in China triggered the most recent successful alarm in 2014. The alarm provided 10 to 57 s warning to the nearby Shao Ton region and Kun Ming city respectively, including 26 schools in two provinces.

## Summary

Pheromones in nature are external memories projected to the physical world and shared within a group. The dynamics of depositing and vaporizing pheromones are sophisticated processes, inspiring algorithms for routing, optimization, searching, and navigation. The generalized pheromone update model can help us to discover motion patterns in video, which transforms the hidden patterns of moving objects into visible trails that accumulate or decay over time, much like a scent. The Motion Energy Image (MEI) is an envelope of motion pixels. It is a special case of pheromone updating that assumes no pheromone decay. The output is a binary image, either 0 or 1. On the other hand, the Motion History Image (MHI) represents the decay of motion pheromones over time.

Sequential patterns of pheromone deposition can be represented in a flocking behavioral model – a set of bird-like objects that can project approximate feature vector distances onto a visual space.

Social media collects digital pheromones in the form of posts, messages, photos, videos, events, and "likes" from users. These digital pheromones often form their own networks that are used for spreading posts, recruiting new "friends", creating user profiles, and streaming targeted commercial advertisements. A pheromone network, on the other hand, connects members with special interests, such as news, sports, hobbies, images, videos, music and politics. Pheromone networks are dynamic in nature. Like living organisms, they are born; they grow; they transform; they merge, and oftentimes; they die. This unique phenomenon fits Zipf's Law, which typically refers to the "size" of an occurrence of an event relative to its rank. Certain clever social insects may discover shorter paths on their way to finding food. Those "elite" members create network hubs to spread pheromones more efficiently. Following the trail of the elites attracts more visitors and more pheromones are deposited.

Social media allows us to perceive visual and textual pheromones across social networks. The "light painting" method allows us to have a rough impression of the "average" color distribution of a user's visual posts. Psycholinguistic analysis reveals the personality traits of an online user. The studies from large social media groups show correlations between Linguistic Inquiry and Word Count (LIWC) and personality, as well as emotional state.

Social media can also alert others of dangerous situations, such as depression and suicide alerts, ad-hoc safety checks, and Earthquake Early Warning Systems. When an alert is time-sensitive or life-threatening, automation of the event detection and alert dissemination is necessary.

CHAPTER 4

# Foraging Behaviors

Foraging is an instinctual behavior for living. It is a continuous activity to search surrounding areas for food either passively or actively. This requires the actions of hunting and gathering. Foraging is a broad search over an area in order to obtain something, whether it is ants looking for green leaves, bees scouting for flower pollen, or even people seeking specific information. Foraging describes how animals behave to satisfy their life-sustaining needs. In the cyberspace, foraging means to actively seek, gather, share, and consume information.

In this chapter, we elaborate on strategies exhibited by the foraging routines of animals and humans, from passive hunting to active and social foraging. These foraging algorithms are tightly connected with other instinctual algorithms, such as pheromone depositing, spontaneous alternation, and learning.

## Collective Foraging

Honeybees communicate the distance and location of resource patches through bee dances. It has been shown that dance activity increases with the quality of a food patch and decreases with the distance from it.[1] Ants use pheromones and "tandem walking" to guide partners during exploitive trips. Recruiting is essential for group hunting. Since worker ants are blind, they rely on tactile and olfactory sensation to forage.

### *Pheromone Paths*

Foraging ants do not want to get lost on their way home. Hence, whenever they travel, they deposit pheromones that help them to return to the nest after they have

[1] Seeley TD (1997) Honighienen im Mikrosmos des Bienenstocks. Basel: Birkhäuser.

© Springer-Verlag London 2016

Y. Cai, *Instinctive Computing*, DOI 10.1007/978-1-4471-7278-9_4

foraged. This helps ants return home quickly, and also enables other ants to reach food spots more efficiently. When an obstacle is encountered on a path, an ant must make the decision whether to go left or right or walk around the obstacle. See Fig. 4.1. The ant will most likely take the path where the intensity of the pheromone deposit is higher, indicating that more ants have followed this path before. The probability of a path chosen depends on the intensity of the pheromone deposit. If there were no pheromone trail ahead, then the chances for selecting one over two paths would be 50/50. The ant may choose either one randomly in this case. Shorter paths have a higher density of accumulation because ants can facilitate more efficient round trips.[2] As a result, the collective foraging trail with a higher concentration of pheromones normally indicates a shorter path. This is only a generalization of ant behavior in nature. Different ant species often vary in their exact use of pheromones.

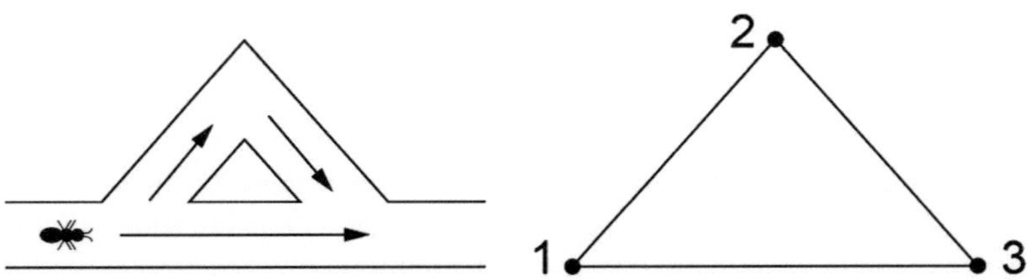


**Fig. 4.1** Ant foraging and ant colony optimization (ACO)

The ant's foraging path can be formulated as a graph problem that contains nodes and the links between them. The pseudo code of the Ant Colony Optimization algorithm is:

```
While (not termination)
   Generate routing solutions
   Update pheromone
End While
```

In order to make a move for the ant, we need to construct ant routing solutions based on the probability of the pheromone on the link, including the influence of the intensity of the pheromone and the desirability of the link (edge or segment of a path) – typically correlating to the distance of the link. The amount of pheromone is updated according to the amount of pheromone deposited on the given link and the rate of pheromone evaporation. Thus, if the ant has a sinus infection, or if the trail

[2]Eswarann (2012) Ant Colony Optimization. https://algorithmsinreallife.wordpress.com/2012/02/20/ant-colony-optimization/

is so old that the deposited pheromones have long since faded away, then the ant's foraging path would change accordingly.[3]

One major application of the ACO algorithm is the Traveling Salesman Problem (TSP) that assumes the starting point is the same as the ending point. This ensures that the ants take the shortest foraging route on their way home at the end of the work day, while they travel to all the food locations once and only once, not missing or repeating any food spot visits. Bonabeau, Dorigo, and Theraulaz's book goes into more detail about Ant Colony Optimization algorithms.[4]

## Quorum Recruiting

Foraging means exploration of new places to find more food or a new diet. In addition to pheromones, many biological organisms can sense and respond to their own *population density*. Bacteria use quorum sensing to coordinate gene expression. Social insects use quorum sensing to determine where to forage or nest. Let us again take ant foraging for example. Assume we have a colony of ants and two diets – red leaves and green leaves. Some ants take only red leaves or only green leaves; some can take both. Within a total of 5 trips, the green leaf area has been visited 4 times and the red leaf area has been visited 3 times. The area where the red and green leaves overlap has been visited 2 times. Table 4.1 shows the ants' foraging trips and food contents.

**Table 4.1** Ant's field trips and diets

| Trip order | Diet |
|---|---|
| 1 | Green |
| 2 | Green, Red |
| 3 | Green, Red |
| 4 | Green |
| 5 | Red |

What would the chances be for an ant to visit the common food area where it takes red leaves and then green leaves? There are two factors to consider: *frequency* and *confidence*. Frequency, or support, is the percentage of trips to the area. In the ant foraging case, frequency is equivalent to a pheromone deposit on the path. The more frequent the visits, the more pheromones are deposited, assuming pheromone decay is negligible in a short period. Confidence is the number of the visits to another dietary area divided by the frequency (support) of the original area – the

[3]Shekhawat A, Poddar P, Boswal D (2009) Ant colony Optimization Algorithms: Introduction and Beyond. http://mat.uab.cat/~alseda/MasterOpt/ACO_Intro.pdf

[4]Bonabeau E, Dorigo M and Theraulaz G (1999) Swarm Intelligence: From Natural to Artificial Systems. Oxford University Press

percentage of foragers that actually choose to that combo diet. For ants, they prefer to follow the trails that have high pheromone deposits because they are the most frequently visited places. When a young ant tries to learn foraging patterns, it would be told to follow the path with minimal pheromone deposits, say, at least 5 % of the time. As the young ant forages, it gains more confidence when it finds the pheromone deposit intensity increasing as it moves from a lower population density area to an area of higher population density.

In our case, the frequencies for the foraging dietary areas: red = 3/5 = 60 %, green = 4/5 = 80 %, both red and green = 2/5 = 40 %. The confidence for the ant to take red and then also green = 40 % ÷ 60 % = 0.67, and the confidence for the ant to take green and then also red = 40 % ÷ 80 % = 0.5. Thus, the ant that takes red leaves and also likely would take green leaves with a confidence of 0.67. Consequently, the "rule" or the foraging paths are not symmetrical. The confidence of the rule that ants who take green leaves and also take red leaves is only 0.5. The pheromone deposit along that direction decreases, so the confidence level decreases as well. Figure 4.2 shows an illustration of the foraging map.

Similar to the recruiting behavior in swarm intelligence of social insects, online shoppers leave their digital pheromones to be used to attract other shoppers. Items previously purchased, viewed, or placed in an electronic shopping cart by the user can be stored in a database of an online merchant, such as Amazon. A computer-assisted system generates recommendations using a previously generated table that maps product items to respective lists of similar items. Personal recommendations can be generated by accessing the table to identify items similar to those previously selected by the user, for example, the current contents of a user's shopping cart.[5] I recently visited Amazon.com to order the autobiographical book "Surely You Are Joking, Mr. Feynman!" by Richard P. Feynman. The bottom of the web page displayed its recommendation message: "Frequently Bought Together." The top one was the sci-fi novel "Snow Crash" by Neal Stephenson. Both books share the

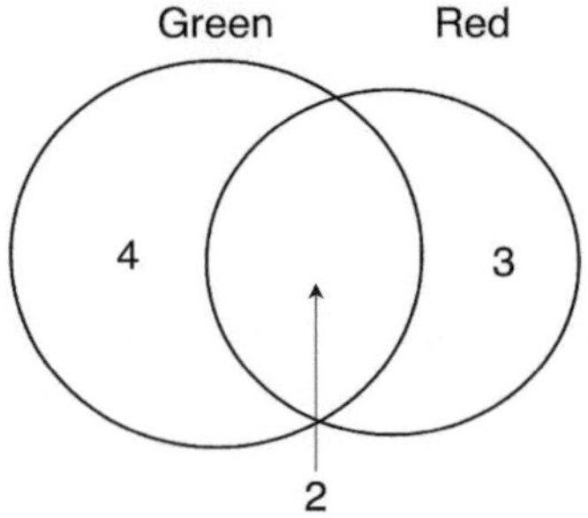


**Fig. 4.2** Ant foraging map of trips and diets

[5]Jacob JA, Benson EA, Linden GD, McKenna P, Chen P (2013) Method and system to connect consumers to information. US Patent, US8468050 B2

common theme of science. How the two were listed together is actually no mystery. The Google co-founder Sergey Brin told the magazine *Business Insider* that these two books changed his life. The article was widely circulated on social media.[6] People who read the article would likely order both books on Amazon, sending the quorum signal to potential readers.

Quorum recruiting behavior can be formulated as an Association Learning algorithm. It is a classic machine-learning model that is aimed at discovering rules from multiple datasets containing common elements – in our case, the common items in shopping carts. The pseudo code of the algorithm is as follows:

```
Generate a list of frequently viewed items
Filter out the items that have less than a minimal popularity level
Generate a list of possible rules based on frequently viewed items
Sort the rules by confidence level
Filter out rules that are below the minimum confidence value
```

The possible association rules can increase exponentially when the number of items in the shopping cart increases. Imagine the number of rules if we have a list of ten items. It would be enormous. We can use an *Apriori* algorithm to reduce the number of rules that we need to test. The principle of *Apriori* is surprisingly simple – if the rule does not meet the minimum confidence requirement, then the subsets of that rule will also not meet the minimum confidence requirement. Going back to the example of ant leaf foraging, assume that the rule Green → Red does not meet the minimum confidence requirement, so any rules within the subset of that rule will also not meet the minimum confidence requirement. This is often true in the foraging practice of social insects whenever they encounter a pheromone deposit level that is less than the minimum requirement. They will not follow that path, or its branches, saving their energy for the promising paths.

## *Tactile Communication*

As we mentioned earlier, worker ants are blind and rely on an olfactory and tactile antenna. Tactile sensing is a way for insects to communicate with each other over a very short range to avoid traffic jams on a narrow path. When a group of worker ants carries a heavy leaf to their home, the ants travelling in the opposite direction will stop and yield their path. By contacting each other in traffic, the neighborhood ants exchange information about their movements and traffic conditions. As a result, a worker ant will decide its next move: Go, Stop, Yield, or Retreat. This metaphor can be applied to traffic signaling in the modern era. There is a standardized protocol in America for wireless Vehicle-to-Vehicle (V2V) communication by broadcasting real-time messages, including short range within 100 m. For short range messages,

---

[6]Feloni R (2015) Google cofounder Sergey Brin says these 2 books changed his life. http://www.businessinsider.com/google-cofounder-sergey-brins-favorite-books-2015-7

the metadata format will be consistent with the V2V protocol SAE J2735 for the DSRC Message Set Dictionary with basic safety messages such as location, direction, and speed.[7] The data set will include messages transmitted from vehicles that are defined in the SAE J2735-200911 Basic Safety Message. These messages are transmitted every 100 ms by each vehicle to make sure the brakes have enough time to stop the vehicle.

The Basic Safety Messages will, at the very least, contain this set of data elements: message information, including temporary message identification and message count; position information, including latitude, longitude, elevation, and position accuracy; motion information, including speed, heading, angle, and four-way acceleration rate; and vehicle basics such as vehicle size. For example, a family van broadcasts its message to neighboring vehicles: "Hello, I am a middle-sized vehicle traveling at 60 mph, heading north, and I am not pressing down on the brakes..." In an ideal world, the short range V2V communication might eliminate the need for traffic lights. However, V2V is not a foolproof solution. Some vehicles may not be equipped with wireless devices, and pedestrians or animals would not wear the devices needed for detection. Possible remedies include autonomous driving, or combining V2V with an infrastructure to vehicle (I2V) communication protocol. For example, highway surveillance cameras would detect vehicles, pedestrians, and other objects at the intersection, and broadcast a short message to vehicles near that intersection.[8]

## Serendipitous Foraging

The abundance of data and spare time often allow us to forage information serendipitously. Like the behavior of beavers who build dams for collecting fish in a river, valuable information can be extracted from intercepting or fishing, which often leads to free or low-cost intelligence collection.

### *Crawlers*

Crawlers are rudimentary search agents that explore the World Wide Web. Like a spider, a crawler explores every link it can find on a web page and stores the data in its database. They are essentially an Internet robot assigned the task of opening all hyperlinks presented on a webpage and creating a database of text and metadata from all the links. A typical crawler starts with an initial set of links to visit. As soon as it visits those links, it adds new links to the existing list of URLs to visit. As the

[7] SAE (2011). US Department of Transportation. SAE J2735 – Dedicated Short Range Communications (DSRC) Message Set Dictionary.

[8] Cai Y, Bunn A, Liang P and Yang B (2013). Adaptive feature annotation for large video sensor networks, Journal of Electronic Imaging 22(4), Oct–Dec. 2013

crawler traverses through the links, it downloads information from those web pages to be viewed later in form of snapshots.[9]

Downloading webpages requires a lot of data bandwidth and it is expensive to the search engine owner and server. In order to make it computationally feasible, crawlers typically follow common policies. For example, a crawler will focus on downloading only the most relevant content of a web page rather than some unimportant data. A crawler schedules the time when it should revisit a web page to update its database. Crawlers often work in parallel with one another, and they may download the same page, so the crawler maintains communication with others to eliminate any chance of duplication. Amazingly, in order to avoid increasing the load on the webserver hosting the website, a "polite crawler" must wait a few seconds after it downloads some data from a webserver (Fig. 4.3).

Web crawlers can not only download data and images for search indexes, but can also download audio and video files. In the past, people used a VCR (video cassette recorder) to record video from a TV channel. One would turn on the video recorder, select a channel, push the "Record" button, stop it after some time, and after rewinding the tape, one could play the recorded program on the screen. In the age of digital media, we can use media crawlers to capture video data. RTPDump is an open source software program for RTMP (Real Time Messaging Protocol) audio, video, and data streams.[10] It is used to connect to RTMP servers like a normal video player client and capture online TV or on-demand video streams from the network to save them as files. For example, there are over 41 cameras along the I-495 expressway in Virginia. Each camera has an ID, (e.g. S-48.8-1). The traffic management center streams real-time videos online

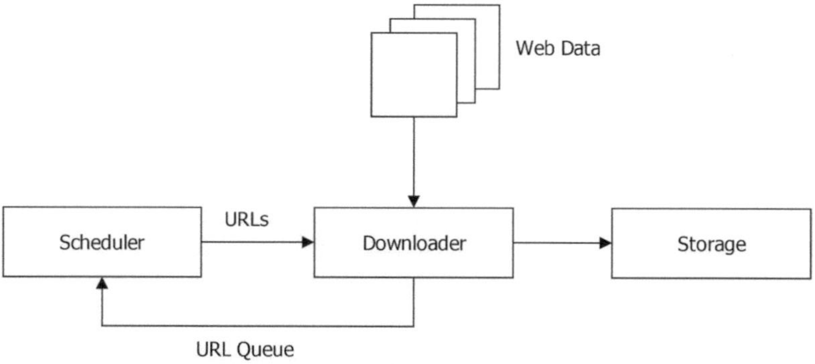


**Fig. 4.3** A spider crawls through the web to forage for food

[9]Tiwari A (2015) How Search Engine Works and Makes Your Life Easier? http://fossbytes.com/search-engine-works-makes-life-easier/

[10]RTMPDump. WikiPedia. https://en.wikipedia.org/wiki/RTMPDump

with an RTMP server. In negotiating a connection, an RTMP client sends and receives a data stream containing elements in command lines. Below is a pseudo code for crawling and downloading the video streams from online traffic cameras.

```
Set recording time
Find video server address
Add camera IDs to the crawling list
Open processors for all cameras
Execute all the recording processors until the session ends
```

RTMPDump has been widely used in academic research on video streaming. However, it can also be used to infringe copyrights due to its ability to decrypt both Encrypted RTMP and Secure RTMP Digital Rights Management technologies. Like other software tools, crawlers can be adversarial too.

## *Interception*

Interception uses sensors set up along a data stream in order to quietly listen to related information. Open-air wireless communication creates opportunities for signal interception, or in some cases, free sensors. For example, many vehicle drivers wear Bluetooth wireless headsets to connect to a mobile phone while driving. The wireless signal strength changes as the vehicle approaches or drives away from the receiver along a highway. By setting up a Bluetooth signal receiver near a highway, we can passively measure the approximate speed of the traffic flow on the highway. This is a low-cost measurement alternative to using expensive radar sensors. Wireless signals can be intercepted for nefarious purposes as well.[11]

In the digital world, foraging is about valuable information such as keywords, relationships, timing, places, events, and changes. Online stores look into customer's shopping carts, browsing histories, and wish lists. Social media looks into users' posts, messages, friendships, relationship statuses, ages, events, and changes. It is all about matching related links to the customer's interests.

The simplest algorithm perhaps is keyword matching. The system scans the keywords in a user's contents and tries to match the keywords with their synonyms in their product database. Like when a shark smells blood, the predator follows the trace of the prey quietly. For example, when I chatted with a friend on Facebook about "black gaffer's tape" for an art project in the comments of a post, Facebook immediately followed up with ads for "black duct tape" on the right hand side commercial column and in the main post area as a "sponsored post" with an image of the black tape in order to catch my attention. As we know, duct tape and gaffer's

[11]Dwulf (2015) Hackers can use RFID readers to steal payment card numbers while you are in public. http://www.techworm.net/2015/02/hackers-can-use-rfid-readers-steal-payment-card-numbers-public.html

tape are two different types of utility tape, and have subtle differences in attributes such as stickiness, texture, and strength.

How do we measure similarities between the intercepted product attributes and the attributes of a product in a database? Given two attribute vectors $A = \{a_1, a_2, \ldots, a_n\}$ and $B = \{b_1, b_2, \ldots, b_n\}$, we can measure the similarity using the cosine angle cos($\theta$) between the two vectors. It is equal to their dot product divided by the product of their norms. The dot product is just the sum of element-by-element multiplication. The norm is the magnitude of a vector.

$$c(A, B) = \frac{A^T B}{|A| \cdot |B|} \tag{4.1}$$

Since the angle $\theta$ is in the range of $[0, \pi]$, the resulting similarity will yield a value of one of the following: $+1$ meaning exactly the same, $-1$ as meaning exactly the opposite, 0 meaning independent, and any values in-between indicating intermediate similarity or dissimilarity. The cosine angle distance is $d_c(A, B) = 1 - \cos\theta$. Amazon in fact filed a patent for the consumer product recommendation algorithm that uses Cosine Angle for the similarity measurement.[12] This was discussed in the section "Quorum recruiting".

In case our feature vectors have binary values, we may use the Jaccard similarity coefficient. Given two feature vectors, A and B, each with $n$ binary attributes, the Jaccard coefficient is a useful measure of the overlap that A and B share. Each attribute of A and B can either be 0 or 1. The total number of each combination of attributes for both A and B are specified as follows: $M_{11}$ is the number of attributes where A and B both have a value of 1, $M_{01}$ is the number of attributes where the attribute of A equals 0 and the attribute of B is 1. $M_{10}$ is the number of attributes where the attribute of A equals 1 and the attribute of B is 0. The Jaccard similarity coefficient is given as:

$$J(A, B) = \frac{M_{11}}{M_{01} + M_{10} + M_{11}} \tag{4.2}$$

The Jaccard distance is $d_j(A, B) = 1 - J(A, B)$.

## Fishing

Fishing is a metaphor for obtaining valuable information using bait. Many department stores offer "advantage cards" with the bait of discounts and gasoline perks in exchange for the customer's personal information such as

[12] Jacob JA, Benson EA, Linden GD, McKenna P, Chen P (2013) Method and system to connect consumers to information. US Patent, US8468050 B2

home address and shopping records, which enable the supermarket to send targeted information to customers, mine spatiotemporal data, and predict consumer demand.

Phishing has been often associated with malicious emails that try to obtain a user's sensitive information such as usernames, passwords, credit card details, and sometimes, even money. Below is an example of a typical phishing message:

> From: "Blackboard" <en-notification@blackboard.com>
> Subject: E-mail Account Update
> Date: August 31, 2015 9:22:30 AM EDT
> To: Recipients <en-notification@blackboard.com>
>
> Dear [Member],
>
> This email has been sent to you because you need to update your E-mail Account within the next 48 hrs.To access your E-mail Account.
>
> Click here to re-login
>
> Update Your Profile Account
>
> Thanks For Your Co-operation.
> Blackboard | Technology Services.

To convince the receiver the message is legitimate, a phishing email normally uses address *spoofing*, in which one person or program pretends to be another by falsifying data and thereby gaining illegitimate advantage. The sender information shown in emails (the "From" field) can be easily spoofed. This technique is commonly used by spammers to hide the origins of their emails, and can lead to problems such as misdirected web links. In the case above, Blackboard.com is a legitimate, popular website for college students and faculty. Although the phishing message appears to come from Blackboard.com, the original headers indicate that the email was sent from another organization and that the "From" address had been spoofed. Sloppy phishing emails are rather easy to detect, for example, UNITED NATIONS <mr.jameshbeljr51@gmail.com>. Many users would simply look at the highlighted subject line, "United Nations," rather than check the full address of the sender. This is especially true when a user browses emails on a mobile phone where full email addresses are often hidden. The "payload" line in the message stating "Click here to re-login" would take the user to an external site not affiliated with Blackboard, where a malicious entity has established a web form to collect user credentials. The email may have appeared legitimate, but a detailed look at it indicates that it is not.

## *Mechanical Turk*

Mechanical Turk, on the other hand, represents the bright side of fishing. It is to attract online users to complete so-called "Human Intelligence Tasks (HIT)," in which humans out-perform machines (e.g. pattern recognition). In order to teach a computer to recognize objects, one must manually annotate massive training samples frame by frame, which is labor intensive, time-consuming, and expensive. Because of this, scientists are foraging for free or low-cost human intelligence resources online. This originated with CAPTCHA, a program that protects websites against bots by generating and grading tests that humans can pass but current computer programs cannot. For example, humans can read distorted text, but current computer programs cannot. In order to prevent spam emails, CAPTCHA generates distorted text or numbers to check if input is indeed from a human. The original distorted images are generated by an algorithm that rotates and bends random text and numbers. The current algorithm, RECAPTCHA, goes a step further, displaying text or numbers from digitized books or street names and numbers from Google street car cameras. The idea is to harvest users' pattern recognition knowledge to help machine vision algorithms *at no cost*. This concept has been implemented in the cloud at Amazon which has established a hub called Amazon Mechanical Turk. As a Mechanical Turk worker, one can work from home, choose their own work hours, and get paid for doing something useful for a research, analysis, or development project. Payment for these tasks is small – usually between zero and 0.20 USD.

Behind every machine-learning algorithm, there is human intelligence. Coding human intelligence into the machine has proven difficult and expensive. Decades ago, this process was called Knowledge Engineering, which was mainly used in semantic coding. It had not become affordable until the booming of social media, online shopping, and cloud-based computing. The strategy of a Turk is to divide and conquer a complex task into simple mechanical steps so that anyone with minimal training can perform them for free or for a few cents.

Taking "Image Regions with Text Snippets" HIT as an example, the task was created by the requester Visual Genome. This HIT requires a qualification test. Time is allocated from 5 min to 24 h. The award level is $0.20 per HIT. The HIT creator wanted to collect human knowledge of the text descriptions of visual objects in images. In order to enable machine vision algorithms to learn how to recognize objects, the developers must manually annotate or label objects (e.g. a car, a person, or a box) in an image with detailed descriptions. The "Image Regions with Text Snippets" HIT is actually a small script that enables the Turk worker to draw a bounding box around each object in an image and write 4–10 words in a text input box to describe the objects. Imagine we have a color photo of a single elephant drinking from a puddle in the middle of a forest. The annotated parts could be: "an adult elephant in a puddle, the ear of an adult elephant, the tusk of an adult elephant, the tail of an adult elephant, the legs of an adult elephant, the eyes of an adult elephant, the ears of an adult elephant, a still puddle without any ripples, a small shrub in the middle of the puddle, a blue sky on a clear day." Essentially, anything can be written into the descriptions, and Turk workers must be creative. Typically, a worker can do the HIT in a little under 2 min. The pseudo code of the HIT is:

```
If the worker draws a box around an object, then assign an object ID and
increase the box counter, and let the worker fill the description for
the object box;

If the worker presses button + or -, then display the next object box ID
and descriptions, or the previous ones;

If the Submission button is pressed, then check errors such as missing
descriptions or less than n boxes.
```

After the Submission button is pressed, the HIT script takes the data and stores it in its database in the cloud of Amazon Mechanical Turk, in the format of: image ID, object ID, location coordinates of the box, and text description of the box. Given the Mechanical Turk results, a machine learning algorithm can build the model for a specific object, such as the tusk of an adult elephant. The system is able to search for all the boxes containing those keywords from all images in the database. This is a tremendous time-saver for research teams because typically, college graduate or post-doctoral students have to conduct manual object annotations. In order to ensure the quality of the mechanical Turk, most of HITs provide examples and training sessions. The popular social media website Reddit created a special subgroup for Amazon Mechanical Turk under the discussion group Artificial Intelligence,[13] which transformed the Mechanical Turk into a social foraging activity.

Human vision is the most sophisticated sensing and computing system. In order to harvest human visual intelligence, the HIT foraging algorithms themselves have become more and more complicated. Many of them need GUIs and sophisticated interaction design. For example, the HIT "Trace Contour of Object" created by Vocds asks Turk workers to use a mouse to trace the boundary of an object.[14] This HIT is valuable for object recognition for complex shapes such as chairs in daily life. How to segment a chair from its background often poses a great challenge to machine vision, because of obscurity and viewpoint. The instruction says: "Please draw the shape around regions that have a single chair. Make the region as tight as possible. Your shape should closely follow the object boundary. Draw multiple regions within the visible parts of objects that are disjointed." The instruction also gives good examples and bad examples, and detailed drawing steps. In this case, a set of graphical editing and navigation utilities is necessary, such as undo, redo, close, delete, move to target, zoom in, zoom out, reset zoom, next. After the Turk worker presses the Submission button, the HIT is saved to the database in the format

[13]NhvK (2015) Any tips for completing "Decribe image regions with text snippets" visual genome hits? reddit.com. https://www.reddit.com/r/mturk/comments/3gd249/any_tips_for_completing_describe_image_regions/

[14]Amazon Mechanical Turk (2016) https://www.mturk.com/mturk/preview?groupId=355YH9BKNIVAGO2MR6LXHR30C01FVF

of image ID, polygon ID, and coordinates of the polygon. Because the HIT assumes that there is only one chair or no chair in any given image, the analysis software can easily identify the human-traced chairs from images. In this process, it also removes all background pixels. The tightly-traced contours or multiple polygons of a chair can then be inputted into the machine-learning algorithm as training samples. The HIT indeed can increase the machine vision accuracy because it helps to avoid background noise frequently found in training samples.

## Scanning and Tracking

Given limited resources, time, and attention, a foraging problem in general is to maximize gains of valuable intelligence per unit cost.[15] For all foragers, including humans and animals, it is a tradeoff between scanning (exploring) and tracking (exploiting). Scanning in breadth is an activity whereby a forager broadly examines a wide variety of information that may or may not be relevant to the goal. The benefit to scanning is that when someone considers a broad variety and volume of data, there is less chance of missing "nutritious" intelligence.[16] In contrast, tracking is an in-depth activity involving zooming into a highly-targeted area, thoroughly examining what is found, extracting information, and closely monitoring and tracking down the target as needed. The human body is designed to scan and track in parallel: our hearing, olfaction, and peripheral vision scan the environment simultaneously so that we will not miss any intrusive event. When we find anomalous events or a moving target, we exploit it and track it down continuously, while virtually ignoring other ongoing events or objects. Scanning is a parallel process, but tracking is mainly a sequential process, as we have a very limited attention span.

In order to maximize the chances of finding something, a scanning path can be arranged systematically around the perimeter of a space. A scanning path may begin at a random position so that no scan will repeat the same trajectory. This is one example of an instinctual default operation, "spontaneous alternation" in order to avoid deadlocks. The iRobot's cleaning robot Roomba™ adopts both scanning strategies. First, it embarks upon a "routine" path by moving in a spiral formation from its origin outward. When it hits a wall or obstacle, it bounces back at a random angle and tries to follow the wall or go back to the spiral mode. When the robot encounters a very dirty spot, it switches from scanning to an intense cleaning (tracking) mode.

[15]Pirolli P and Card S (1999) Information Foraging. Psychological Review. Vol. 106, No. 4 (Oct. 1999), 643.

[16]Pirolli P and Card S (2005) The Sensemaking Process and Leverage Points for Analyst Technology as Identified Through Cognitive Task Analysis. 2005 International Conference on Intelligence Analysis, Vienna, Virgina, May 2–6, 2005

**Fig. 4.4** The video wall of the traffic management center in Pittsburgh, PA, USA

Like the robot Roomba™, scanning and tracking are used in alternation during foraging. The appropriate amount of scanning depends on the context. It is a balance between detection rate and accuracy. In order to increase the detection rate, more scanning is needed. To increase the accuracy, in-depth tracking and thorough exploitation of high-value areas is needed. Figure 4.4 shows the video wall of the traffic management center in Pittsburgh, Pennsylvania. Operators monitor approximately 30 screens from over 200 surveillance cameras installed on highways, bridges, and tunnels. As a routine, the operator scans the 30 screens in front of him and switches to other cameras from time to time. However, most video feeds are unseen by the operator. Whenever something out of the ordinary occurs, the operator can switch the video feed to display the event on large screens and use pan, tilt, and zoom functions to track the event or view details. For example, when the operator discovered a traffic accident, his gaze was continuously focused on this event, while other events received less attention. The danger here is that perhaps an even more serious accident could be hidden from view and potentially missed by the preoccupied operator.

## *Saccadic Gazing*

How do we scan a video wall around us? There are too many objects to be processed by our limited attention spans. A saccade is the rapid movement of an eye. It contains a quick fixation in between movements. Our eye movement is a sequence of saccadic motions that forms a zigzag trajectory, jumping from one point to another. The human eye fixates mainly on certain elements of salient objects. The eye rests much longer on some of these than on others, while many elements receive little or no attention. What distinguishes attractive elements from those that do not draw attention? In general, objects with the brightest or darkest elements attract more attention. Human eyes and lips also attract attention.

Saccadic gaze is actually a spatial and temporal compression algorithm for visual information processing. In fact, the human eye can only acutely see a tiny area the size of a thumb. The vision in-between fixations is entirely from our imaginations. Every day, we are literally blind for 4 hours. In order to get a clear picture of the world, our eyes move saccadically to cover large areas. Saccadic gazing is like taking still photographs to sample the dynamic world. Because the sampling rate is not fast (less than 30 Hz), it compresses the "video" temporally. In addition, saccadic gazing enables humans to measure the size, shape, and relationship of objects in a scene. In order to measure these characteristics, the eye must stop at the fixation spots and bounce back and forth iteratively, forming a tracking trajectory around the features.

Psychology studies reveal that human saccadic motion is actually driven by instincts and purpose. In the 1960s, Russian psychologist Alfred L. Yarbus published his seminal work, "Eye Movements and Vision," wherein he describes an eye tracking experiment with famous Russian artist Ilya E. Repin's oil painting "An Unexpected Visitor."[17] In the painting, an exiled revolutionist man walks into his home and his wife and children appear surprised. Yarbus recorded the eye movements of seven different subjects. First, each subject examined the picture freely without instruction with both eyes for 3 minutes. Then, Yarbus recorded eye movements by the same subject who was asked questions before the subsequent recording sessions: "estimate the material circumstances of the family in the picture"; "provide the ages of the people"; "surmise what the family had been doing before the arrival of the unexpected visitor"; "remember the clothes worn by the people"; "remember the position of the people and objects in the room"; and "estimate how long the unexpected visitor had been away from the family." The observers formed different eye movement patterns according to particular questions. The eye gaze focused on the people in the room when the observer was prompted to figure out their ages, remember their clothing, their positions, or how long the man had been away from his family. On the other hand, the eye gaze fixated on the environment of the room when the observer was trying to figure out the material circumstances of the family and the objects in the room.

Yarbus' experiments reveal that saccadic eye movements not only are the traces of scanning, but are also used in tracking. Scanning and tracking can work *simultaneously* by moving the eye's gaze around an object or increasing the fixation time on an object while paying little or no attention to the rest of the objects. Today, social media are much like a video wall that demands attention from online users. Figure 4.5 shows a sample of the eye movements of a user browsing Facebook. As you can see, Facebook divides the screen into four columns: friends, main content,

[17] Yarbus AL (1967) Eye movements and vision. pp. 171–185. Plenum Press, New York

Fig. 4.5 The eye movements of a user browsing Facebook

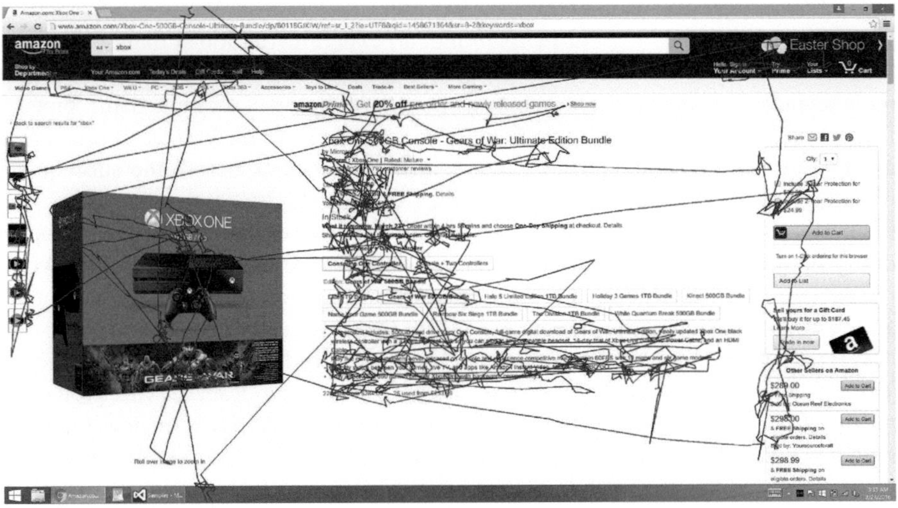

Fig. 4.6 The eye movements of a user browsing a page of Amazon

commercials, and activity status. The eye movements were typically concentrated on the main content column in the middle, while sparsely scanning the other columns. Figure 4.6 shows the eye movements of a user browsing a page of Amazon. It contains rough scanning patterns between the menu and products. As Facebook and Amazon collect user data (e.g. mouse clicks and pages or items in shopping carts), it is plausible for them to also collect a user's eye movements and subconscious interests by continuously measuring pupil sizes from the eye tracker.

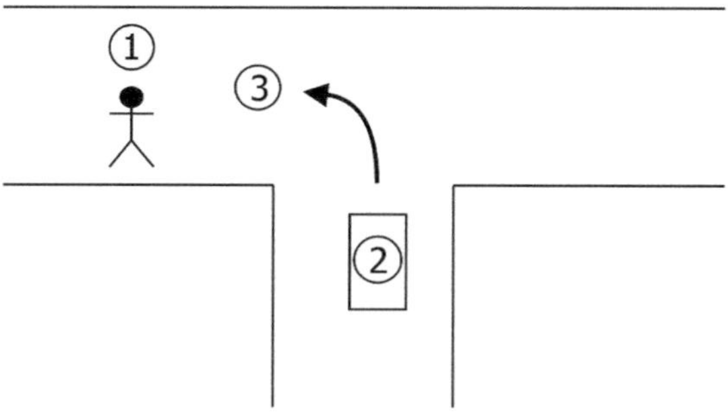


**Fig. 4.7** Saccade and saliency of traffic monitoring: (*1*) the walking pedestrian in the distance; (*2*) the moving car; and (*3*) the projected collision course of the pedestrian and the car

A series of biologically-inspired algorithms have been developed to create "virtual gazing" by simulating human attention. Itti's Surprising Model is based on the prior probability of events.[18] However, the algorithm focuses on low-level features such as texture and color. The Motion-Context Attention Shift (MCAS) algorithm, on the other hand, is a higher-level gazing simulation model that focuses on attention shifts during the tracking of multiple objects in a video. The model assumes that a system can scan a limited number of salient moving objects simultaneously, such as pedestrians and cars. We can assign saliency to events semantically. For example, in Fig. 4.7, we have an intersection scenario. The gaze remains upon moving objects with higher saliency levels. When the saliency value creeps above the threshold, sparse scanning mode is shifted to intense tracking mode for closer examination. The model assigns saliency 1 to the walking pedestrian in the distance; saliency 2 to the moving car; and saliency 3 to the pedestrian and the car that are on a collision course.

While most classic eye tracking experiments are based on still images, our eye tracking studies show that animated images or videos make a significant difference in scanning and tracking patterns. Animated pictures normally attract more attention, if even for a brief period. Our eyes instinctively pay attention to moving objects. Perhaps this explains why Facebook recently added animated GIF and video preview functions, aiming to shift users' attention from scanning to tracking, and to spend more time on the page. The pseudo code of a saccade:

```
Find salient object or feature
While tracking
    If an object has a salient motion
       track it
    Else
        find the next salient object or feature nearby
        track it
    End If
    find the next salient object or feature nearby
End While
```

[18]Itti L (2005). Bayesian surprise attracts human attention. Information Processing Systems (NIPS), 2005.

Assume we can harvest human eye movement data. How do we teach a computer to adopt the human attention model of scanning and tracking? We need to first encode eye movements into a digital format. A saccadic trajectory is normally a three-dimensional data entity, including x and y coordinates, and time stamps. A saccadic trajectory can be simplified as a *chain code* that converts a curve to a sequence of numbers. Each line segment in a chain code is represented by its length and a unique code.[19] This transforms two-dimensional features into one-dimensional numbers or letters. Chain code can be defined as either four-directional or eight-directional. For example, if we define moving to north a unit as 1, west as 2, south as 3 and east as 4, then the sequence of movements: north, east, south and west can be represented as a string: "1432." Next, we can implement a fast text search algorithm to match the patterns with defined errors, for example, using Levenshtein distance, in which an error can be defined as insertion, deletion, or replacement of a letter or number in the chain code. For example, comparing the string "1234" with "1342", the Levenshtein distance is 3. It is worth to note that chain code has its weaknesses. It reduces the angular resolution to limited numbers and it is neither rotation-invariant nor scale-invariant.

## Foveal Vision

Saccadic gazing is a series of sequential movements. As you can see, gazing at a particular area may cause one to miss an event happening simultaneously in another area. We wonder if it is possible to perform scanning and tracking in parallel? One possibility is using foveal vision. As we are primitive hunters, human vision is a multi-resolution perceptual system. The most acute vision in the human eye is located at the fovea of the retina with a line of sight connecting to the outside world. The discovery of the line of sight can be attributed to the Italian renaissance painter, scientist, and inventor Leonardo da Vinci from his experiment in which he has found the clearest vision to be at the fovea. According to his experiment, the eye has a central line and everything that reaches the eye through this line can be seen distinctly. Foveal vision, or central vision, has a meager 3° angle of acute vision. The rest of the field of view is peripheral vision (about 60 to 120° wide).

Economically, we only need to display a significantly small region with high acuity if a system can track a viewer's eye movement. In light of this, Gaze-Contingent Displays (GCDs) have been developed to degrade the level of detail (LOD) of peripheral image regions in order to reduce the computational workload

[19] Yang Y and Cai Y (2010) Virtual gazing in video surveillance. ACM Multimedia, SMVC'10 October 29, 2010, Firenze, Italy.

during image transmission, retrieval, or display. It aims to match the amount of information displayed against human visual processing capacity through real time eye movement sensing.[20]

To create a GCD system, we diminish spatial resolution (pixels), temporal resolution (frame rate), and color depth (number of colors) for the peripheral or unattended regions for better network data traffic management. This is an appealing option for transferring video over low-bandwidth networks. The key concept is that we only pass the video information when we need to. This is similar to human's *sensory inhibition* behavior. For example, we often focus on interesting scenes or objects while ignoring the background or uninteresting things. For video stream networks, there are two scenarios of such sensory inhibition: (1) the network transmits the video channel in the highest resolution whenever the operator is watching, while setting other channels to lower resolutions; and (2) the network transmits the video channel that contains objects of interest at a high resolution and the rest in low resolutions. Figure 4.8 shows the architecture of a video network system that incorporates gaze tracking and object detection input to set the multiple resolutions at the camera side.[21] This design optimizes the network traffic so that the important objects or camera channels get high resolution.

Gaze tracking and automated object detection can also enhance the objects of interest on the sensor side. This is called sensory enhancement, in contrast to sensory inhibition. For example, when the operator of a highway traffic management center watches the scene of a traffic accident, the camera can zoom in for more close-up shots and the video processing computer can further enhance details with shadow illumination, video stabilization, and super-resolution.

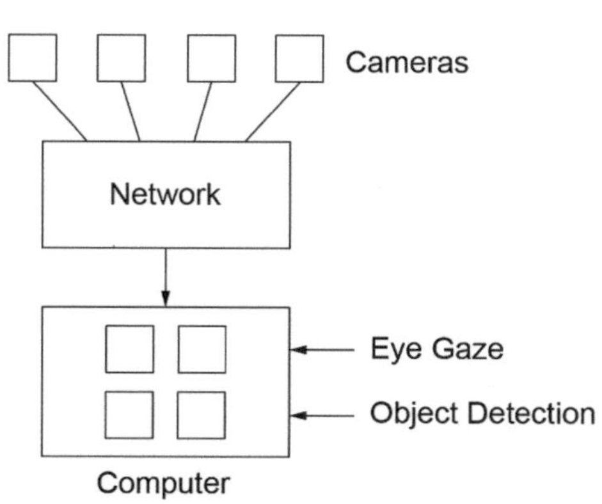


**Fig. 4.8** The gaze-contingent display system for multiple resolution video streaming (Cai Y (2010) US 20100283843 A1. Multiple resolution video network with eye tracking based control. 2010)

[20]Duchowski AT, et al. (2004) Gaze-Contingent Displays: A Review, Cyber-Psychology and Behavior, Vol. 7, Num. 6, 2004

[21]Cai Y (2008) Visual digest networks. In Cai Y (ed.) Digital Human Modeling, LNAI 4650, pp. 28–46, 2008

## Summary

Foraging algorithms have emerged from biological systems at many levels. Ant colony optimization algorithms are based on the pheromone tracing and updating models that enable collective and quorum foraging and tactile navigation. Ironically, these simple strategies have been adopted into advanced computing systems today, ranging from network flow optimization to Amazon product recommendation and vehicle-to-vehicle wireless communication.

More advanced foraging algorithms such as interception, fishing, and mechanical turk add a level of serendipity into the foraging world, which appears to be low-cost and ubiquitous.

Finally, more sophisticated human foraging strategies such as *scanning* and *tracking* are introduced here, which have primitive but applicable broader applications. These applications range from information foraging to video digesting. Two human vision algorithms, saccadic gazing and foveal vision, enable humans to manage limited attention spans, multiple resolutions of representation, and complexity of visual perception, which is carried out by half of the neurons in the human brain.

# CHAPTER 5

# Primitive Learning

Natural selection has produced instinctive behaviors through the slow and gradual accumulation of numerous slight, yet advantageous variations. The result is automation, a behavior that is second nature. This is called the Baldwin Effect. This principle can be applied to both humans and machines. In order to enable a machine to learn as efficiently as a human or animal, we need to understand primitive learning processes within humans, animals, and even insects. In this chapter, we explore primitive learning behaviors, including perceptual adaptive learning, peak-shift, lateral learning, learn-by-tapping, and indirect learning from virtual experiences, such as playing or gaming.

## Perceptual Adaption

A primitive vision system emerged from a simple light sensor in order to perceive brightness or darkness. Then, it evolved into a pinhole focusing mechanism – an eye with no cornea or lens, but a very tiny hole for the pupil. The light that passes through this hole forms an image on the retina. Leonardo da Vinci discovered the line-of-sight ray-tracing model to illustrate the image formation process. Starting with a point in the world, the light will be reflected in many different directions. There is only a single ray of light from each point in the scene that makes it into the pinhole, while most of the rays of light are irrelevant to our eyes. As a result, the image on the retina appears upside-down.

The visual cortex in the human brain takes what our eyes perceive and flips it right-side up, so everything we see is actually an altered image – a virtual reality! This seems crazy, but we indeed live in a virtual reality world. How does this principle operate in physical reality? An object standing straight up is actually straight up, but the lenses in our eyes refract the light of the object upon seeing it, and the visual cortex then has to then flip that image upside-down to make sense

© Springer-Verlag London 2016
Y. Cai, *Instinctive Computing*, DOI 10.1007/978-1-4471-7278-9_5

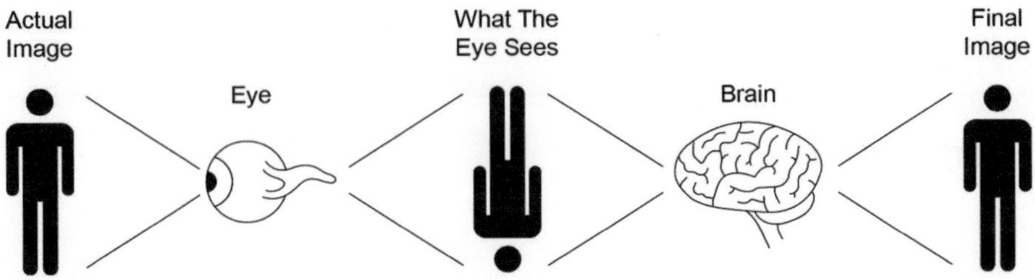


**Fig. 5.1** The images of the world that fall on our retina are upside-down, yet we see the world right-side up

of it. In other words, if our visual cortex did not flip everything we see upside-down, we could not operate in physical reality. See Fig. 5.1.[1]

Since the time of Leonardo da Vinci, people have been puzzled by this upside down imaging phenomenon. In the 1890s, American scientist George Stratton conducted an experiment to see what would happen if he viewed the world through an inverting prism.[2] He found that after 4 days, his brain started to compensate, and he could see the world the right way up again.

There are many reverse goggles on the market that are available for similar experimentation. Take, for example, the ULINE™ goggles that flip everything we see and interact with upside-down. Within a few days, our visual cortex will have compensated for this so that it begins to see things right-side-up again. This is an evidence of the plasticity of the human mind. Everything is relative, and our minds can be easily reprogrammed. We can be told absolutely that black is white, up is down, left is right, etc., and the mind will conform to this reality to enable us to keep operating in this dimension.[3] When we view the world normally, our brain reverses the images on the retina in both the up-down and right-left directions. These aforementioned prisms reverse the image in the up-down direction only. However, if we were to rotate the prism by 90°, it would then reverse from left-to-right instead. Furthermore, with today's video camera technology, we can flip images both vertically and horizontally.

Out of sheer curiosity, I bought the reverse ULINE™ goggles and started to wander around my office building. At first sight, the goggles have a rather narrow view – about a half of the height. Nonetheless, the narrow view was compensated by actively pitching my head up and down quickly to cover the full view of the upside-down world. I found it difficult to wear the goggles and try to shake hands with someone, to pick up an object, to pour water from a jug into a glass, to type, or to find the way around an unknown area. These skills would have required 4 days or more to adapt.

---

[1]Heeger D (2006) Perception Lecture Notes: The Eye and Image Formation. http://www.cns.nyu.edu/~david/courses/perception/lecturenotes/eye/eye.html

[2]Stratton G (1896) Some preliminary experiments on vision without inversion of the retinal image. Psychological Review, 3, 611–617

[3]Ick D (2009) Fact: Eveything you see is BACKWARDS. www.davidicke.com. http://forum.davidicke.com/showthread.php?t=95746

**Fig. 5.2** The views from the upside-down goggles: the empty hallway (*left*), the hallway with obstacles (*middle*), and the stairs (*right*). The images are superimposed with edges based on a simple Canny edge detection algorithm to highlight the boundaries between areas

However, it took me *only a few hours rather than days* to navigate through the environments that I am familiar with, such as my office building and home. Figure 5.2 shows three typical scenes in my office building: a straight, empty hallway, the hallway with obstacles along the wall, and stairs. Walking in an empty hallway is the easiest task. I simply walked along one side of the wall and paid close attention to any landmarks or obstacles in order to avoid bumping into them. The floor, at which people normally look while walking, was where the ceiling should be. I needed to keep looking up to check the floor, since this is what I was trying to safely navigate. The vertical edges became critical boundaries between walls, floor, ceiling, and obstacles. These outlines helped define my position and the locations of obstacles. For example, at the corner of the hallway, there were some obstacles along the wall. Knowing the position of the obstacles on the right, I was able to move forward by walking along the left wall (see middle scene in Fig. 5.2). The hardest task is to walk on the stairs because the motion seen through the goggles is the total opposite. This challenge can be met by touching the railings on the side or on the wall. Using tactile sensing and limited boundaries, I was able to walk through a hallway with obstacles, go upstairs, open the door, and access the mailbox.

This empirical experiment shows that we actually do not rely too much on visual information for navigating a familiar environment. Our instinct prioritizes critical information pertaining to safety: preventing a fall, avoiding obstacles, and monitoring the position of the trip. We can almost walk blindly in a familiar environment, simply using our tactile sensing capacity. Of course, I tested myself walking blindly at home and in my office building, assuming there were no unpredicted intruders in the way. It was not efficient, but it was smooth enough to move from point A to point B. Does this mimic the behavior of low-vision mice running around the house? They normally scurry along the wall and use tactile sensing. This experiment also shows that if we are familiar with the environment, or perform simple tasks such as indoor navigation, perceptual adaption time could be much shorter.

Basic perceptual adaption functions can also be found in digital cameras in functions such as auto-focus, auto-intensity adjustment, auto-orientation, and so on. Machine vision algorithms push the envelope further to more advanced applications, for example, image stitching. Given a set of randomly-oriented photos from a scene, one can stitch them together to form a panoramic picture. This requires computing a homographic transformation to rotate, scale, and project an image for aligning with other images. A homographic matrix can be calculated by matching two sets of feature points: one in the selected image, and another in the neighboring image.[4] These feature points can be extracted from the maxima of the Difference of Gaussian (DoG), and filtering out points at the edge or low-contrast points.[5]

The reverse goggle experiment also demonstrates that our brain circuits can adapt to a new environment from old experiences. For a mobile robot, the alignment of a previous visual world in their memory can be represented as a homographic transformation of the new virtual world. However, finding a homographic transformation matrix from feature points is a computationally expensive optimization process. There are two ways to improve the performance of perceptual adaptation: (1) fusing with other sensory data, and (2) prioritizing feature points. Sensory data such as gyroscopic and magnetic data on a mobile phone can derive information about the orientation of the phone. Therefore, the orientation of the camera can be embedded into the metadata of the photo, which can be retrieved for displaying or stitching. Prioritizing extracted feature points is essential for real-time data processing that filters out non-essential data, only focusing on immediate needs – a common process in nature.

The reverse goggle experiment indicates that the perceptual adaptation process can be much faster if we only pay attention to survival variables such as cliffs and obstacles, while ignoring non-essential details, such as ceilings and obstacles behind others, or obstacles that are far away. This is vital to the survival of a living creature because processing time is very limited, and response must be immediate.

## Peak Shift

To capture the essence of an object, our instinct often amplifies the differences and unique features of that object, highlighting key features and reducing redundant information.[6] This process in the brain has evolved to do naturally by more powerfully activating the same neural mechanisms activated by the original object.[7] This psychological phenomenon is called the "peak shift" effect. Peak shift has been most widely understood in regards to its applications in animal discrimination

[4]Wikipedia: Image Stitching. https://en.wikipedia.org/wiki/Image_stitching

[5]Lowe DG (1999) Object recognition from local scale-invariant features. Proceedings of the International Conference on Computer Vision. pp. 1150–1157.

[6]Zeki S (1999) Inner Vision: an exploration of art and the brain. Oxford University Press

[7]Ramachandran VS, Hirstein W (1999) The Science of Art: A Neurological Theory of Aesthetic Experience. Journal of Consciousness Studies 6 (6–7): 15–51.

learning. In the peak shift effect, animals sometimes respond more strongly to exaggerated versions of training stimuli. For instance, a rat is trained to differentiate a square from a rectangle by being rewarded for recognizing the rectangle. The rat will then respond more frequently to the object for which it is being rewarded, so much so that it will respond more frequently to a longer, narrower rectangle than to the original with which it was trained. Cartoon artists often exaggerate figurative features to enhance visual effects, for example, making American comedian Joe Leno's distinctively large chin even larger. Exaggeration of distinctive features helps people to memorize and recognize a figure in a picture. It especially helps children because their cognitive skills are still developing.

We can apply the peak shift principle to sound recognition. For example, we may crop out the background noise and only store representative signals in the training data for machine learning. Figure 5.3 shows an example of the peak shift process for training a computer to learn asthma sound patterns. The peak shift sampling windows crop out the background noise for maximal discrimination in learning. The asthma sound patterns contain a wheezing sound by its unique, sharp, high-pitched musical tone. By highlighting the horizontal bright patterns, we can threshold ranges of frequency and magnitude, thereby removing non-significant background sound patterns. Our experiments show that cropping the feature-rich patterns from raw data improves recognition accuracy. Based on forty training samples for wheezing sounds and normal sounds, after applying the peak shift preprocessing, the positive detection rate of the k-NN classifier increases from 33 % to 89 %. Furthermore, if samples have distinguishable patterns and low noise, one can use simple classifiers such as k-NN for recognition instead of more sophisticated ones such as the Gaussian Mixture Model, which requires more specific training data to run than k-NN.[8]

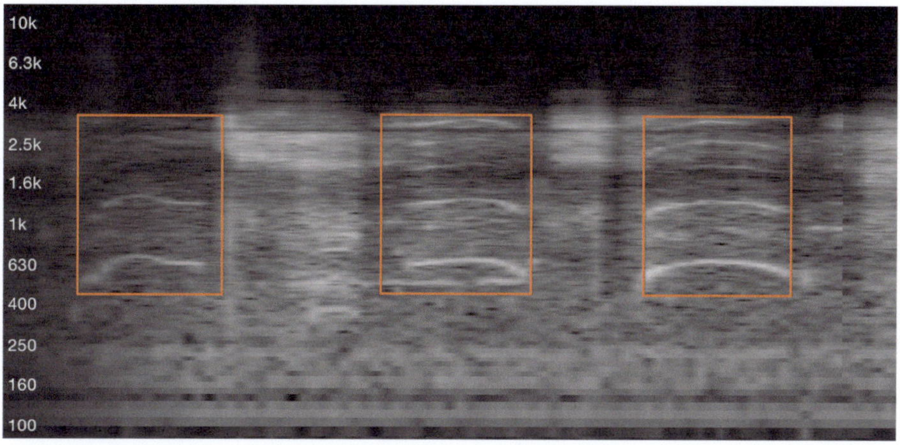


**Fig. 5.3** Peak-shift processing of the wheezing sound signal

[8]Cai Y (2014) Ambient Diagnostics, CRC Press, 2014.

## Lateral Learning

In addition to peak shift processing distinctive features in the data, we often learn patterns and trends from its environment. We can sense that something is wrong in a car by smelling certain odors or listening to sounds from the engine. We call this process "lateral learning," which derives patterns from the same source but different channels, which are sometimes hidden and ignorable.

Digital images normally contain metadata from the camera used. The camera metadata typically include cues: subject distance (SD), focal length (FL), exposure time (ET), aperture value (AV), f-number (FN), and flash fired (FF). Assume that we want to classify a collection of digital photographs into two classes: indoor or outdoor. We may look at the low-level cues such as color and texture features and train a classification model, say, a Support Vector Machine (SVM). This method can only achieve about 81 % of total classification accuracy.[9] On the other hand, we can do data mining from the semantic cues from the camera metadata and feed them to a similar classification model, or an even simpler model like a Bayesian Network. Amazingly, using metadata alone ET + FF + SD can achieve 91.5 % total accuracy for the classification of indoor or outdoor scenes. This seems to make sense to us: we often turn on flash (FF), use longer exposure time (ET), and shorter subject distance (SD) while taking indoor photos, and we tend to do the opposite in outdoor settings. Combining cues from camera metadata and low-level image features in a Bayesian Network, the accuracy can reach a 94.1 %, whereas the low-level image feature-based learning only achieves 3.6 % total accuracy. Relatively, image processing is more computationally expensive than numerical processing in orders of magnitude. This case study indicates the great potential of lateral learning.

Lateral learning can help determine the correlations of multiple sensors and compensate for the weakness of individual sensors in different situations. Take for example, detecting vehicles from visual or auditory channels. It is very easy to detect a motorcycle or truck by the sound because of the unique acoustic signatures of such vehicles. This is beneficial to the overall detection at nighttime, when the detection accuracy of regular color cameras often drops significantly.

## Learn-by-Probing

Learn-by-probing is an active learning process that explores an unknown environment by sampling a small set of essential data along the path. Imagine a blind person walking around a city with a probe stick. Using a tapping technique, which can be either tactile feedback or auditory alert, the blind person can detect cliffs, walls, and

[9]Boutell M and Lou J (2005) Beyond pixels: exploiting camera metadata for photo classification. Pattern Recognition. 38 (2005) 935–946

other obstacles in the path, and move forward with amazing speed. The amount of data collected from the probing device is much smaller than the data perceived from a visual system.

## *Hill-Climbing and Randomization*

Learn-by-probing is to learn by doing on the fly. While executing the task, the system adapts itself for better performance the next time. Imagine you are wearing a pair of upside-down goggles and you attempt to walk down a hallway. Your visual perception is compromised, and your brain and limbs need to rewire. You have a few options to move forward: moving alongside the wall, walking in the center of the hallway, using a stick to tap the floor, looking up to see the floor, and so on. You would likely try several options to find a safer and more efficient moving strategy. Typically, you would keep using one strategy until it fails, resulting in a collision. Sometimes you must improve the strategy along the way, even without a collision, for example, on a different floor of the building, in different locations, or with different arrangements of obstacles in the hallway.

This process is similar to the "Hill-Climbing" algorithm – the analogy of a blind person trying to find the highest point on the ground. If the blind person probes four directions at each step, he or she may find a satisfactory solution in the end. However, this does not guarantee the best solution because the hill the person climbs may not have the tallest peak. In order to avoid the local optima, the blind person needs to randomize the order of probing choices, or randomize the initial starting point. This is similar to the Spontaneous Alternation Behavior (SAB) discussed in the chapter about Default Mode. The anecdotes of the upside-down goggles and the blind person walking illustrate that when visual signals are not available or are significantly reduced, we fall back into acting instinctually, using strategies that are common in nature. For instance, social insects like ants use the hill-climbing strategy for detecting pheromone trails, foraging for food, and learning to navigate new environments. They also apply spontaneous alternation to avoid deadlock. Amazingly, ants use the "parallel computing" technique to implement the randomization operation by teamwork of connected ants at random initial starting positions. They gradually learn their optimal path through their shared external memory – the emission of pheromones along their routes.

## *Scanning and Tracking in Learning*

As we discussed in the chapter about Foraging Behaviors, scanning is to search a large area in order to detect potential targets. Tracking is to further recognize or classify the target. Like foraging, primitive learning often alternates between these two modes. Take the example of a blind man using a stick to scan the road in front of him. From the reverse goggle experiment, we discovered that the perceptual adaption process can be divided roughly into two stages: (1) scanning for dangerous situations to ensure survival, e.g. to avoid falling down the stairs, or to keep from colliding into the walls or obstacles, and (2) tracking for controlling our hands in order to accomplish more complex tasks, such as fetching a pen or moving a mouse. The scanning process aims to search in a systematic trajectory in order to avoid missing a target. On the other hand, the tracking process focuses on individual targets in a small area, which requires subtle maneuvering and cognitive processes. Adapting scanning skills in a new environment is faster. In the reverse goggle experiment, it only took a few hours. However, adapting tracking skills is a relatively longer process. It may take 4–5 days to learn daily activities beyond simple navigation. The following case studies introduce how to apply scanning and tracking strategies in real-world applications for autonomous robot design.

## *Pedestrian Detection Case*

As it is with human drivers, safety should be the ultimate instinct for autonomous driving vehicles. For decades, it has been a challenge for a machine vision system to detect pedestrians on the road. LIDAR (Light Detection and Ranging), or laser imaging, is a critical device for autonomous or semi-automatic vehicles to detect obstacles on the road. It scans the environment with rotating near-infrared laser beams that measure the line-of-sight distances between the obstacles and the sensor, collecting a ring of points for each layer of the laser beam. Four or more layers of beams can create a three-dimensional point cloud model of the vehicle's surrounding environment. LIDAR is an ideal device for implementing the systematic scanning of pedestrians and other obstacles. Its region of interest (ROI) can even be designed "instinctively" for better response time. For example, if the vehicle is moving forward at a certain speed, the pedestrians on the sidewalks or behind the vehicle will not be a problem. LIDAR can also help to segment targets from the background based on their 3D profiles. However, sparse laser beams, e.g. only four layers of contours, are sometimes not enough to recognize the shape of a pedestrian. It may have a high false detection rate, for example, interpreting a utility pole as a pedestrian, or vice versa. See Fig. 5.4.

In order to classify whether the target is human or not, we need a tracking and recognition model. The LIDAR scanner needs more vertical resolution, or the system needs to employ additional sensors, such as a visible wavelength imaging camera with a shape recognition algorithm like Histogram of Oriented Gradient (HOG). This algorithm enables individual image-based detection based on the "cell-like" histogram of oriented gradients. The recognition model can then be trained with samples. However, the algorithm has its weaknesses. For instance, occlusion handling is a serious problem for pedestrian detection algorithms, which

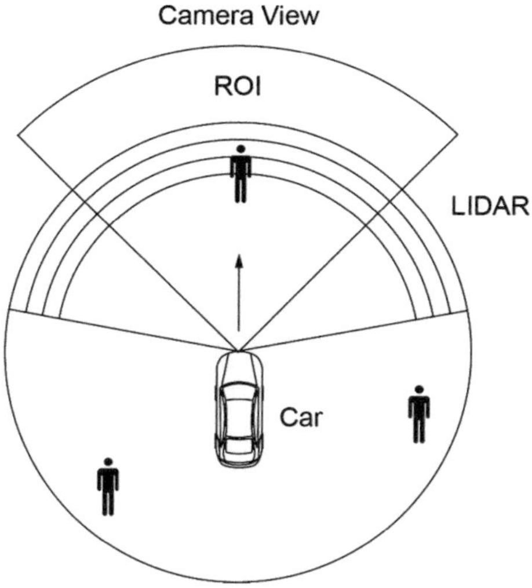


**Fig. 5.4** The scanning and tracking processing for pedestrian detection with LIDAR and camera data. When the car travels forward, the region of interest (RIO) is the pedestrians or other obstacles in front of the car

may contribute to a higher rate of missed detections.[10] Therefore, it makes sense to combine the LIDAR and camera data into a sequential method: LIDAR does the scanning to locate potential targets, and the camera tracks each target to identify whether or not the target is a pedestrian. The LIDAR's scanning process behaves like a filter that removes unnecessary computation, including the potential for false positives in the recognition stage. According to tests in the author's lab based on the LIPD dataset, fusing LIDAR and visual camera data can greatly improve the reliability and accuracy of the overall detection rate. The LIDAR sensor alone has a missed detection rate of 1.83 % and a false detection rate as high as 21 %. The camera data based HOG algorithm yields a missed detection rate of 4.9 % and a false detection rate of 1.26 %. The fused method reduces the missed detection rate to 0.87 % and the false detection rate to 2.16 %. Fusing LIDAR and camera data not only helps to improve classification accuracy, but can also be beneficial in extracting additional information, such as distance of the pedestrian from the vehicle, time of impact, and approximating the intent of the pedestrian.

## Landmine Detection Case

Landmine detection is an extreme case where primitive instincts come into play. The human operator behind a metal detector often relies on a strict procedure to make

[10]Deshpande S and Cai Y (2016). Pedestrian detection. In Loce R, et al. (eds.) Computer vision and imaging in intelligent transportation systems. IEEE and Wiley, to be published in May 2016.

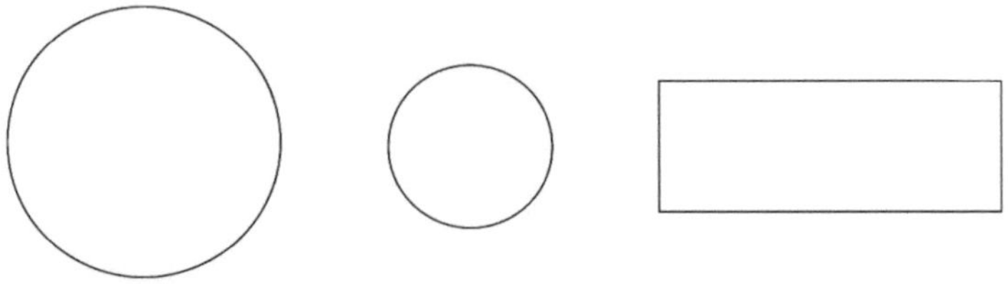


**Fig. 5.5** The three common types of mines used in this game. From *left to right*; an anti-tank mine, and two anti-personnel mines and their relative shapes and sizes

the next move. A landmine simulator enables us to safely investigate the learning experience for scanning and tracking landmines.

In the simulator, the landmine field is an empty screen where an unknown number of mines are hidden in random locations and in random orientations, e.g. rectangular mines. The mouse is used to simulate the metal detector. Whenever the mouse touches the mine, an alarm sound will beep. Using a metal detector is similar to blind people searching for something by tapping around with a probe stick. It is a means of tactile sensing. In the simulator, we consider three types of conventional landmines: large circular anti-tank mines (e.g. TM-46), small circular anti-personnel mines, e.g. MD-82B, and rectangular anti-personnel mines based on the catalogue of mines from Freedom Fields USA.[11] See Fig. 5.5.

First, let us look at how human landmine detection works in an experiment setting. Human subjects started by scanning through the open field row by row until the metal detector alarm sound triggered. Then, the players probed the metal detector around this spot to track the shape of the landmine using sound in order to predict the outline and type of landmine. After multiple trials, two heuristic strategies emerged: (1) cross tapping and (2) edge tracing. Cross tapping probes the farthest spot to determine the rough central axes of the shape. Edge tracing probes the contour of the shape. We call these methods "Human Heuristics." See Fig. 5.6.

Second, let us look at how a computer performs this task. As with the human subjects, the computer starts with scanning the open field until the alarm from the metal detector is triggered. Next, it tries to recognize the shape by converting the points on the edge into a string of numbers or labels. A common technique is the chain code algorithm, which computes the angle between adjacent pairs of points in a sequence. This angle is then stored as values based on a series of direction-value pairs. Based on the deviation in values, the computer can differentiate between a circle and a rectangle, and determine the size. Figure 5.7 illustrates an example of the Chain Code.

[11] Freedom Fields USA: Landmines Types (2010). http://www.ffusa.org/landmine_types_md_82b.html

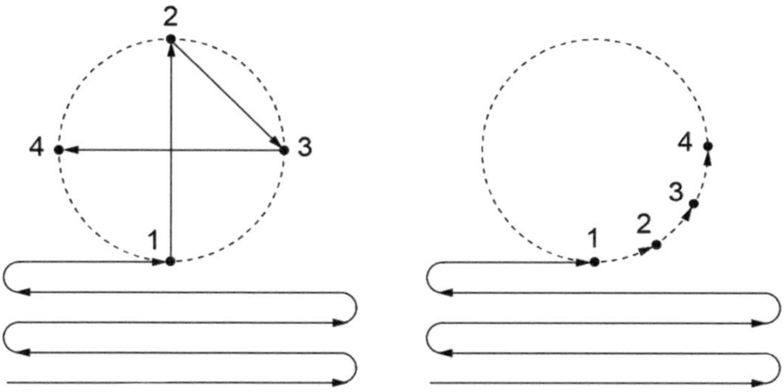


**Fig. 5.6** The "Cross Tapping" and "Edge Tracing" strategies

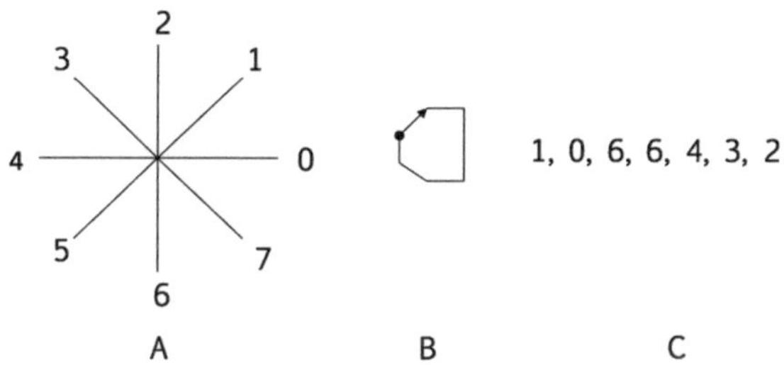


**Fig. 5.7** The Chain Code directional definition and sample shape, along with its Chain Code: *A* is the definition of the code associated with eight directions. *B* is a sample shape. *C* is the Chain Code for the shape in *B*

The computer is able to quickly perform an exhaustive and complete search of the grid to recognize the shapes of mines. We call this the "brute force" method. Still, human heuristics can make accurate judgements in fewer steps. The computer can certainly learn tracking strategies from human heuristics. By simulating human heuristics in this simulation, the computer processing time decreases significantly. See Fig. 5.8.

## Projected Memory

In a real-world landmine field, human mine detection experts often spray bright marks or add grid wires to assist in the dangerous task. These can be viewed as projected memory on the ground because of the limited capacity of short-term memory in our brains – especially trying to remember imaginary spots on the ground. Psychological experiments show that the number of objects an average human can hold in short-term memory is in a size of 7 plus or minus 2 "chunks." For imaginary objects in an open field, the number of chunks could be even smaller. Thus, we need a form of external memory, like the pheromones of ants on a

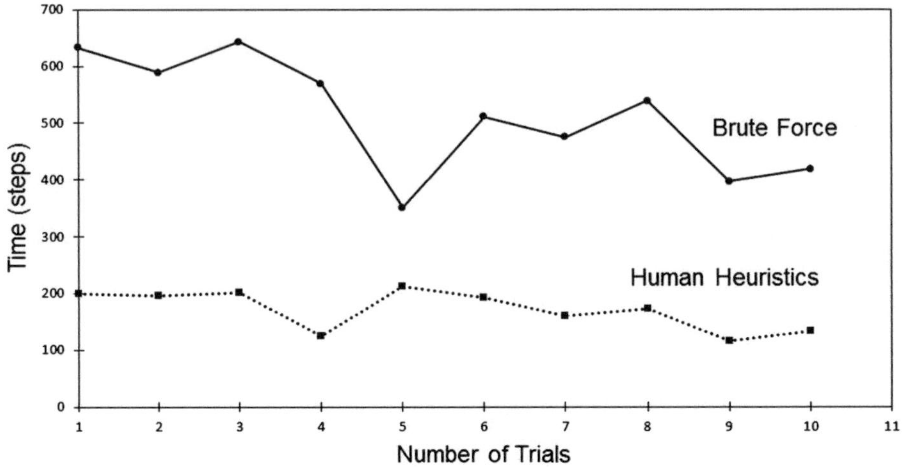


**Fig. 5.8** Runtime results for computerized shape recognition: the brute force algorithm versus the human heuristics algorithm

**Fig. 5.9** Screen display of the landmine detection game with assisting tools of markers (*left*), highlights (*middle*) and grid (*right*)

path. Here, we want to generalize the projected memory with physical markers, highlights, and a grid on the ground to represent the inspected areas. Figure 5.9 shows three tools to help the user to memorize the inspected areas, including markers, highlights and grid.

With all of the tools in place, several different variations of the game were developed. The first, labelled "No Help", involved no tools at all. The variations labelled "Grid," "Unlimited Markers," and "Highlight" involved the provision of tools as their names suggest. Each variation of the game was assigned to four individuals who proceeded to play it ten times in a row, restarting after each round ended. To compute their score, we used the following formula weighing the number of mines counted greater than the shape of mines determined accurately: if the

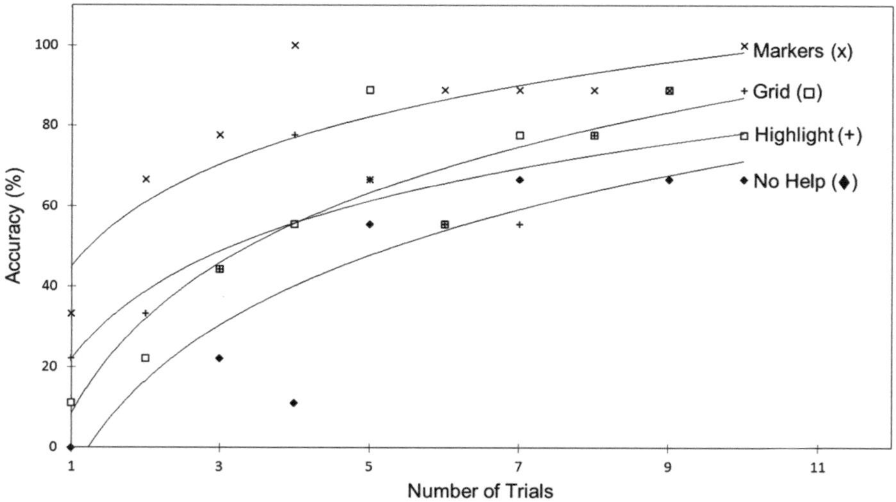


**Fig. 5.10** Plots of accuracy against number of tries for various players with a logarithmic best-fit. Some players were given tools to improve performance

number of mines was wrong, 0 points were given; if the number of points were right, the mines were given in proportion to the percentage of shapes correct, with 0 the minimum and 1 the maximum score. Figure 5.10 shows the data points of average accuracy against number of attempts with a logarithmic best fit line.

From the graph, we see the accuracy increases with the number of trials, suggesting that humans naturally tend to learn from their initial mistakes, and improve along the way as more attempts are made.

The presence of a superimposed grid, markers, and highlights significantly helped human mine detection accuracy. This implies that the computer's memory capacity aids the human subject in developing more informed decisions about the locations and shapes of hidden mines. To elaborate, the superimposition of a grid assists the human by showing a relative size and displaying straight lines for comparison. Superimposing a grid also helps the human determine the size of the mine by giving the subject a one-centimeter reference point. The markers provide a mechanism to remember all previously-hit mines, benefitting the human greatly since he or she can now paint a visual picture of the marked areas. Without the markers, the human can only remember the locations of possibly the previous 5–10 mine locations. Finally, highlighting serves a similar purpose to using markers by enabling the human to keep track of previously-scanned areas. Therefore, despite the human's effective potential to learn, the input from a fast processing computer expedites this learning process tremendously. This allows the human to form more accurate predictions, and reach the goal of perfect accuracy quickly.

However, we note that even with many tools, the human subjects performed poorly on their first attempts. An important issue here was the time limit. Although the access to all helping tools should have allowed the players to easily gain points, many players did not have enough time to accurately estimate all of the shapes. This problem was quickly rectified in the latter stages, as players displayed much better time management. Primitive learning is seen here – humans are extremely fast learners and tend to rely on instinctive deductions based on past mistakes, improving their present performance. Even without being instructed to learn from their mistakes, humans can perform so instinctively.

## Learn from Virtual Experience

A significant portion of our knowledge is learned from a virtual experience through playing, gaming, and analogy. Universal concepts such as death cannot usually be experienced directly. Such concepts can only be learned in virtual ways. See Fig. 5.11. Simulated games play an important role in learning. According to Constructivism, we do not simply learn things but rather construct them. A prime example is the understanding and solving of biomedical problems. This requires insight into the complex interactions between the components of biomedical systems by both domain and non-domain experts. This is challenging due to the enormous amount of data and knowledge in this field. Therefore, non-traditional educational tools have been developed such as biological storytelling systems, animations of biomedical processes and concepts, and interactive virtual laboratories.

**Fig. 5.11** The destructive learning experience

## *Virtual Crash Dummies*

Since the 1930s, the automobile industry has been using human bodies for studying the physical impacts on the body during a vehicle accident. One of the fathers of the crash test dummy, a biomechanics professor Lawrence Patrick,[12] first tested the physical impacts on his own body, including riding the crash rocket sled himself, experiencing crushing blows to the head and body, and other forms of physical abuse like a "mad scientist". In one study, he subjected himself to a 23 kg pendulum to the chest plate in order to test the effects of a steering wheel column on a human body during collision. The research goal was to collect data about how the human body responded in a vehicle accident. Patrick quickly realized those tests were not sufficient for testing the limits of human endurance, because if a test had reached his physical limits, he would have been dead already. He then switched to using human cadavers for extreme tests, for example, he placed a cadaver at the bottom of an elevator shaft and dropped heavy objects onto the head to test the strength of a human skull.[13] Since then, cadavers have been used to observe multiple impact scenarios and their effect on different parts of the human body. Using cadavers has its drawbacks: different bodies yield variable impact data because no two cadavers are alike, and each can only be used once in testing. Cadavers donated for scientific research also tend to be older and more fragile, e.g. the average age of cadavers used was 72, which would have an impact on the objectivity of studies because car accidents are leading cause of death in younger people.

One of Lawrence Patrick's students, Harold Mertz, started to develop the first mechanical dummy named Hybrid III[14] – the current worldwide standard frontal crash test dummy. It was initially only a 50th percentile male dummy, which was 180 cm tall and had a mass of approximately 78 kg. He sat on the driver's seat in 65 km/h offset frontal crash tests. Now, this dummy has a big brother – the 95th percentile Hybrid III, a 188 cm and 100 kg male model. Other members of the crash test dummy family include the 5th percentile female, plus two child dummies representing a 21 kg six-year-old, and a 15 kg three-year-old. Leading companies such as Ford Motors developed their own pediatric crash test dummies. During the tests, calibration marks are fastened to the side of the head and knees to aid in measurements taken from slow-motion video footage. Up to 58 data channels located in all parts of the dummy from the head to the ankles, record between 30,000 and 35,000 data items in a typical 100–150 millisecond crash. However, mechanical dummies are not perfect. First of all, a fully-articulated dummy is rather expensive. On average, these cost anywhere from $300,000 to $400,000 USD.[15]

---

[12]Wikipedia: Lawrence Patrik. https://en.wikipedia.org/wiki/Lawrence_Patrick

[13]Marquis E (2013) The Driving Dead: Human Cadavers Still Used In Car Crash Testing. http://www.autoblog.com/2013/10/24/the-driving-dead-human-cadavers-still-used-in-car-crash-testing/

[14]Wikipedia: Hybrid III. https://en.wikipedia.org/wiki/Hybrid_III

[15]Wikepedia: Crash Test Dummy. https://en.wikipedia.org/wiki/Crash_test_dummy

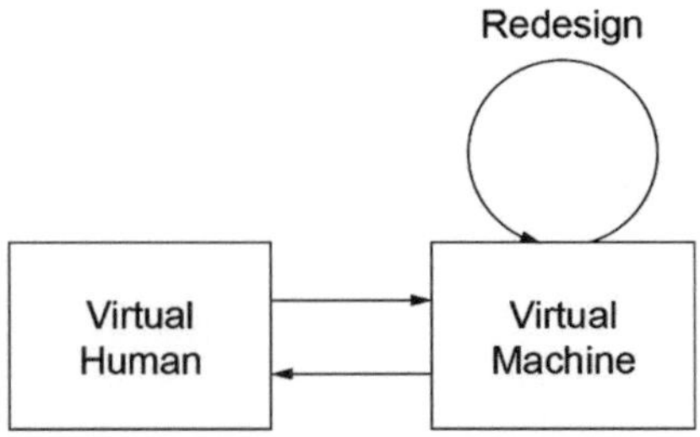


**Fig. 5.12** The learning system based on automated virtual experiments

Second, current mechanical dummies still do not display enough detail on what happens to internal organs during a crash.

The computer simulations and technologies of today have become affordable for advanced crash test dummies. So far, eleven automobile and parts companies have decided to put their competitive differences aside and create a consortium of Virtual Humans for virtual vehicle test standards. The consortium, "Global Human Body Models," initially contained six models, including large, medium, and small male and female body types. These models simulate how the human body responds to vehicle accidents. The models include finite element models of muscles, and respiratory and circulatory systems based on biomechanical and impact dynamics data. It is expected that this kind of simulation will replace the use of mechanical dummy models because it saves time and is more economically affordable.

Furthermore, it is worth noting that finite element vehicle simulation models have been developed for decades, and are well-tested and mature enough to yield realistic test data results. The next breakthrough would be the connection between the virtual human models and the virtual vehicle models to form a closed-loop automatic computational system in which no mechanical or human intervention is needed. This would enable vehicle safety designers to invent new structures or optimize existing design parameters (Fig. 5.12).

## *Virtual Reality Training for Robots*

The feasibility of using virtual reality to train nonhumans is demonstrated in studies that tested virtual reality environments on rats. Hölscher et al. found that rats are capable of navigating in virtual environments, while Nekovarova and Klement discovered that rats could learn to press levers for rewards when certain configurations appear on a computer screen. While previously, only humans and primates were thought to have been able to navigate virtual worlds, these rat studies show that non-primates can be trained to navigate them as well.

Then comes an interesting question: If rats can navigate in virtual reality and humans can train in virtual reality, can robots also train in virtual reality? Since humans who are trained in virtual reality may perform equal to or better than humans who are trained in real-life, the possibility exists that training a robot in virtual reality may result in improved (faster or more accurate) performance than training a robot in real-life. In VRMC's DARWARS Student State Assessment study, military personnel learned to navigate a virtual shoot house before running through a real-life shoot house that had a different layout. Their improved speed in the shoot house demonstrates that spatial skills learned in virtual reality were transferred to real-life experience. Therefore, there is reason to believe that if a robot were trained in navigational skills in a virtual reality setting, it would learn improved spatial skills that transfer to unfamiliar real-life environments.

Virtual reality can be applied to simulate operations in extreme environments that are not feasible or are too expensive in conventional training systems, such as a weightless environment. For example, NASA has worked with SONY to turn the game engine PlayStation into a virtual reality platform to enable astronauts to experience the operations of the humanoid robot Robonaut 2 in a weightless environment. Through virtual cameras in the simulation model, the operator can manipulate Robonaut 2's two arms using the handheld controller to accomplish planned tasks in a space station.

The current affordable virtual reality training systems are mainly based on the classic stereoscopic goggle design. The goggle feeds left and right channel screens directly to the user's eyes. Lenses are added to allow the eyes to focus on the screens in the goggles. Control buttons are attached for interactive input. Signals from head movement sensors are used for tracking the head motion for synchronizing the point of view in the video. Prevailing systems can be divided into three groups: client-server based, smartphone-based, and augmented reality-based. The client-server-based systems such as Oculus Rift mainly do all the calculations on the computer server and communicate to the goggles as a client. This architecture enables heavy-duty computation for high-fidelity rendering and multi-player synchronization on the server.

Smartphone-based systems such as Google Cardboard and Samsung VR conduct all of the computation on the phone, and have minimal to no communication with the computer, merely mirroring the data from the phone to the computer screen. Given the limitations of the computing power of smartphones compared to the average full-sized laptop or desktop computer, the size of a VR model appears to be restricted. The higher-end augmented reality devices like Microsoft's HoloLens work by projecting left and right channel images onto holographic film in the goggles that superimpose the projected 3D model onto the background. The holographic film is almost transparent, but reflective enough to project images. For better 3D appearance, the background usually needs to be darker than the projected images so that the objects may appear to float around in the background. Holographic architecture enables users to see 3D models on top of the background. This is useful for surgeries, navigation, real-time control, and education, where users need to see fused images from multiple channels.

The greatest advantage of these consumer-grade game platforms is that they are much more economically affordable compared to the traditional custom immersive simulators. Users of these consumer-grade platforms can also rapidly prototype simulation models using off-the-shelf software and hardware components, as well as by using developer toolkits. However, current consumer product-based systems continue to face challenges in providing high-quality images, fast interfaces, and rich virtual reality content. The interoperability between the VR systems on the market still needs to be improved.

## *Role-Play Games*

How do you teach medical students empathy toward elderly patients? Trainees may know how to diagnose diseases, but they usually find it difficult to understand the physical challenges faced by elderly patients daily. A new simulation system enables medical students to personally experience the effects of aging. The medical student wears a suit that strains the muscles along with a pair of goggles that reduces eyesight by 80 %. This way, the student can truly understand their elderly patients' physical problems better, and provide them with a higher level of care. This technology not only helps medical students hone their knowledge and skills, but impart students with emotional qualities such as empathy. The aforementioned example facilitates mental preparation in terms of experiencing emotions while communicating with the patient[16] (Fig. 5.13).

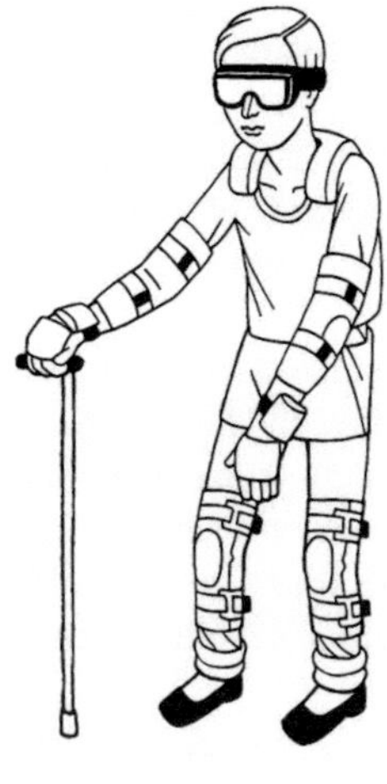

**Fig. 5.13** The suit and goggles that enable the in-vivo elderly learning experience for medical students

The idea of role-play game-based learning is to allow the user to gain knowledge while playing a simulation game. We cannot learn the dynamics of the human body by dissecting a cadaver. One must use virtual experimenting. Learning the principles of biology is different from learning engineering because biological systems are

[16]Linssen D (2015) Age simulation suits are teaching medical students empathy. http://www.sciencedump.com/content/age-simulation-suits-are-teaching-medical-students-empathy

more complex. The main learning strategy is hacking rather than systematic design and implementation. This is similar to how a software engineer keeps a program running and modifying small portions gradually. The author's team developed a computer game that enables children to explore biomedical knowledge. It is a biological world model in which users can explore biological interactions by role-playing "characters," such as cells and molecules, or as an observer in a "shielded vessel", both with the option of networked collaboration between simultaneous users. The system architecture of these characters contains four main components; (1) bio-behavior, modeled using cellular automata, (2) biomorphic, using vision-based shape tracking techniques to learn from recordings of real, biology dynamics, (3) bio-sensing, based on molecular principles of recognition to identify objects, environmental conditions, and progression in a process, and (4) bio-dynamics which implement mathematical models of cell growth and fluid-dynamic properties of biological solutions. Principles are demonstrated in a simple world model of the human vascular system through a biomedical problem. The problem involves an infection by Neisseria meningitides, where the biological characters are white and red blood cells and Neisseria cells. Our case studies show that the problem-solving environment can inspire strategic, creative, and innovative thinking in the user.

## Summary

Primitive learning consists of simple yet powerful strategies that may inspire future development of learning systems. The explorations in this chapter began with perceptual adaption for walking while wearing reverse goggles, where visual perception functions are compromised. In this case, the adaption process forces the user to switch to instinctual learning models. This includes utilizing the hill-climbing method to search for survival navigation, spontaneous alternation behavior for collision avoidance, and the use of scanning and tracking processes.

Scanning mainly deals with searching and locating, whereas tracking focuses on object recognition and measurement. In the two real-world case studies, the scanning and tracking model helps to design a pedestrian detection algorithm that fuses three-dimensional laser range finder data with two-dimensional shape recognition models for better detection accuracy, including lower missed detection rates and false positive detection rates. In the landmine detection simulation study, human-like shape tracking models are developed to improve accuracy of the recognition of mine type. The minefield experiments also show that projected external memory such as painted marks, highlighted traces, and grids can help human operators improve their detection performance.

Peak-shift strategy exaggerates the features of an object over background for better learning results. Lateral learning is to find the correlations of multiple sensors and to compensate for the weakness of individual sensors in different situations.

A significant portion of our knowledge is learned from virtual experiences, such as playing or gaming. In these processes, we are not only learning, but also constructing cognitively. Simulation-based learning is a primitive learning strategy but is also a contemporary approach in the digital age. Virtual reality models such as digital humans, virtual space robots, and role-playing simulation games have quickly become important and valuable tools for scientific exploration, product design, and education.

As the great physicist Linus Pauling once predicted, simulation will become the third scientific discovery method, in addition to the age-old induction and deduction methods.

# CHAPTER 6

# Tool Making

One quality that defines human instinct is not our compulsion to survive, but our ability to create things in the most unpredictable ways. *Hacking* or *improvisation*, in a technical context, is to solve a problem with the tools and materials immediately at hand, in ways the materials were not originally designed for. It is problem solving on-the-fly in spontaneous moments of sudden inventiveness.

In this Chapter, we will explore tool making processes. How do we adapt ourselves as tools? How does a robot extend its limb to reach a banana like a chimpanzee? How do we build physical and mental scaffolding for problem solving? How do we use visual language to substitute for functions or concepts? How do humans or robots make tools collectively? Is it possible for a machine to self-replicate? What would be the consequences of such tool-making behavior?

## Self-Tooling

Self-tooling is a primitive way to adapt oneself as a tool to accomplish a new task that is not anticipated. It is a self-sufficient approach to problem solving that requires no external resources. Self-tooling behavior has two aspects: adapting the human body or adapting a machine as a tool for various new tasks.

### *Adaptive Design for the Human Body*

For tens of thousands of years, human body parts have been used as tools for endless purposes. When taking measurements, for example, certain legacy units are still in use today. At one time, the inch was figured as the span of the knuckle of King

© Springer-Verlag London 2016

Y. Cai, *Instinctive Computing*, DOI 10.1007/978-1-4471-7278-9_6

Edgar's thumb. Charlemagne decreed one foot to be the length of his foot. A yard was the distance measured on King Henry I, from his nose to the fingertip of his outstretched arm.[1]

In the digital era, human finger gestures have been studied intensively for human-computer interaction design, specifically for mobile computing input. For thousands of years, our opposable thumbs have been used for holding items, or in the expression of the "thumbs-up" gesture. The repertoire of the human thumb is humble compared to the highly notable index finger. Since the invention of the typewriter 148 years ago, human thumbs have been used for typing a few limited letters and pressing the spacebar on the typewriter. However, within only the last decade, the human thumb has "evolved" from limited typing to full alphabetic text typing while other fingers just hold the mobile phone. This is attributed to the emergence of touch screens on mobile devices.

The touch screen on mobile devices is a matrix of pressure sensors beneath the screen that detects the coordinates of pressed areas. It is virtually a playground for designers to invent a whole new language of hand gestures for many functions. Readers may use a deck of index cards to write down the potential functions and possible hand gesture categories to see how many combinations are possible. For example, we can first write down the following functions:

```
typing
drawing
cutting
slicing
flipping
rolling
rotating
resizing
stretching
navigating
zooming
cropping
erasing
controlling (volume)
shooting (photos)
poking (friends)
```

And we can write down following hand gestures on cards:

```
single tapping
double tapping
sweeping
holding
dragging
two-finger rotating
two-finger stretching or narrowing
```

So far, we have 16 cards for functions and 7 cards for hand gestures. The possible combinations are: $16 \times 7 = 112$. We can also incorporate more dimensions in the design space, such as 3D touch sensing. The random combination of cards is a way of generating design ideas. However, not all the combinations are feasible. A

[1]Fairbanks AT and Fairbanks E (2005) Human proportions for artists. Fairbanks Art and Books.

knowledge-based filter is needed to trim the alternatives down to a manageable size. In the upcoming chapter, "Creativity," we will explore this method further.

Microsoft's Xbox Kinect interface is another example of self-tooling design by adapting human body gestures, including the movement of arms, hands, legs, and the body as a whole. The Kinect enables gesture control for games, without direct control devices such as a game controller or a motion sensor (Nintendo Wii).

Steve Jobs took self-tooling seriously in the Apple products he was responsible for. Jobs insisted on not having a large screen for the iPhone, because it prevented users from typing words and holding the phone with just one hand. He also opposed the use of a stylus, favoring the use of his fingers for navigating the touch screen of the iPad. Simply put, fingers are self-sufficient – one never has to worry about losing the stylus. After Jobs' death, Apple released the 5.5-in. large screen iPhone 6+ similar to what its competitor did years ago. However, appreciating the existing market for 4.7-in. smaller screen phones, Apple decided to continue the production of the smaller phones into near future. In addition, Apple released the stylus iPen for iPad Pro, but they targeted professional groups such as architects, artists, animators, and designers who need drawing precision in their workplaces. Behind every market, human instincts are the invisible hands that drive consumer behavior. Self-tooling is one of the biggest factors in product design.

## *Adaptive Design for Machines*

We often try to retool existing machines for new tasks. The more comprehensive the systems are, the more spin-off applications there can be. Taking a car for example, we can deconstruct a car into a few major components for self-tooling usages:

```
lights
seats
pumps
engine
generator
batteries
wheels
sound system
wireless connections
```

Lights and seats have commonly been used for camping outdoors. The windshield water pump has been used by the well-known comedian Mr. Bean for brushing his teeth in one of his movies. Some cars are powerful enough to provide towing force for pulling other cars out of the mud. The electrical generators and batteries of a car can provide 24 V of electricity for appliances. Tesla started to sell the batteries for home power storage for off-the-grid living. Ford Global technologies filed a patent in 2015 for transforming a car wheel into a self-propelled electric unicycle. Users would simply detach the back wheel of the car and then attach the kit stored in the trunk that provides the controls and seat for riding the electric unicycle. A contemporary car-tooling example is a smartphone or tablet that can be used as a dashboard. This enables the user to park farther from their final destination and to safely maneuver in a crowded downtown area. Modern vehicle users also enjoy advanced sound systems for music that is downloadable from onboard wireless systems. However, smart systems allow for adversary attacks. For example, computer viruses may be able to access the backdoor of the sound system network and enter the vehicle's signaling and control systems. One man's channel is another man's tunnel.

Similarly, smartphones can be used for applications far beyond their original purpose. A phone can be used as a flashlight. The touch screen of a smartphone with three-dimensional pressure sensors can be used for weight measurement. A phone can measure your heart rate by detecting the color changes of your finger as pressure is applied on the phone's camera. The smartphone can also be a sound decibel meter using its microphone input. Every year, thousands of apps are developed for mobile platforms. Because of their mobility, smartphones have more potential to be adapted to new situations than desktop computers.

A more comprehensive case study is how to enable an autonomous mobile robot to adapt itself to challenging situations, such as opening a door in a hospital, without having to reconfigure itself. Assume a mobile robot has a basic architecture – an arm and gripper that can reach the height of any door accessory, e.g. doorknob, door handle, door press latch, card reader, or combination keypad. See Fig. 6.1. For turning a doorknob, it is possible for the robot to use its gripper to rotate the knob with enough friction. A sliding sensor would be needed to detect sliding in order to apply sufficient pressure on the gripper. In addition to rotating the door handle, the gripper must exert some force in the downward direction. The gripper may push the door press latch forward in order to open the door. When additional force is needed, the robot may back up and use momentum to add thrust. The gripper may also slide a card through the slot in a vertical card reader. However, picking up a card from a flat surface is rather challenging, so an alternative is to hang the card from the robot's neck on a lanyard. An even more challenging task is to open a door with a combination keypad. The robot may require more than two fingers because some keys must be pressed simultaneously, and the fingers must be small enough to press individual keys without accidentally pressing surrounding keys. If the robot forgot

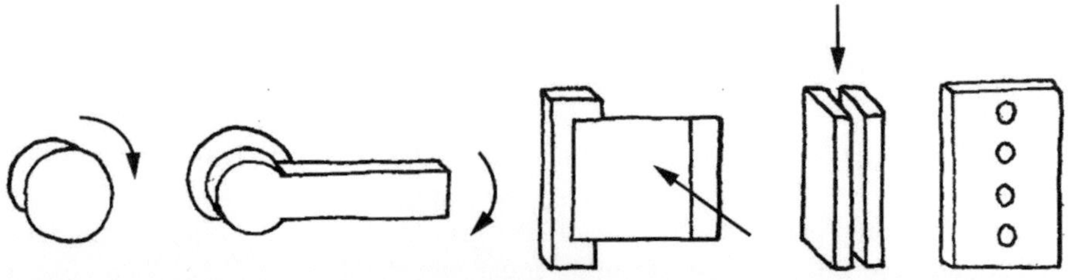

**Fig. 6.1** Various door opening challenges for a mobile robot from *left* to *right*: door knob, door handle, door press latch, card reader, and combination keypad

the password combination, it would have to run a lengthy password cracking script. Some robots are specifically built for lock-picking, but in this case it would not be necessary because of the flexibility of the robotic fingers.

This "Gedanken Experiment," or thinking lab, shows that a human-like robotic hand can be adapted for many tasks that we take for granted. Precision and coordination between fingers are needed to process the subtle, spatiotemporal details in challenging tasks such as door opening. Like human systems, modern mechanical systems such as smartphones, robotic hands, and connected vehicles, contain abundant recourses for self-tooling. Many of these functions can be improvised through additional programming for good or for bad purposes.

## Limb Extension

In some cases, self-tooling is not enough to solve the problem at hand. For instance, to reach areas that are too high for arms to reach, or too small for fingers to poke. Many ethological studies about animal tooling behaviors have focused in this realm, which we call *limb extension* behaviors.

Fifty years ago, psychologist Wolfgang Köhler conducted the famous "banana test" for chimpanzees. The idea was to test how the animals would solve the problem of reaching for food by making tools. Köhler hung a basket of bananas upon the ceiling where the chimp's arms could not reach. The lab provided a few materials for the chimps: multiple wood sticks and boxes. The sticks were hollow and had different diameters. The chimps were able to insert the small sticks into the backs of the larger ones to make a longer stick. In other groups, chimps stacked the boxes and climbed on top of them to reach the basket. Köhler dubbed the moment when the chimp suddenly found the solution, "insight." Insight is part of an animal's natural instincts.

It is worth noting that tool making is often limited by the *availability* of materials and/or the *manufacturability* of the process. It is common practice to make tools from the available materials within a close range, e.g. crows can pick up a leaf stem near a tree to make a stick. *Manufacturability* is referred to as the capacity to perform actions such as grabbing, aligning, assembling, and cutting. Humans, chimps, and monkeys have the most flexible, versatile hands in nature, leading to the ultimate level of manufacturability for instinctive tool making. Here, we may hypothesize that as actuators, our hands enable us to make tools – not our brain. Evolving tools have evolved human intelligence.

Let us look at how a robot makes tools in the "banana test". Similar to Köhler's experiment, a basket is hung high from the ceiling and the robot cannot reach it with its arm. Assume there are different sizes of hollow sticks and boxes. The robot may pick up one or more to see if the single stick can reach the basket. If not, then it must connect multiple sticks together. To accomplish this, two arms are needed – one for holding and another for alignment, insertion, and rotation. This scenario seems quite staged because it anticipates that the robot would try to connect the

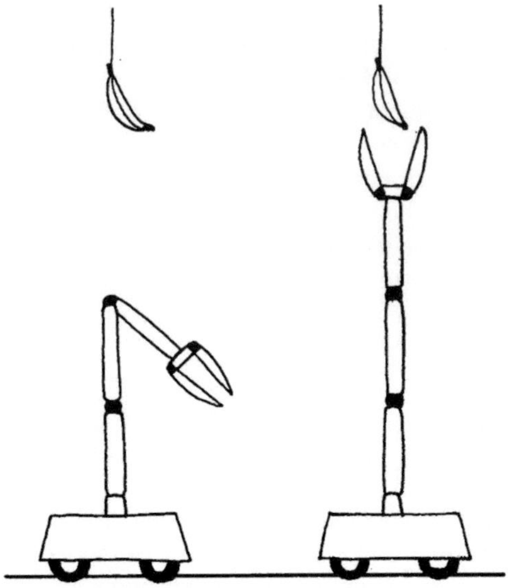

**Fig. 6.2** The "Banana Test" for a robot

sticks. The robot should have the knowledge of how to connect things for length or height adjustment. As the room also has boxes, there is nothing to prevent the robot from exploring the boxes. The robot can push one or more boxes under the basket and try to climb on top. If the stack of boxes is not high enough, the robot is expected to add more boxes on top. This would also require two arms, or two one-armed robots to stack boxes. To make matters worse, unless the robot is a legged robot, climbing on top of a rectangular box is challenging. It appears that the staged materials for the "banana test" were straightforward enough for chimpanzee subjects, but uncertain for a wheeled mobile robot. In desperation, the robot may improvise a way of "climbing" the boxes by lifting its base to the edge of the box with its arm(s) and then pushing itself to the top of the box. This mimics the action of a long arm earthmover loading itself onto a truck (Fig. 6.2).

Self-loading is a process of a self-tooling behavior. It is a plausible, but not ideal solution. This is because a robotic arm and gripper can be damaged while exerting or sustaining extra force. Now, let us assume the robot can 3D print items with one of its arms. The high level of manufacturing capability would overcome the limited availability of materials. Assume the robot can 3D print a stick with a hook or knife attachment to cut the rope. See Fig. 6.3. With these special tools, the robot can easily take down the basket. The printed stick alternatives may work alone, which reduces the assembly process, or in conjunction with other existing materials. For example, the robot may 3D print a ramp for climbing to the top of a stack of boxes. The second alternative seems more plausible because it requires minimal printing while also maximizing the usage of the available materials.

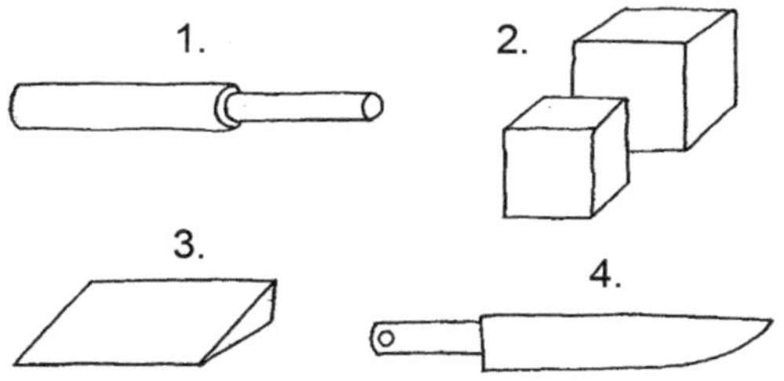


**Fig. 6.3** Extension alternatives for a mobile robot: (*1*) stick, (*2*) boxes, (*3*) 3D printed ramp, and (*4*) assembled stick with a knife for cutting the rope

Let us also explore how to reach a tiny spot that is too small to reach with a finger. Crows use thin wooden sticks to pick up worms inside holes that are too small for their beaks to reach. Wooden sticks are the simplest solution for birds to reach into small holes because that material is widely available, and the size and shapes of the sticks allow crows to hold them by their beaks. Crows learn how to manufacture wooden sticks into different shapes for different situations. As a part of group survival instincts, crows share tool-making skills amongst their flock. Therefore, crows from different flocks may have different strategies. Monkeys also use wooden sticks to poke at food in small holes. They can use their hands instead of their mouths to hold sticks, giving them more flexibility and precision for controlling the tool.

The medical robot Da Vinci is an excellent example of the limb extension needed for human doctors to perform minimally invasive surgery. Take for example keyhole gastric bypass and brain surgeries. Rather than having to open the patient's torso or cranium, robotic arms can enter the body through a small incision to reach the targeted area. Using this technology, there is no need for traditional, invasive open chest or open skull procedures. Furthermore, endoscopy technologies enable doctors to observe, take biopsies, or operate in digestive systems such as the colon or the stomach. The emerging technology of capsule, or pill cameras can reach even the tiniest of areas such as the small intestine, and send back exam images from front-view and rear-view cameras at a rate of at least 4 frames per second. The biggest advantages of capsule cameras are that they are non-invasive, no anesthesia is necessary, and they are low-cost – only a fraction of the price of a conventional endoscopy procedure.

However, like any mobile device, pill cameras have to have batteries. The camera battery lasts approximately 8 hours, but it takes 24 hours for the camera to exit the body. The battery is turned on only during the last 8 hours of the exam. In order to extend their lifespan, in-vivo robots can harvest energy from the biomass resources inside of the body using specialized bio-fuel cells. To perform routine procedures like endoscopies, the pill camera must also have arms and legs for taking biopsies, as well as for maneuvering inside of the body. Capsule cameras are tiny robots inside of our bodies that can do exams while we work, play, or sleep. It is only a matter of time before we have nano-sized robots travelling through our bloodstream for medical diagnoses or operations.

## Scaffolding

Scaffolding is a temporary structure used to support a work crew and materials to aid in the construction, maintenance, and repair of buildings, bridges, and other man-made structures. The use of scaffolding can be traced back as far as 17,000 years. Archeologists found sockets in the walls of Paleolithic cave paintings at Lascaux, suggesting that a scaffolding system was used for painting the ceiling. As a tool, scaffolding has unique characteristics. It provides support for accessing a work area without damaging it, and it can be made out of inexpensive materials. After work is complete, the scaffolding is removed, leaving virtually no trace on the finished product. The idea is to provide a temporary support for the tool-making or problem-solving processes and the support can be removed afterwards.

### *Physical Scaffolding*

Scaffolding has been physically applied to 3D additive manufacturing using a technique known as fused filament fabrication (FFF). Materials including plastic filament, metal wire, concrete, or chocolate are melted and pushed through the printer's extruder, and laid down layer by layer to produce a part. FFF uses the molten filament material itself as scaffolding. However, the toothpaste-like filament is too elastic to support itself across wide spaces. Also, the part will not be stable if the base of the object is too small, such as a sphere. 3D software can convert a 3D model into machine instructions known as "G-Code." Within a G-Code file, the software can identify areas where scaffolding is needed, and insert the related printing commands into the machine code. To print a ball for example, the machine prints out scaffolding columns around the base until the print reaches the halfway mark of the height of the ball where the upper part of the ball can then be supported. 3D printed scaffolding shapes are designed to be thin, sparse, and easily breakable, resembling forms such as hollow columns, wafers, zigzag sheets, and honey combs (Figs. 6.4 and 6.5).

Scaffolding is important when 3D printing biomedical tissues and organs. The choice of materials, cell types, growth and differentiation factors, and technical challenges related to the sensitivities of living cells and the construction of tissues

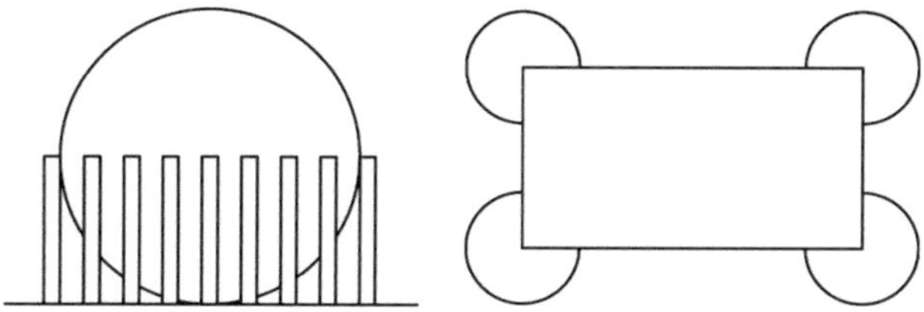

**Fig. 6.4** The scaffolding structure for printing a ball and the Mickey Mouse ear-shaped rafts

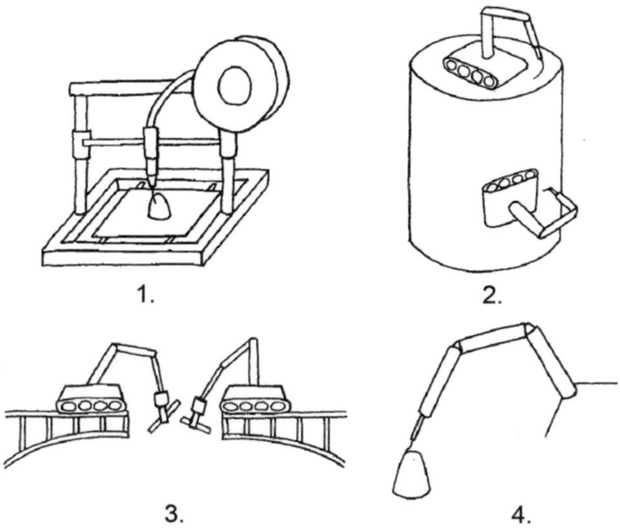


**Fig. 6.5** 3D printing architectures: (*1*) three-axis desktop printer, (*2*) climbing robots, (*3*) welding robots, and (*4*) 3D printing robotic arm

make bio-printing much more complicated than non-biological 3D printing. In bio-printing, scaffolding materials can be organic or inorganic. For example, by removing cells from a donated heart, the remaining structure is ideal for providing the scaffolding for printing and growing new cells onto it. Tissue can easily be printed on artificial scaffolding structures, allowing new cells to propagate and grow.[2]

[2]Murphy SV and Atala A (2014) 3D bioprinting of tissues and organs. Nature Biotechnology, 32, 773–785, 2014.

## *Cognitive Scaffolding*

Scaffolding can also be used as a metaphor for cognitive supports to facilitate a learning process. Take, for example, instructional scaffolding, where support materials such as templates and models are introduced to students and then gradually removed as students develop autonomous learning strategies.

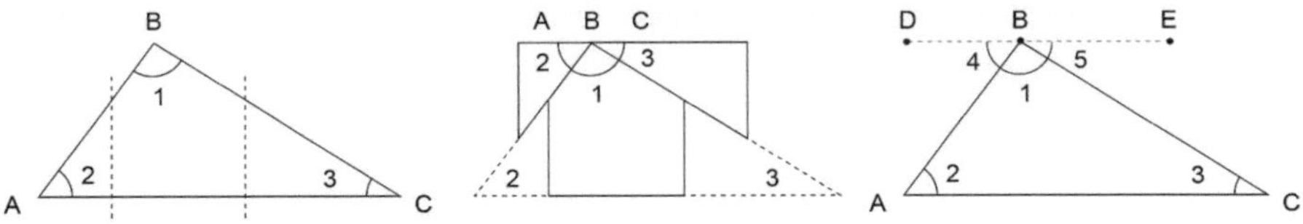


**Fig. 6.6** Auxiliary line for proofing a trigonometry problem

Scaffolding has been widely used in mathematical learning processes. In trigonometry, *auxiliary lines* are used for problem solving. They are imaginary lines that help to solve the problem, but do not alter the shape. For example, given a triangle ABC, prove the sum of the angles of triangle ABC, angles 1, 2, and 3 is equal to 180° (Fig. 6.6). A cave man's proof is to cut the triangle ABC corners and put them together to form a straight line. Since a straight line is 180°, it proves this hypothesis. However, this method destroyed the shape. A more elegant method would be to draw an auxiliary line DE such that line DE is parallel to the line segment AC. According to the parallel postulate, when a transversal cuts parallel lines, alternate interior angles are congruent. Thus, angle 4 is congruent to angle 2. Similarly, angle 5 is congruent to angle 3. Angles 1, 4, and 5 were defined by a straight line. We know that the measure of a straight angle is 180°. The sum of angles 1, 4, and 5 form a straight line. Therefore, based on substitution, the sum of angles 1, 2, and 3 equals 180°.

## Substitution

From the trigonometry problem above, we find that substitution of equivalent concepts leads to logical reasoning towards the proof. Substitution is to replace one item with another based on similar functions, values, or concepts, but with advantages such as abstraction or improvement. Mental imagery or visual diagrams can be viewed as the substitution of physical objects. Those visual representations are not totally abstract as with numbers, but intuitive enough for humans to associate with their physical properties.

Bees build up the same home from generation to generation. Humans on the other hand often have a mental image or an external drawing or physical model of the home before they build it. For decades, researchers have studied representations of mental images and diagrams and the functions they play in problem solving in diverse multidisciplinary areas. Philosophers have long been interested in the nature of mental imagery and have debated the reality and nature of mental imagery and visual representations. Design theorists have always been interested in the role of sketches and diagrams as design aids. Artificial Intelligence researchers have been interested in diagrammatic reasoning, especially in the early work on geometric theorem proving.[3] Here, we look into evolving engineering diagrams as tool-making languages.

## *Diagram Languages*

Although computer programming languages have been developed for half a century, programming in C, C++, Java, Python, or Ruby is still challenging to the majority of people. Conversely, engineers who were familiar with hardwired relay circuits and critical control systems, invented their own diagram languages to enable field engineers to "wire" the mechanical and electrical or computational components together on a computer, similar to how they would wire hardware systems in the physical world. Wiring rather than coding allows factory engineers and technicians to develop new software without additional training to learn a formal programming language. Development and maintenance have been simplified because of the resemblance to familiar relay hardware systems. Diagrams can substitute physical systems with off-line simulators, therefore debugging becomes simulation. From a cognitive perspective, diagrams are substitutions for physical systems.

Ladder logic was originally a symbolic notation system for designing relay circuits for process control. It is named as such because the layouts resemble ladders. Each device in the relay rack is represented by a symbol on the ladder diagram with connections between those devices shown. Ladder logic has evolved into a language that represents a program with a graphical diagram based on relay logic hardware. It has been used to develop software for programmable logic controllers (PLC) for industrial control applications.

Figure 6.7 shows an example of the common industrial latching start/stop logic circuit, where [ ] means open and [\] means closed. We have a "Start" button to turn on a motor, and a "Stop" button to turn off the motor. When the "Start" button is pressed, the input goes true, it activates "Run" to be true, and keeps the mechanism latched even though the "Start" button may be released. When the "Stop" or "Emergency Stop" button is pushed, the current is cut off and the motor will stop. Consequently, the "Run" circuit logic returns to its inactive state.[4]

---

[3]Chandrasekaran B, Glasgow J and Narayanan NH (1995) Diagrammatic Reasoning – Cognitive and Computational Perspectives. The AAAI Press and The MIT Press

[4]Wikipedia: Laddler Logic. https://en.wikipedia.org/wiki/Ladder_logic

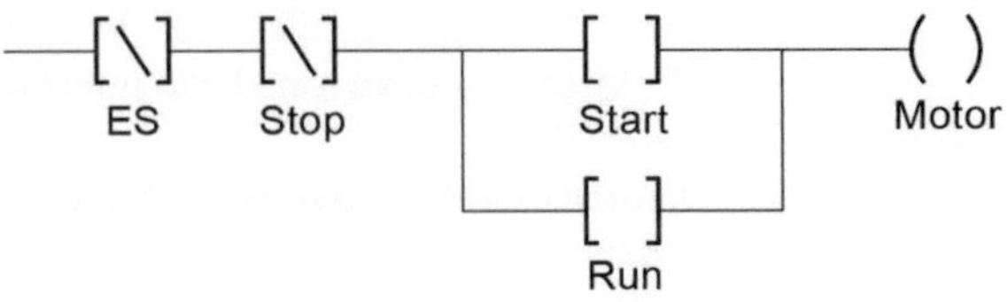


**Fig. 6.7** Ladder Logic diagram of a motor control system

Like music notations to composers, ladder logic diagrams enable factory engineers to compose and perform sequential control of a process or manufacturing operation. In ladder logic, engineers may substitute components (e.g. replacing a motor with a scanner). Additional functionality can be added to ladder logic as a special block. When the special block is powered up, it executes code contained in the block, similar to adding a new chip to the circuit board, and additional wiring is required.

LabVIEW™ is a symbolic design software package made by National Instruments.[5] It has been widely used in the development of scientific and engineering instruments and systems. It is essentially a modular simulation block language with which engineers can "wire" the software blocks together to get offline simulation or real-time online results. LabVIEW has gradually become a visual programming language for special components such as an FPGA (Field Programmable Gate Array), a large-scale integrated circuit of primitive gate-based computer chips, which is conventionally programmed using logic design languages such as Verilog. Because LabVIEW has the pre-fabricated code in blocks, it increases programming productivity by orders of magnitude.

The Video and Image Processing (VIP) Blockset is a tool used for rapid design, prototyping, graphical simulation and to effect code generation of video processing algorithms. It was developed by the Mathworks as a component of its original SimuLink language. Like LabVIEW, it is a high-level visual language with pre-fabricated code in each block. It can also automatically generate the code in other languages, such as embeddable C for Digital Signal Processing (DSP) chips.

The diagram languages above contain a set of pre-fabricated blocks that resemble hardware components such as computer chips. Typically, a chip contains a data sheet that specifies its input, output, function, and other specifications, as well as sample circuits. Engineers often search online for specific modules that are interchangeable. This creates a new paradigm of software development. Instead of coding from scratch, the developer may search for a specific block or module online and "substitute" the descriptive block with the pre-fabricated code. Figure 6.8 illustrates the concept of this process: a diagram of an edge detection function containing blocks with snippets of code for converting color images to grayscale (A), the edge detection filter (B), and thresholding (C). The developer may search

[5] National Instrument: http://www.ni.com/labview/

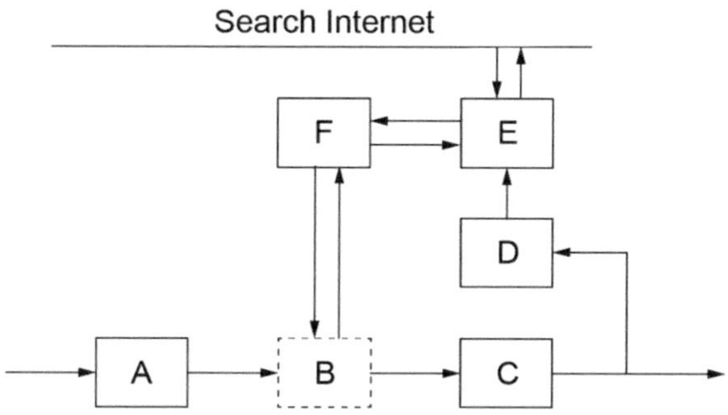


**Fig. 6.8** Substituting a software block from online repositories

online repositories for appropriate algorithm blocks, e.g. Canny filter, Roberts filter, or Prewitt filter, and so on. Block B can be substituted with any of the filters to accomplish the edge detection task.

Diagram languages align mental imagery with visual representations so that developers may focus on function rather than low-level code. In fact, many diagram languages automatically generate code. However, they have certain limitations. Normally, diagram representations work well if the functions can be clearly divided into blocks and there are as few interactions as possible between blocks. They are not flexible compared to conventional programming languages, but they allow programmers to make fewer errors because the blocks have been pre-fabricated and pre-tested for quality assurance. Visual representations may also have scalability problems. For example, if we want to have 100 multipliers in a LabVIEW diagram, we may have to copy the multiplier 100 times onto 10 pages, but in a conventional programming language such as Verilog, we may need just one line of code.

## *Substitution in Machine Learning*

The advantage of machine learning algorithms perhaps lies in the "black box" approach, in which developers need little knowledge about the structure of data. In contrast, the "white box" approach is to model everything in the data with a full understanding of the relationships and dynamics within it. In reality, we often fall in between those two approaches and use a "grey box" approach, in which we only partially have explicit semantic or mathematical knowledge about the data, but we do not know how these pieces of knowledge work together to get satisfactory results. In many cases, we can design a layer of feature blocks followed by a layer of learning algorithms, wherein feature blocks generate input data for learning blocks. Feature blocks can be added or removed based on their contributions to the goal function.

Let us look at a medical video analytics problem. Assume we have a facial expression video, in which we want to detect whether the eyes are open or closed. The level of distention is an important measure for affection analysis. Given samples of open and closed images, we can represent a layer of semantic feature blocks: edges, length of edges, areas, shapes, as well as other subtle features like texture, and high or low frequency distributions of pixel variations, which we do not have words to describe. We can then construct a classification layer that takes combinations of the feature descriptors as input and generates classification results as output. The classification model can be a learning model such as Bayesian Network or AdaBoost. This "grey box" approach incorporates high-level human descriptions (edges and curvature) and the "hidden," low-level imagery descriptions like texture. In contrast to the prevailing learning algorithms that assume all the feature descriptors, disregarding if semantic or non-semantic descriptors should be included in the model, the semantic descriptor blocks can be substituted with other blocks. Furthermore, they can be eliminated if their contribution is below the threshold. As a result, it expedites the learning process because fewer parameters are involved. This simulates the human learning process in that we are inclined to keep optimizing, or to describe the object using fewer parameters. The substitution process can keep going throughout the life of the software system, which is also different from prevailing machine learning strategies that normally only learn once and use the same parameters forever (Fig. 6.9).

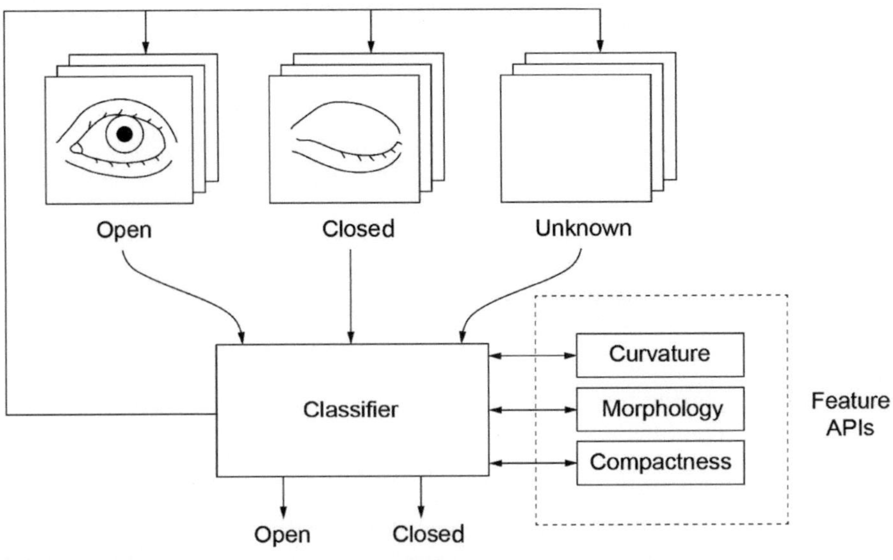


**Fig. 6.9** The "grey box" approach incorporates high-level human descriptions and the "hidden," low-level imagery descriptions as well as weights

A well-trained machine learning system is a knowledge system that contains fine-tuned values or parameter weights based on a massive number of training samples. When we have a new sample dataset that is significantly different from the previous ones, then we have to retrain the model. For example, a popular face detection algorithm is based on wavelet features and machine learning classification (e.g. AdaBoost models). It can only detect frontal-view faces. If we want the algorithm to detect profile-view faces, then we must collect profile-view face samples for training. If we want to detect both frontal-view and profile-view faces, then we must train the algorithm with both sample sets, which can be time-consuming. A more efficient way of training would be to add the profile-view face detection block to the existing frontal face detection block, and take an OR output from the two blocks.

A machine learning system can be divided into three general blocks: the shell of a learning algorithm, the trained model, and the original training sample data. These three blocks can be substituted with other blocks serving similar functions.

Massive original training data is critical to the efficacy of any learning system, however, it is physically challenging to move it along. Modeling is necessary to simulate the distributions and dynamics of raw data so that the model can "substitute" raw data, even fill in missing data that is not available in the original dataset. For example, medical imaging studies often use dummy "phantoms" to test medical imaging systems. Another example is mapping the epidemic spread of the Ebola virus. Software agents can be used for generating data of household members, school-age children, shopping trips, etc. The nearest neighbor data can be used to estimate any missing data in rural areas.

Machine learning algorithms can be hardened into a chip or standalone board that can be implanted into a system. In fact, several neural network algorithms have been implemented on chips to be faster and cheaper than implementations on a computer. For example, Cognimem's CM1K chip[6] has 1024 neurons using the Radial Basis Function neural network learning algorithm. The board is 5 cm $\times$ 5 cm with a $640 \times 480$ pixel black and white camera. It can be used for simple pattern recognition tasks such as sign language recognition, face recognition, gender classification, and object tracking.

## Self-Assembling

Social insects, such as ants, are well known for working together as a whole unit. Ants carry food and protect their nest together. In flooded forests, fire ants swim by interlocking their legs together to form a living raft. A group of ants can form a bridge to cross any gaps on the way home. Collective tooling is the making of

[6]Cognimen Technologies: Chips and Modules. http://www.cognimem.com/products/chips-and-modules/CM1K-Chip/index.html

a tool by connecting a group of individuals. Collective tooling mechanisms can be implemented in hardware such as robotics or software such as online games.

In robotics, collective tooling is achieved by shape formation and inter-robot communication to achieve *self-assembling*. M-Block is a $15 \times 15 \times 15$ cm cubic magnetic robot designed to mimic social insects that can climb, tumble, roll about the ground, leap through the air, and move while suspended upside-down.[7] These mobility functions enable multiple cubic robots to work together to self-assemble into different types of tools or functions. This could be to arrange themselves into scaffolding for building projects, or they could tumble into disaster areas and reconfigure themselves to survey the area. Shape formation of the M-Block is based on the robot's inner kinetic energy. Each robot has an energy-conserving flywheel inside that spins at up to 20,000 rpm. When it stops, the rotational force transfers to the cube, causing it to roll. If the flywheel spins fast enough, the cube will jump. Each cube has magnets on the faces and corners. The face magnets join cubes together, while the magnets on the corners help keep the cubes in contact as they experience changes in position, like the connected legs of fire ants. The magnets on the faces of the cubes can be turned on or off. Because the robots are cube-shaped, they are stable and relatively easy to stack into multiple configurations, such as a tower, a bridge, or a wall. Cubes can also slide alongside each other for alignment without any external forces. As a result, M-Block can form almost any desired shape.[8] (Fig. 6.10)

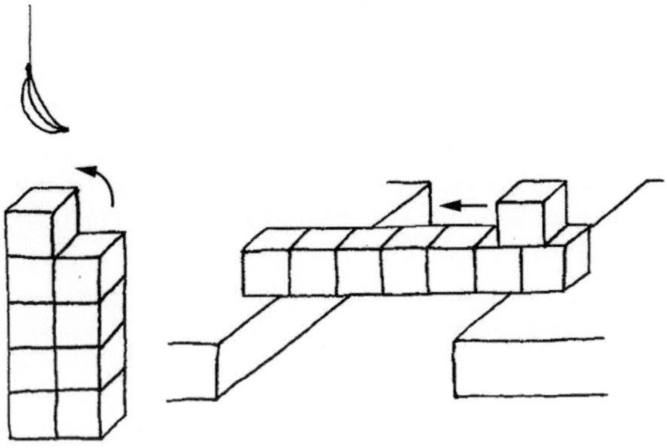

**Fig. 6.10** Self-assembling M-Block robots

Collective tooling can be found in software development. The Open Source movement has introduced crowdsourcing platforms such as GitHub, Mathworks

[7]Emspak J, Cubic Robots Build Themselves. http://news.discovery.com/tech/robotics/cubic-robots-build-themselves-131004.htm

[8]Romanishin J, M-Blocks Modular Robots, YouTube. https://www.youtube.com/watch?v=mOqjFa4RskA

Central, and Kickstarter, and thousands of sites for sharing, such as Instructables and Pinterest. However, usually collective tooling is more about sharing than purposeful development.

Mechanical Turks collect massive individual intelligence in order to improve software tools, ranging from emotion recognition, to disease diagnostic systems. Online games consume roughly 200 million minutes of human attention each day, which is a lot of cognitive surplus. Here, collective tooling is a process that connects many volunteers to solve a complex problem. An example of crowdsourcing is to divide a puzzle into many small pieces and allow thousands of online users to work together to solve it. Collective tooling can be used in gaming platforms so that users are entertained, and also trained or educated. Additionally, collective tooling games can serve as an important resource for scientific research. "Gamification" of certain tasks makes users more motivated and engaged in the problem at hand.[9,10,11,12]

A video game is a real-time database that continuously updates key variables. The core functionality provided by a game engine includes the input process, the rendering engine for 2D or 3D graphics, and a physical engine for simulating the game world. The feasibility of a collective tooling approach has been tested in a game-based approach to malaria image analysis. In particular, this approach investigated whether anonymous volunteers with no prior experience would be able to count malaria parasites in digitized images of thick blood smears by playing a web-based game.[13]

Foldit is a crowdsourcing game for 3D protein folding.[14] As we know, proteins can be involved in diseases like AIDS, cancer, and Alzheimer's in many different ways. The number of different ways a small protein can fold is astronomical because there are so many degrees of freedom. In Foldit, the players manipulate different lengths of side-chains and backbones until they fit into the smallest volume possible. Game scores are awarded by how much energy would be needed to hold a protein in the shape the player has created. The real challenge comes from competing against other players to make the highest-point version of a specific protein. So far, more than 100,000 people have downloaded Foldit, turning the game into a massively multiplayer competition.

---

[9] Good B, Su A. (2011) Games with a scientific purpose. Genome Biol 2011(12):135

[10] Nielsen M (2011) Reinventing Discovery: The New Era of Networked Science. Princeton, NJ: Princeton University Press.

[11] McGonigal J (2011) Realty Is Broken: Why Games Make Us Better and How They Can Change the World. New York: Penguin (Classics)

[12] Swan M (2012) Crowdsourced health research studies: an important emerging complement to clinical trials in the public health research ecosystem. J Med Internet Res 2012; 14(2):e46

[13] Luengo-Oroz MA et al. (2012) Crowdsourcing Malaria parasite quantification: an online game for analyzing images of infected thick blood smears. J Med Internet Res 2012:14(6)e167. http://www.jmir.org/2012/6/e167/

[14] FoldIt. http://fold.it/portal/

## Self-Replicating

In nature, many animals can reproduce without sex, from the smallest insects to the biggest self-fertilizing sharks, budding jellyfish, and all-female reptiles. Asexual reproduction comes in many forms too, from basic cell reproduction by mitosis, to complex cloning cycles.[15] Computational self-reproduction is straightforward. Computer viruses can reproduce themselves across connected media. However, self-reproduction in hardware is rather challenging due to the complexities of materials and assembly.

Taking a self-replicating 3D printer for example, it involves reproducing mechanical and electrical components from the printer and assembling them together. The goals of self-replication are: to make a 3D printer that can produce as many of its own parts as possible, to make it easy to assemble, and to use parts that are easy to find. The first self-replicating machine started with an open source system called RepRap (short for replicating rapid prototype) invented by Adrian Bowyer from the UK in 2005. RepRap uses fused filament fabrication (FFF) to lay down material in layers. A plastic filament or metal wire is unwound from a coil and supplies the material to produce a part. RepRap has used its own printed parts, including plastic gears and joints. Today's affordable FFF 3D printers contain components and mechanics that were originally designed for RepRap printers. The recently-developed self-replicating printers Snappy[16] and Dollo 3D[17] consist of over 70 % 3D printable parts. These designs require no belts, pulleys, or metal rails, and almost no screws outside of the motors or extruder hot-end. The entire system can be snapped together within a couple of hours.

The non-printed parts of Snappy present major obstacles for self-replicating an FFF 3D printer. These parts usually include electronic components (stepper motors, cooling fans, controller circuit boards, and power supplies), and metal components (steel threaded rods, bearings, copper extruder hot end, steel extruder drive gear, mechanical limit switches, wires, and glass build platform). In theory, these "non-printed" parts are could be printed at a certain level.

3D printed stepper motors, or brushless pulse motors, have been prototyped.[18] Stepper motors usually contain a rotor with a permanent magnet and a motor base with copper coils around arms. The coils are synchronously powered by a switching power supply that produces an electric pulse signal to drive the rotor in a rotating magnetic field. For example, a motor prototype may have ten magnets. All of the magnets face outwards. Ten coils alternate between push and pull mechanisms to

[15]Foden S (2016) Animal that Don't Have a Male for Reproduction. http://animals.mom.me/animals-dont-male-reproduction-9838.html

[16]Reprap: Snappy. http://reprap.org/wiki/Snappy

[17]BenjaminE4, Dollo 3D: Self replicating 3D printer. http://www.instructables.com/id/Dollo-3D/

[18]Bell D (2015) 3D Print Your Own Stepper Motor. http://makezine.com/2015/08/18/3d-print-stepper-motor/

spin the rotor. At the current time, the coils are wound by hand.[19] However, it is also possible to print a copper coil around the plastic arm. In order to increase the turning angle resolution, a gearbox can also be printed.

On the other hand, high-end metal additive manufacturing has been applied in industry for two decades, with applications ranging from dental crowns and kneecap implants, to jet and rocket engines. This technology continues to expand rapidly.[20] 3D printed metal objects often outperform those produced with traditional casting and milling techniques in terms of accuracy and complexity. A typical metal printing machine uses a high-powered laser to melt metal powder in a box layer by layer, similar to other powder-based 3D printing processes. The difference is in the powder. A metal printer can print objects in aluminum, stainless steel, copper, or alloys. Therefore, it is possible to print the metal parts needed for a self-replicating FFF printer, such as steel threaded rods, micro switches, and copper wires. The biggest problem with metal 3D printers is their high cost – they are only affordable to large companies like General Electric or SpaceX. However, the situation has changed rapidly as low-cost metal printers such as MatterFab have emerged.

Large-scale integrated circuits can be printed and packaged in smaller forms. However, they are normally specially designed. Generalized circuits such as an FPGA (Field Programmable Gate Array) are essentially computers that can be reconfigured for different purposes. Large-scale integrated circuits can also be integrated with mechanical components such as micro sensors in the form of microelectromechanical systems (MEMS). MEMS are made up of components between 1 and 100 μm in size, and devices generally range from 20 μm to 1 mm in size. Integrated circuits can further scale down into nanoelectromechanical systems (NEMS) and nanotechnology. MEMS became practical once they could be fabricated using modified fabrication techniques in the existing semiconductor industry, including molding, plating, and etching.

Printed circuit boards (PCB) are used to assemble microprocessors, stepper motor controllers, sensors, switches, and power sources. Circuit boards can now be printed in multiple layers, with a printer similar to an inkjet printer, using insulating and conductive nano inks. This brings exciting new possibilities to the field of custom manufacturing for self-replicating systems.[21]

Assume we have all of the electronic components and we must assemble them onto a PCB. For decades, engineers dreamed of better ways to join electronic components together. The traditional way to stick electronic components together

[19]Halterman TE (2015) Designer 3D Prints his Own Brushless Pulse Motor. http://3dprint.com/89359/3d-print-brushless-pulse-motor/

[20]Dunham S (2015) Metal Additive Manufacturing Market Continues Rapid Expansion. http://3dprinting.com/metal/metal-additive-manufacturing-market-continues-rapid-expansion/

[21]Grunewald S J (2015) 3D Print Multi-layer Circuit Boards with the Nano Dimension Dragonfly 2020 3D Printer (2015) http://3dprint.com/59360/dragonfly-2020/

involves soldering and welding with a heater. The new method involves using special hot glue made out of metal that sets at room temperature, and requires very little pressure to seal.[22]

Overall, it is possible to self-replicate most of the components in a 3D printer, say, over 90 % or more. However, the cost and complexity of such a system would be overwhelmingly high, not to mention printing in multiple materials on multiple scales, let alone precisely assembling parts with two robotic arms. Perhaps we are better off with low-cost designs with the maximum amount of feasibly printable parts and the most minimal assembly processing required.

## Adversary Tooling Behaviors

Tool-making behaviors show great promise in making intelligent systems. However, they can have adverse impact as well. Similar to our immune systems that can detect and terminate intruders, killing tools can do harm to our bodies if left unchecked. In order to prevent immune cells from overtaking our bodies and attacking healthy cells, immune cells stop self-replicating when their time has expired and instructions are received to discontinue reproducing.

Computer viruses are self-replicating chunks of code that spread across networks. They can make tools from hosts. A polymorphic program exhibits self-tooling by *substituting* code in different forms, yet operating with the same functionality. For example, equations $3 + 1$ and $5 - 1$ both achieve the same result, yet use completely different code. This can be exploited by computer viruses and malware, making it difficult for the anti-virus software to detect them.

Adversary code may also exploit vulnerabilities in the Internet of Things (IoT) where many IP addresses are attached to the online devices, such as surveillance cameras. Hackers can make tools like Shodan[23] browse online devices and hijack the cameras' IP addresses, infecting devices with malware scripts. Then, they can launch spontaneous attacks from a massive number of infected devices to a targeted web site. This is called a Distributed Denial of Service (DDOS) attack. Because there are many infected devices with authentic IP addresses worldwide, it is difficult to trace back the origin of the attacker.

Adversary tool-making behavior perhaps poses the most dangerous threat to our connected world. In the following chapter Self-Awareness, we will explore further about how to develop artificial immune systems to detect adversary behaviors and dissolve malicious code.

[22]Stagon SA, Elliott KP and Huang H (2016). Metallic Glue for Ambient Environments Making Strides. Advanced Materials & Processes, January 2016

[23]Shodan.io. https://www.shodan.io/

## Summary

Humans, animals, and machines can make tools. We often improvise tools when given unexpected situations such as having to survive in the wild. When we make tools, we explore the potential of existing knowledge or systems for new applications. In many cases, we must make tools for solving problems, using limited resources to adapt to an infinite number of scenarios.

Self-tooling is a primitive way to adapt oneself as a tool to accomplish a new task that is not anticipated. This includes adaptive designs for the human body and machines to do things that were not planned before.

Limb extension is a common experiment to test animals' tooling behavior. Consider, for example, the banana test for chimpanzees. The chimpanzees' success in reaching the fruit depends on the availability of materials and manufacturability of the processes. For individual robots, it is possible to extend its limb with sticks, boxes, and 3D printed tools that can be found or made in the field.

Scaffolding is a temporary structure used to support construction, which includes physical scaffolding such as support mechanisms for 3D printed objects, as well as cognitive scaffolding, such as auxiliary lines in trigonometry.

Substitution is the replacement of one item with another based on a similar function, value, or concept. Our mental imagery or visual diagrams can be viewed as a substitution of physical objects. This implies that visual representation is important to learning at any level. For example, engineering diagram design languages have replaced abstract programming languages in many industries. Substitution can also be used for optimizing software in online repositories and machine learning algorithms.

Like social insects, robots can make tools collectively in self-assembling processes. They can form different shapes and move in different directions in order to form a bridge, ladder, and so on. Collective tooling can also be found in software development, as with Mechanical Turks and crowdsourced online games.

Complete self-replication of a machine such as the RepRap 3D printer is possible but expensive in terms of printing and assembly. Additive manufacturing processes lay materials down layer by layer. In contrast, biological self-replicating processes use growing and dividing cells that can potentially be useful in the field of biomedical engineering and beyond.

Tool making can also be used for adversary purposes such as computer viruses, creating threats to not only the cyber universe, but also the physical world. It is an arms race in tool making. Tools are us.

# Part II
# Instinctive Cognition

CHAPTER 7

# Perceiving Motion Patterns

Humans are visual explorers. Human vision is a highly dynamic process. Consider our instinctive reaction to every sudden change. Each change may be inconsequential; but it may also be vitally important. We inspect the world of objects and events with rapid eye movements, which consist of frequent gaze shifts at two to four fixations per second. During a fixation period, the underlying neural processes are updated in less than half a second.[1] Subconscious motion perception enables us to perceive motion patterns, correlated or anomalous events in our surrounding environment. Human vision is by far the best computer in our everyday life. In this chapter, we explore how to map spatiotemporal data into motion pictures, and how to incorporate human perceptions in pattern discovery in motion and events.

## Discovery of the Planet Pluto

Astronomy has been built upon massive amount of observational data. It is like finding a needle in a haystack. How big is the "haystack"? On a clear, moonless night sky, people can see up to 2500 stars with their naked eyes from Earth's night side. With a good pair of binoculars, the number jumps to about 200,000. A small telescope will let us count up to 15 million stars. Large observatories can perceive billions of stars. As early as 100 years ago, astronomers started to record star locations with photographs taken over the course of a few days or a few months apart, resulting in tens of thousands of images. Rapidly moving objects such as asteroids and comets would change their relative positions between photographs. Since planetary objects orbit the sun, their relative movements would be detected in

[1] Öğman H and Breitmeyer BG, eds. (2006) The First Half Second. The MIT Press.

© Springer-Verlag London 2016 
Y. Cai, *Instinctive Computing*, DOI 10.1007/978-1-4471-7278-9_7

these photographs as they shifted their positions relative to the surrounding stars. Astronomers had to comb through massive archives of collected images to see changes.

However, human eyes alone could not reliably detect changes from one photographic plate to another. There were simply too many stars on each plate. As a remedy, an analog device called a "blink comparator" was invented. This device mechanically alternated a projection of one photographic plate with another, presenting the astronomers with a single location to view the two plates, rather than having to move one's head and eyes from left to right to compare photographs. Using the blink comparator, any difference in the location of an object on the two plates would "blink" in front of the astronomers, drawing immediate attention to the difference between the two images. A "blinking comparator" is also called a "blink microscope," and has been used in medical radiology, with X-Ray photographs, in which doctors compare two images to assess changes.

About 100 years ago, an amateur astronomer named Clyde Tombaugh, a farmer in Kansas in the US built his own powerful 33-centimeter lens telescope with a long exposure camera to study the stars. The photographic plates Tombaugh used were $36 \times 43$ cm in size. For 20 years, he took telescopic photos at night, then developed the photographic plates and compared them during the day using a blink comparator device. Tombaugh spent weeks and months searching the vast depths of space looking for moving objects amongst the thousands of stars exposed on those plates. Finally, in February 1930, he found something that he could not explain right away as a nearby asteroid or some other form of space debris. It moved too slowly to be an asteroid. If this were a planet, it was farther away from Earth than Neptune. Tombaugh had discovered the dwarf planet Pluto! Tombaugh was blinking between the two plates taken on January 23 and 29, 1930, when he found something strange. Figure 7.1 shows the two famous plates. There is an animation available at the web link.[2] It is not easy to spot Pluto. The two plates were not perfectly aligned, the time exposures did not yield the same brightness, and the orientation of the two fields were not exact. But with a bit of training, it is possible to spot moving patterns. The essential purpose of the blink comparator is to project two images onto one spot alternately. Thus, the alignment of the two images is key.

Today, most modern astronomy and radiology analyses no longer use films for discovery or diagnosis. Images are now routinely obtained, evaluated, interpreted and stored in a digital fashion using a computerized display and storage system. However, despite the advancement of digital media and vision algorithms, it is still a challenge to detect anomalous motion and events autonomously. In many cases, we still need visualization techniques to discover hidden patterns in massive data. In the following sections, we will discuss how to map, align, interface, and filter raw

[2]The National Air and Space Museum (2015) Finding Pluto with the Blink Comparator. http://blog.nasm.si.edu/planetary-science/blink-comparator/

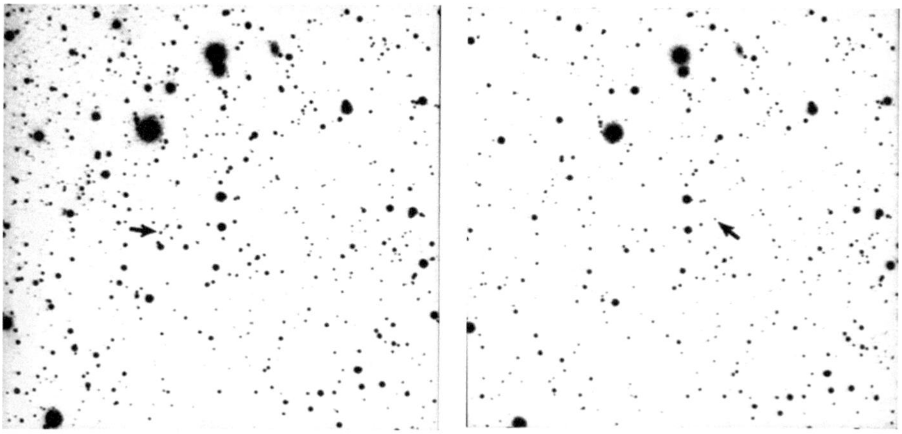

**Fig. 7.1** Discovery of the planet Pluto: January 23, 1930 (*left*) and January 29, 1930 (*right*) (Images courtesy of Lowell Observatory Archives)

data in order to discover patterns in motion and events. In contrast to conventional visualization techniques, the methods in this chapter focus on data animation that match human perception instinctively.

## *Spatial Mapping*

To animate a dataset, we must setup a visual space first. For example, if we have geographical data such as flight records and credit card transaction records, the latitude and longitude coordinates could be natural location attributes. In more general cases, we have non-geographical attributes such as life expectancy versus national income per capita. We can select two or three attributes from a dataset to configure the visual space, using the following variables:

- Position of geometric coordinates
- Position of non-geometric coordinates
- Shape with multiple attributes
- Length
- Volume
- Area
- Intensity

Geometric mapping with GPS coordinates is relatively straightforward. In some cases, only a postal code (zip code) is available. A table look-up method can be used to derive the GPS coordinates from each postal code. For non-geometric data such as life expectancy and GDP data, each variable is a dimension of the visual space. For more than three-dimensional variables, it is possible to convert the multiple attributes to a glyph shape in which the radii of its splines represent values. Finally, the values of one- or two-dimensional variables can be mapped to the intensity of pixels in the form of a heat map.[3]

A principle of spatial data visualization is to enable users to perceive patterns accurately. The data scientist William S. Cleveland conducted experiments on human responses to graphs.[4] He summarized basic graphical perception tasks from the most accurate to the least accurate: (1) position, (2) length, (3) angle, (4) area, (5) volume, and (6) color. The height judgment is the most accurate when the objects are positioned along a common scale. Color, saturation and density have been put in the last place in the ordering because they can hardly convey the relative magnitude of values quantitatively.

## *Temporal Scaling*

In Max Wertheimer's seminal paper in 1912, "Experimental Studies of the Perception of Movement,"[5] he discovered that animation contains more information than still images. If we turn on a string of holiday blinking lights in an order, we would perceive the light in motion, even though individual lights are stationary. This is because the whole sequence contains relationships among the individual lights. This effect is called *apparent motion*, and it is actually the basic principle behind motion pictures.

In the phenomenon of apparent motion, two or more static objects appear to be moving in one stream as they are placed at different times next to each other. Assume $d$ is the perceived distance between stimuli, and $t$ is the time from the start of one event to the start of the next. Korte's velocity is the speed of apparent motion:

$$v = \frac{d}{t} \tag{7.1}$$

Korte also examined the quality of apparent motion as a function of the relationship among three parameters: (1) the alternation rate, (2) the separation between the two points of light, and (3) the intensity of the two lights. Korte discovered that larger separations require higher intensities; slower rates require higher intensities; and larger separations require slower rates.

[3]Zheng Y (2015) Trajectory Data Mining: An Overview. ACM Transactions on Intelligent Systems and Technology. http://research.microsoft.com/pubs/241453/TrajectoryDataMining-tist-yuzheng_published.pdf

[4]Cleveland, WS (1985). The elements of graphing data. Wadsworth, Inc. 1985

[5]Wertheimer M (1912) Experimentelle Studien über das Sehen von Bewegung [Experimental Studies on Motion Vision]. Zeitschrift für Psychologie 61 (1): 161–265. Retrieved on Feb. 12, 2016: http://gestalttheory.net/download/Wertheimer1912_Sehen_von_Bewegung.pdf

Apparent motion is also the foundation of computer animation. After setting up a visual space to animate a dataset, we need to map the timestamps in the dataset to the video frames. For NTSC format videos, which are common in North America, it is 30 frames per second. For PAL format, which is used in the rest of the world, it is 25 frames per second. To remember these two frame rates, readers may simply divide the local AC frequency by 2. A visual update needs to be close to 30 fps but it is not necessary to exceed this value because human vision can't work faster than that. The human brain also takes 250–500 milliseconds to respond to any visual stimuli. Therefore, it is desirable to display visual stimuli for longer than 250 milliseconds. An example of this can be seen in early radar systems. In order for the operator to see dynamic signals more clearly, a looping circuit was added to slow down the signal update.

## *Looping and Motion Memory*

In order to see moving objects, a vision system needs a memory. Early radar and sonar systems used the blip, a spot of light on the screen, to indicate the position of a detected object, such as an aircraft or a submarine. It was intended to work like a television picture of what was in the sky or ocean. In practice, radar and sonar signals are very noisy. Radar detects things that are not in the sky and sometimes fails to detect things that are in the sky. It can also display one object in multiple places and display an object in the wrong place or in the wrong direction. To improve the detection rate, the Moving Target Indicator (MTI) was devised in the 1950s to remove all stationary targets from the radar screen.[6] To do so, it needed to compare the current blip position with the previous one, thus requiring a memory to store the previous position.

MTI originally worked using a delay line memory invented by American electrical engineer John Adam Presper Eckert, Jr.[7] Coincidentally, Eckert also pioneered the first digital computer ENIAC, where he incorporated his invention of the mercury delay line memory. This was the first digital data storage device. His delay line memory was made up of tubes of liquid mercury that were several feet long. Electrical pulses were converted to sound and back to electrical signals in a continuous loop. We know that sound propagates in a liquid much slower than electricity does, which slowed down the data transmission and served as a data memory or storage. If the current blip position was the same as the stored previous blip position, then the blip was removed. Modern radar systems use computers to accomplish this same task, which is to remove any stationary target from the screen using algorithms such as frame subtraction and background segmentation.

[6]Delay Line and Radar: http://www.computerhistory.org/revolution/memory-storage/8/309

[7]Delay Line Memory: US Patent 2629827: http://www.google.com/patents/US2629827

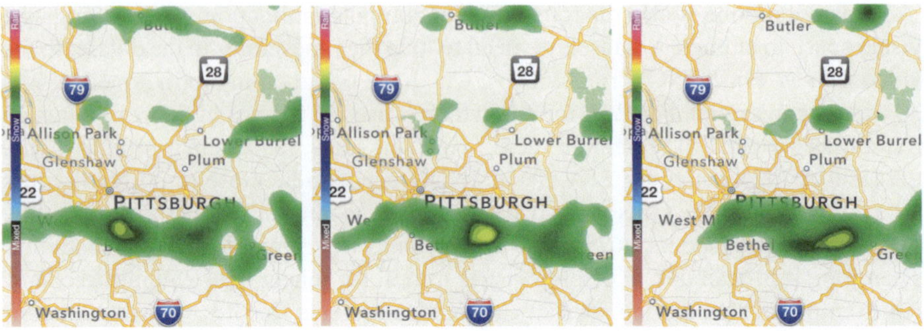


**Fig. 7.2** NOAA weather satellite radar images animated in a loop: 30 min ago (*left*), 20 min ago (*middle*) and 10 min ago (*right*), showing the center of a rain cloud passing through Pittsburgh. Our perception is able to predict the cloud movement in next 10 min

Figure 7.2 shows three animated NOAA weather satellite radar images in a continuous loop: 30 min ago, 20 min ago, and 10 min ago. The sequence of frames illustrates the center of a rain cloud passing through Pittsburgh. Our perception is able to predict the cloud movement in the next 10 min by extrapolating the momentum of the highlighted rain cloud.

## *Alignment for Visualization*

In order to effectively discover motion patterns, we must align data properly in accordance with human visual perception. For example, a fixed camera location helps to align the pixels of an image for users to spot moving patterns. Figure 7.3 shows two images from the same camera viewpoint. By animating the two images back and forth, our eyes will automatically ignore the static dots and perceive only the moving dots that have the highest "motion contrast," or displacement. If we misalign the images, say, by rotating the second image 90°, then we will not see any meaningful patterns.

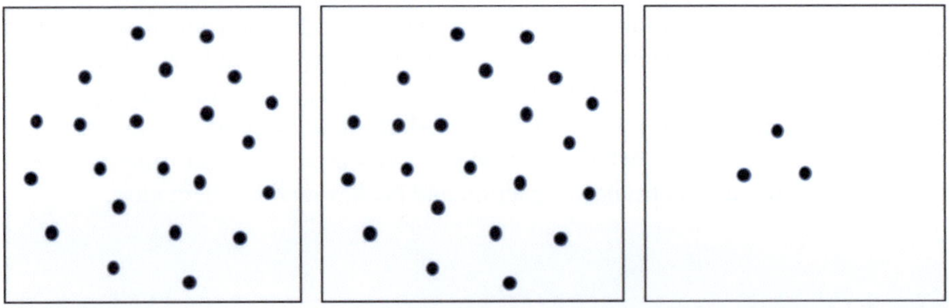

**Fig. 7.3** The alignment of pixels helps to create motion contrast for pattern discovery. By animating the two images (*left and middle*), our eyes will ignore the static dots and perceive motion patterns (*right*)

Alignment helps users to perceive motion patterns. It may even help users discover *missing data* when users don't have any prior knowledge about a dataset. For example, given a set of financial transaction data in ATMs (automated teller machine) in a region that is mapped to geographic locations, glyph sizes and animated frames, we would see the animated shapes that represent cash transaction activities. When some ATMs in one zip code do not have any activity for a day or fail to submit their data, the animated video would show jumpy or motionless patterns.

Geometric landmarks, corners, or gradient features can be used for alignment, by matching corresponding points between two images. To determine whether a pair of points are corresponding or not, we can use a *homography model*, which is a set of geometric transformations including scaling, rotation, shearing, translation, and perspective projection. Homographies between four pairs of points in the two images are computed using *Random Sample Consensus* (RANSAC) for verification. Figure 7.4 shows an example of the homographic transformation where the point $p$ corresponds to $p'$. These two points are referred to as corresponding points.

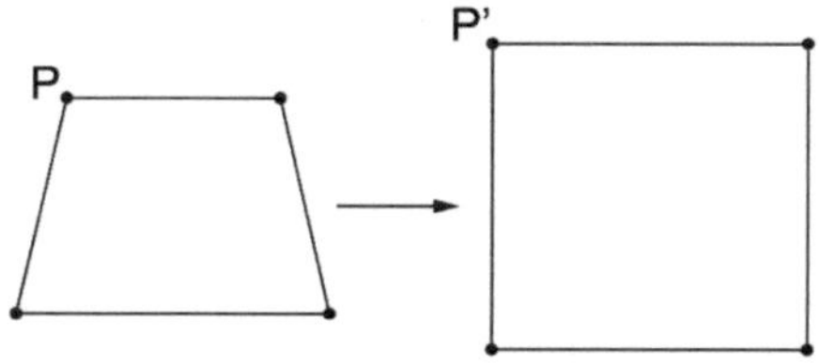


**Fig. 7.4** Homographic transformation

The main function of RANSAC is to identify a "majority vote" or consensus of mapping directions to determine inliers or outliers of the key-points or features of interest. The pseudo code for RANSAC for estimating homography is as follows[8]:

```
Select four pairs of points randomly
Compute homographic transformations
Compute inliers where the differences between the original and mapped
  points are less than a threshold
Repeat until all points are covered
Keep largest set of inliers
Re-compute optimal homographic transformations for all the inliers
```

[8]RANSAC Tutorial from CSAIL, MIT. Retreived on Feb. 12, 2016: http://saturday.csail.mit.edu/fa12/lectures/lecture13ransac/lecture13ransac.pdf

## Motion and Vision

The vision scientist David Marr thought that the fundamental task of vision is to construct a representation of the internal three-dimensional model of the world. The reconstructive process in the brain is visual cognition.[9] Vision evolved in animals not only to enable them to see the world, but also to guide their movement through it. Studies in Neuroscience show that perception and action are intrinsically linked in animal brains. Both perception and action perhaps share some portions of common code in the cognitive process.[10] Vision, for example, guides actions with respect to the world by transforming visual input into the appropriate motor output. This is referred to as *visuomotor* skill, which enables creatures to catch prey and avoid predators, and also enact visual control of barrier avoidance.

Here we want to explore the third aspect: *how motion enhances vision*. We call this "dynamic vision." Because every vision system has its limitations regarding viewing area, distance, and dimensions, the motion of a visual sensor creates multiple views that serve different purposes:

First, motion enables measurement, which is the most important function for scientific visualization. To have a more accurate visual estimation, we often try to find a baseline or reference plane. In many cases, an orthographic projection is preferred, for example, a top view, or a profile view. Figure 7.5 shows a front view, side view and 3D view of a house. In order to estimate the footprint of the house,

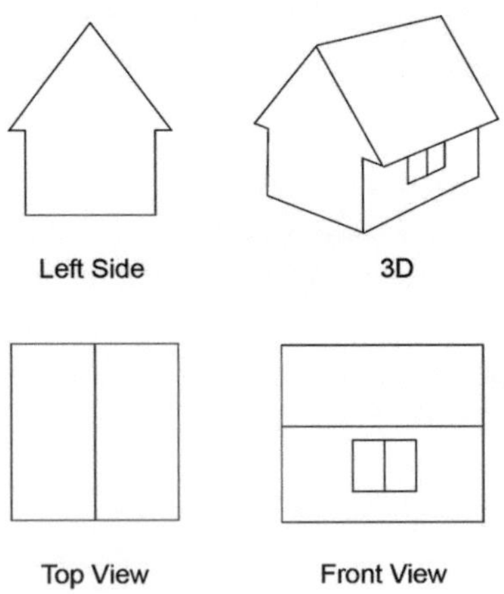


**Fig. 7.5** Multiple views of a house

[9] Marr D (1982) Vision, the MIT Press

[10] Common Code Theory, WikiPedia. Retrieved on Feb. 12, 2016: https://en.wikipedia.org/wiki/Common_coding_theory

the top view is superior. To estimate the height of the house, a front view, or a side view is preferred. Relatively speaking, a top view, front view or side view are better suited for measurement. However, even the 3D representation of the house helps to define the overall layout.

Second, motion enables us to generate an internal three-dimensional model of the world. Moving the camera sideways enables us to sense a three-dimensional world, given the fact that the displacement of distant objects appears to be less than objects that are closer. In its infancy, the digital camera was a very expensive piece of equipment. Some mobile robots could not even afford to have multiple cameras installed. As a result, engineers frequently installed a single camera and moved it between the left position and the right position to simulate stereo vision. This design was a compromise of cost and time, and it proves that motion can create a new basis for vision. Furthermore, multiple views can create multiple baselines for the three-dimensional reconstruction of a scene from two-dimensional images. However, a three-dimensional reconstruction may only model the surfaces that are in the line-of-sight. There are usually many blind spots in a model unless one can spin the object in front of the camera. For example, the popular three-dimensional model of Notre Dame in Paris was reconstructed from millions of tourists' online photos. Because none of the tourists could fly over the roof of the building, the roof details could not be constructed.

Third, motion enables us to have a panoramic view of a large object or scene, particularly when observing things behind obstacles. We cannot move an object as big as a mountain. However, the linear motion of a camera can be used to create panned panoramic images and the rotating motion of a camera can create cylindrical or spherical panoramic images.

Finally, motion is also used to survey location and maps landmarks along the way, a method called spontaneous location and mapping (SLAM). This capacity is found in video odometers that are widely used by mobile robots as an alternative to GPS-based navigation.

In summary, motion enables four valuable vision products: (1) stereo vision, (2) panoramic vision, (3) mapping, and (4) measurement. The "dynamic vision" approach combines navigation with vision in order to view an object in 3D, in a wide range, and for better measurements. This is a biomimetic approach because humans utilize these strategies in daily life. However, adding navigation to vision can create problems. For example, navigating through a visual space can be time consuming and confusing to novel users.

## Gamefication

Gamefication maps a dataset into a virtual world where users can explore the data interactively. A video game is an event-driven program. An event is simply some intended user input, like the push of a button or the click of a mouse. A game must specifically define what actions players can perform and what events those actions should trigger. The following sections describe key components such as the "first-person" view, collaborative intelligence, physics engine, scalability, and randomization.

## *The "First-Person" View*

A world model is the stage on which a game occurs. A dataset can be mapped to a game world model in 2D or 3D space, where data forms actors that come, move, and go. If the data is dynamic, then the world is animated. The advantage of a game engine here is to enable the user to navigate the virtual world through a first-person view. The game engine should allow the user to navigate the virtual world from any viewpoint through zooming, panning, and rotating operations. For data visualization, photorealistic rendering is not necessary. Instead, simple geometric shapes are enough. The challenge is to create an immersive environment for the user to perceive spatiotemporal patterns during the exploration.

In the aviation data visualization study, William Eddy, et al, maps the aircrafts' coordinates to a three-dimensional model where each ball represents an airplane with its size inversely proportional to its depth within the view volume.[11] Users are able to move freely around the dataset looking from different directions with differing view angles. Eddy calls this the "magic carpet", because it moves so fast it appears as if time is frozen. In fact, users can literally stop the clock of the simulation while moving around the data. Users are also able to select any of the planes in the dataset to "ride" on the data from the viewpoint of the pilot seat, creating a surreal experience for data exploration, in which time is compressed by orders of magnitude. To create a virtual image from the aircraft, only the data that are in the line of sight of the aircraft are displayed. The displayed subset of the data changes as the aircraft moves. According to the study, the first-person view indeed captures several air traffic patterns. For example, an airplane from Cleveland cuts in front of the aircraft, requiring the aircraft to slow down. The aircraft from Cleveland precedes the aircraft to Denver. Then the aircraft makes a sharp right turn to avoid another aircraft that crosses its path at the same altitude. Figure 7.6 shows the pilot's view looking towards the airport.

## *Collective Intelligence*

The networking capacity of a game engine enables multiple users to collaborate in data exploration. Most online action games played today are client-server games. In

[11]Eddy W and Oue S (1995) Dynamic three-dimensional display of U.S. Air Traffic. Journal of Computational and Graphical Statitics, Vol.4, No.4, pp. 261–280, 1995

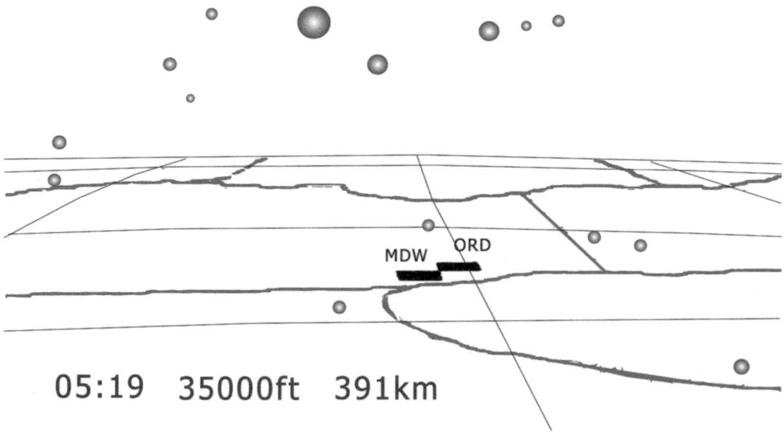


**Fig. 7.6** The pilot's view of the flight record data where each plane's geographical coordinates and time stamps are mapped onto the 3D space so that the viewer can observe the air traffic patterns from a single plane's point of view. Redrawn from the video courtesy of William Eddy

these games, there is a single, authoritative server that is responsible for running the global logic. One or more thin clients are connected to it. Typically, the clients cannot communicate directly with each other, as this is not a peer-to-peer connection. The server controls the global map and the players update local activities. For example, player 1 tells the server, "I have moved one step to the left and my screen shows I am walking to the left." The server would respond, "I received your data and I am updating your position on the global map; and I am going to tell the other players that you have moved to a new position."

In video games, an event can be expressed as a message. Network game clients collect local events and send them to the server. The server updates its real time event database and broadcasts events to related clients. For example, NASA scientists and NOAA scientists often collaborate on oceanographic satellite image datasets. Assume NASA scientists discovered an event in the Gulf of Mexico. They would share their view of the model with the collaborators at NOAA. Both are network clients connected to the same server. The server updates clients with new viewpoint coordinates, in either a synchronized or asynchronized mode. A user may place a tag at the location so that collaborators can see it from their terminals. Figure 7.7 shows a screenshot of a tag in the two networked game windows.

## Physics Engine

A physics engine serves the sole purpose of moving objects around in a simulated physical world. Physics engines take into consideration the physical properties of objects, such as their mass, shape, material, and current trajectory. It constantly

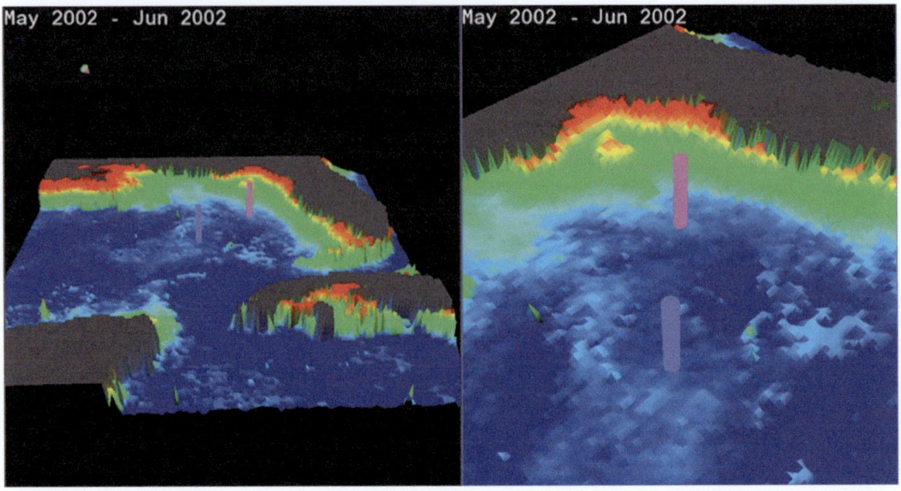


**Fig. 7.7** The cylindrical tag in the two game client windows

calculates new positions and states of those objects. In general, a game engine feeds initial state data to the physics engine, and then responds to the physics engine's feedback.

One of the most important functions of a physics engine is collision detection. Collision detection is an essential part of 3D games. It ensures that the game physics are relatively realistic, so that a man will walk on the road instead of underneath the road, and that he will walk alongside a wall instead of walking through a wall. The believability and enjoyment of a game is very dependent on how well a game can detect collisions. One way to detect a collision in a 3D world is by using the sphere-plane detection method.[12] Detecting collisions with a sphere tends to be easier to calculate because of the symmetry of that object. The entire surface of a sphere is the same distance from the center, so it is easy to determine whether or not an object has intersected with the sphere. If the distance from the center of the sphere to an object is less than or equal to its radius, then a collision has occurred.

## *Scalability*

Video games teach us scalable design strategies for computation-intensive tasks. First of all, consider the viewing scope – only the data that are in the *sight* of the user need to be displayed at any given time. Furthermore, only the data near the user are displayed in high-resolution, while the rest of the data are displayed in

[12]Collision Detection (2003). http://www.edenwaith.com/products/pige/tutorials/collision.php

low-resolution. The multiple resolution approach can also be implemented in the multiple levels of detail. For example, the data close to the user might have a full-representation of all visual variables, whereas the data in the background may show hints of detail. For three-dimensional rendering, the near-field scene is rendered in 3D, but the far-field scene is simply rendered in 2D. Video games also utilize tight collision detection algorithms in near-field objects, while applying the rough collision detection algorithm to far-field objects. To minimize the event handling workload, the game engine usually defines an effective zone for an event.

### *Randomization*

Randomization spontaneously alters control variables in order to maximize the chances of discovering new motion and event patterns. As we have discussed in the previous chapter, "Default Mode," spontaneous alternation behavior (SAB) is essential for survival. In discovery games, one ought to introduce random factors to the game. For example, a game can vary the user's starting position, their selection of a data point to "ride" on, the selection of key variables to display, and so on. A recording mechanism is also necessary to ensure the repeatability of the exploration process so that any experimental sequence can be playback or recovered. This improves learning and sharing potential. Furthermore, randomization reduces boredom in the game because it introduces new scenarios and objects to users. As humans, it is also our instinctual drive to explore new things.

## Filtering

Filtering is critical to a decent visualization. Filtering serves two purposes: enhancing patterns and reducing noise. There are many filtering methods, but here we will focus on two essential cognitively-inspired methods: *continuity* and *curvature* filters.

### *Continuity Filter*

Imagine you manage a visual memory system. What visual samples would you select for representing a variety of shapes and intensity gradients? You would probably pick symmetrical, simple shapes, or smooth gradient images because they require less memory to store and need less processing time to reconstruct a shape or intensity distribution. For example, a circle only needs one parameter: its radius. A two-dimensional Gaussian distribution only needs two parameters. Now, if a hypothetical bad guy breaks the symmetrical shape or for some reason the symmetrical shape is distorted in the real world, then the visual system must come up with more resources to represent that complex shape or distribution. Assume the brain prefers the least effort to represent a shape or a distribution. You might simply ignore any imperfections or distortions and *complete* the shape or intensity gradient distribution in its ideal condition. For instance, everyone knows that the opening of a cup is a circular shape by default, despite the fact that a perspective view of the opening is usually oval-shaped. Automatic completion of a shape or a contour has been widely studied in Gestalt psychology, specifically in terms of continuity and what defines "good shapes."

Continuity filtering can be implemented through spatial and temporal interpolation algorithms. Spatial interpolation methods include nearest neighbor, bilinear and bicubic interpolation, which enable us to determine an unknown value at a given point. The nearest neighbor method sets the value of a point to the value of the nearest known point. It is the simplest way to fill in missing values for visualization. It only requires one value of the nearest point for the interpolation to work properly. In many cases, the resolution is sufficient. However, the nearest neighbor method does not provide subtle gradient information, and the intensity at the borders appears jumpy and unnatural.

Bilinear interpolation is the product of two linear interpolations along the vertical and horizontal axes. The value of a given point is linearly interpolated from the four values of the nearest points. In addition to filling in the missing data, this algorithm can also reduce some of the visual distortion caused by resizing an image. However, it may produce artifacts such as aliasing, blurring and edge halos.

Bicubic interpolation is an extension of bilinear interpolation for much smoother surface reconstruction. In contrast to linear interpolation, which only takes 4 points into account, bicubic interpolation needs 16 points. Images resampled with bicubic interpolation are smoother and have fewer interpolation artifacts. However, bicubic interpolation is computationally expensive. If computational speed is not an issue, then bicubic interpolation is recommended to create the smoothest surface or gradient for animation. Studies show that fluidity of motion in spatiotemporal visualization better enables humans to perceive patterns, compared to instances when the nearest neighbor or bilinear interpolation methods are used.[13] See Fig. 7.8.

Temporal continuity is also necessary for animated movies. For example, many closed-circuit surveillance camera videos have a low frame rate, creating the jiggling artifacts that prevent users from perceiving continuously moving patterns or events. In order to provide temporal continuity, we must interpolate the frames in a video. The simplest way to do this is to perform bidirectional, bilinear interpolation for inserted frames. However, it would produce blurriness in the video. Fortunately, video compression algorithms facilitate more sophisticated temporal

[13]Eddy FW and Mockus (1994) Anexample of the estimation and display of a smoothy varying function of time and space – The incidence of the disease mumps. Journal of the American Society for Information Science. 45(9):686–693, 1994

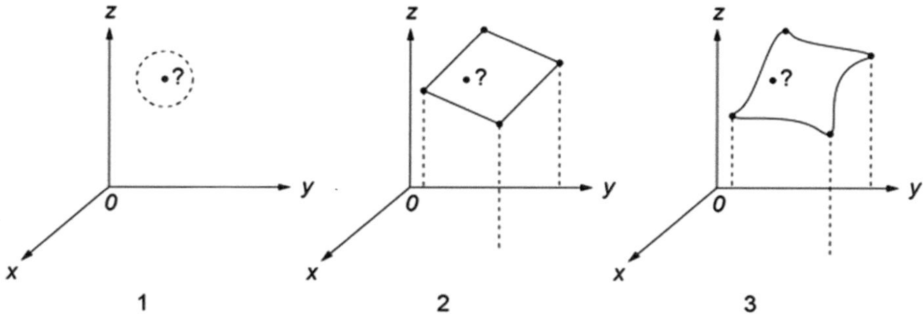


**Fig. 7.8** Illustration of (*1*) nearest neighbor, (*2*) bilinear interpolation, and (*3*) bicubic interpolation. The question marks indicate the interpolated values

interpolation methods. Most online videos are extremely compressed, because it is not computationally economical to send each uncompressed frame across the Internet, when the human eye can only capture a small portion of the content. A typical compression algorithm uses microblocks to represent moving objects. These microblocks can be in $4 \times 4$, $8 \times 8$, or $16 \times 16$ pixels. We can encode the moving objects with microblocks and predict their positions in the next few frames based on a linear interpolation model. If we name the original full-resolution frame the I-frame (intra-frame), then we can compute the predicted frame, or P-frame. Now we can send I-frames followed by many P-frames and then I-frames and P-frames again. Sometimes, the prediction goes too far and creates jiggling transitions between P-frame and I-frame. Now, the bidirectional frame, or B-frame, can be added. It is the pixel-based interpolation between an I-frame and a P-frame in both directions. The sequence of the mixed frames looks like this:

I P P P P B I B P P P P B I B P P P B P P B I B P P P B I B P P

The dynamics of the objects and compression level determines how many P- and B-frames to add. For smoother transitions, more bells and whistles can be added, such as adaptive sizes of microblocks for flat areas and edges.

For imagery motion analysis, computer scientist Takeo Kanade discovered the impact of continuity on animating 3D models from multi-view images. For example, he tried to reconstruct a baseball pitcher's rotating movement from multiple cameras. The video appeared noticeably jumpy because of the low frame rate. By interpolating existing frames, the gait motion becomes much more natural and complete. Similarly, we can apply continuity to CCTV to increase the chances for witnesses to recognize criminal suspects caught on camera. Normally, surveillance camera footage has a very low frame rate to conserve storage space or network bandwidth. Consequently, a suspect's movements appear jumpy and blurry. By adding frames, the motion patterns become more obvious.

## *Curvature Filter*

A recent fMRI study by Dayan, et al shows the brain's inherent need for simple, smooth movement that has a low rate of acceleration change. It is no coincidence that this can be related back to Gestalt psychology in the motor areas of the brain, as increasing evidence shows that the brain looks to find the simplest, "whole" of the parts it perceives, in both static imagery and motion events. The preference for smooth movements following physical laws exists widely in nature. For example, the momentum of a running animal prevents it from making a sharp turn at a high speed. A flying airplane normally does not make a sharp zigzag path at a high speed. Our instinct enables us to spot such anomalous movements.

Here we will apply the instinctual curvature model to filter out noise along a physical object's path, for example, the trajectory of an aircraft. In aviation datasets, it is possible that two different flight reports are mixed together due to the proximity of aircraft in time and space. Figure 7.9 shows a saw-toothed flight path caused by mixing reports from two different flights that are overlapped in time. Intuitively, we know that it is impossible for an aircraft to fly like that.[14] How do we then filter out

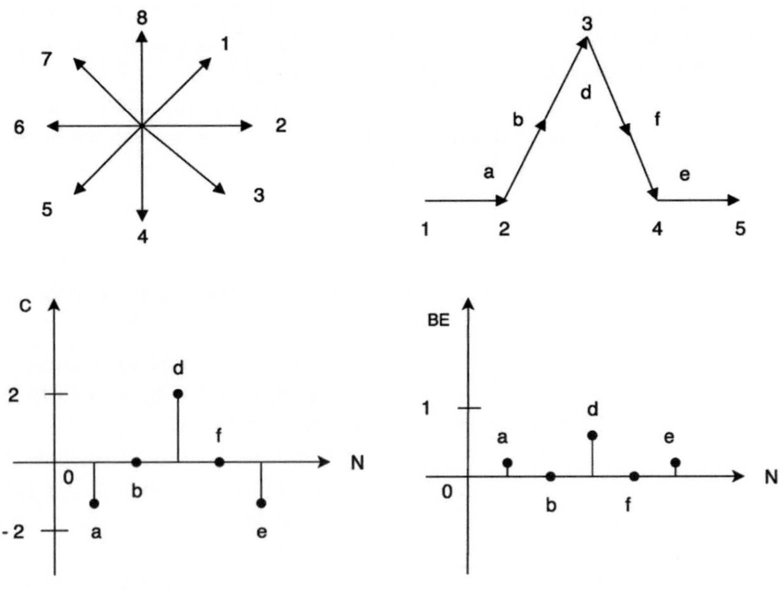


**Fig. 7.9** The definition of a Chain Code (*top left*). A saw-toothed flight path caused by mixing reports from two different flights (*top right*), the curvature histogram (*bottom left*) and the bending energy histogram (*bottom right*)

[14]Eddy FW and Oue S (1995) Dynamic three-dimensional display of U.S. Air Traffic. Journal of Computational and Graphical Statitics, Vol.4, No.4, pp. 261–280. 1995

the points that are not on the true path? We can do this by calculating the curvature and the bend energy.

There are many methods for calculating curvature. For discrete segments of a curve, we can apply the simple method called Chain Code. It encodes the unit line segments into numbers. For example, we can assign 1 of 8 directions to a unit line segment: north (0), northeast (1), east (2), southeast (3), south (4), southwest (5), west (6), and northwest (7). The Chain Code for the flight path in Fig. 7.9 would be:

$$\ldots\ 2,\ 1,\ 1,\ 3,\ 3,\ 2\ldots$$

The curvature of two adjacent segments is the difference of their chain code at the location *a*, *b*, *d*, *f*, *e*.

$$\ldots\ -1,\ 0,\ 2,\ 0, -1\ldots$$

The bending energy (BE) of a curve is the sum of squares of the curvatures over the curve length, similar to the energy necessary to bend a physical rod. The bending energy:

$$BE = \frac{1}{L}\sum_{k=1}^{L} c^2(k) \tag{7.2}$$

where $L$ is the curve length and $c$ is the curvature of the segments. Assume the total curve length is 6. We have the bending energy values at the location *a*, *b*, *d*, *f*, and *e*:

$$\ldots 0.167,\ 0,\ 0.667,\ 0,\ 0.167\ldots$$

Now, we can locate the sharp bending locations. By applying a bending energy threshold, say, 0.5, we can confirm that the observations 1, 2, 4, 5 are in the actual path and observation 3 is the "contaminant."

## Artifacts and Motion Cognition

Missing data, cross-contaminants, and misalignment cause artifacts in visualization such as jiggling and flashing. The dilemma here is that they might reveal actual patterns or they might just be distracting noise. Applying filters may reduce artifacts but it may also be risky to ignore the potential motion patterns within those artifacts. One solution is to adjust the opacity of the filtered areas to see the visual impact. How does one objectively evaluate the cognitive process of a visualization with and without filtering? Let us to consider the fMRI for brain imaging. The fMRI has a high spatial resolution, but relatively lower temporal resolution – typically 10 frames per second – compared to other devices such as eye gaze tracking (60 frames per second) and EEG (over 18 Hz). In the next section, we use an eye gaze tracker and EEG imaging to analyze the impact of spatial filtering.

## *Eye Gaze in Response to the Spatial and Temporal Filtering*

Eye gaze often reveals attention, scanning and tracking patterns in visual exploration. In our experiment, we wanted to compare eye gazing patterns before and after spatial filtering. In NASA's oceanography satellite SeaWiFS image dataset, the chlorophyll channel correlates to the concentration level of harmful algal blooms (HAB). The time-lapse video of the chlorophyll heat map would reveal the HAB occurring, moving, and fading patterns, where the red, orange, and yellow colors indicate a high concentration of HAB. However, there are blank spots in many images caused by missing data hidden under clouds in the sky. We assume that spatial and temporal filtering can fill in the black holes and missing frames in order to reduce the distractive noise so that the observer would pay more attention to key HAB moving patterns. The spatial filtering is implemented with bilinear interpolation. The temporal filtering is based on bilinear interpolation between frames. We have two test video clips:

- rough video – before spatial and temporal filtering, and
- smooth video – after the spatial and temporal filtering.

The smooth video is longer than the rough video after the temporal interpolation. Figure 7.10 shows screenshots of the rough video and the smooth video from NASA SeaWiFS satellite image.

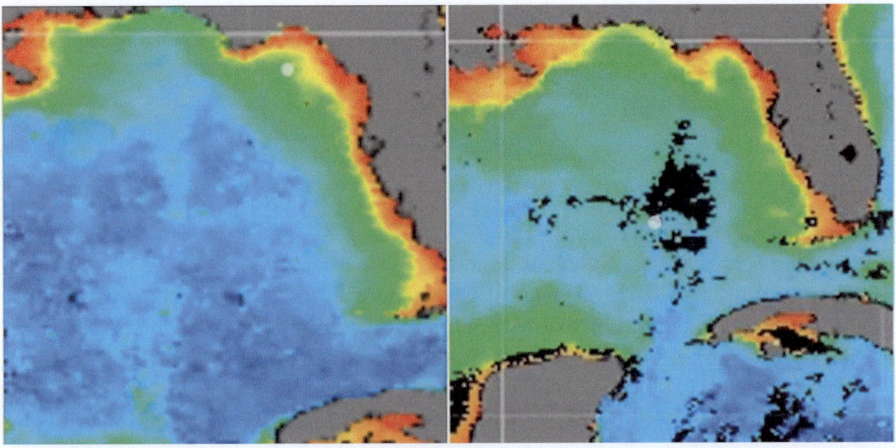

**Fig. 7.10** The SeaWiFS image after the filtering (*left*) and before the filtering (*right*)

We had 40 subjects in 20 groups experiment with eye tracking for the rough video and the smooth video. We used the EyeScience Tracker II, at a camera speed of 60 frames per second. After recording the eye gaze positions with time stamps, we plotted the gaze path over time and overlaid the data on top of the original video.

Prior knowledge and instruction have an impact on the eye gazing paths. When a subject is told to focus on the high HAB area along the Gulf of Mexico coastline, then the subject focuses his or her eyes on that area and completely ignores the rest of the area. To avoid any biased instruction, we kept the instruction neutral and minimal. We only ask subjects to look for patterns of high HAB concentration areas and the color map for the concentration levels.

We found that the majority of the eye gaze paths follow this pattern: Before the filtering, the eye gaze paths are running wild, chasing the irregular black hole patterns. Some move down to the bottom area where no HAB even exist. After the filtering, the eye gaze paths become more consistent, and more focused on the high HAB area along the Gulf of Mexico coastline. Figures 7.11 and 7.12 show the first 20 frames and 90 frames of the eye gazing paths when watching the rough video (unfiltered) and the smooth video (filtered) samples, where the white lines represent eye gazing paths and the size of white dots represents the eye fixation time. Figure 7.13 shows the eye gazing paths in the final frame of the rough video and the smooth video. We observe that the eye gazing paths on the smooth video are much more concentrated on the hot spots of high HAB levels near the coastline, while the eye gazing paths in the rough video are more random. Also, from Fig. 7.11 through Fig. 7.13, we found that the eye fixation times in the smooth video are longer than those in the rough video, allowing the observer to spend more time on visual reasoning.

However, a small number of subjects have the opposite eye gazing patterns. Their eye gazing paths move even more wildly in the smooth video. On-site interviews disclose that smooth areas may cause visual boredom, even fatigue after staring at the screen for too long, due to short visual attention spans. To validate these assumptions, we followed up with EEG brainwave experiments that will be discussed in the next section.

## *Brainwaves (EEG) in Response to the Spatial and Temporal Filtering*

We had 40 subjects in 20 groups watch the rough video (unfiltered time-lapse) and the smooth video (spatial and temporal filtered time-lapse) sequence of the SeaWiFS satellite images. We assumed that the filtered results would have an impact on brain activities in terms of locations on the brain surface and the corresponding moments in time in the video sample. We used the EEG brainwave sensory device Emotiv

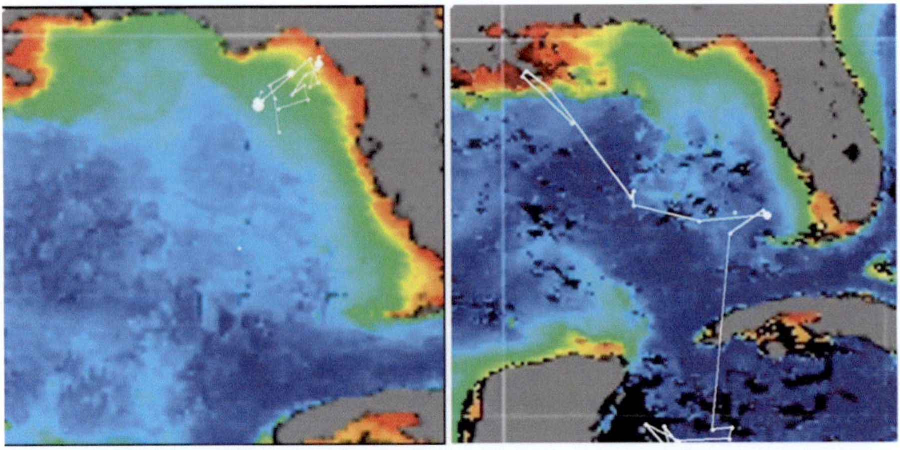

**Fig. 7.11** The 20th frame of the eye gaze of smooth video (*left*) and rough video (*right*)

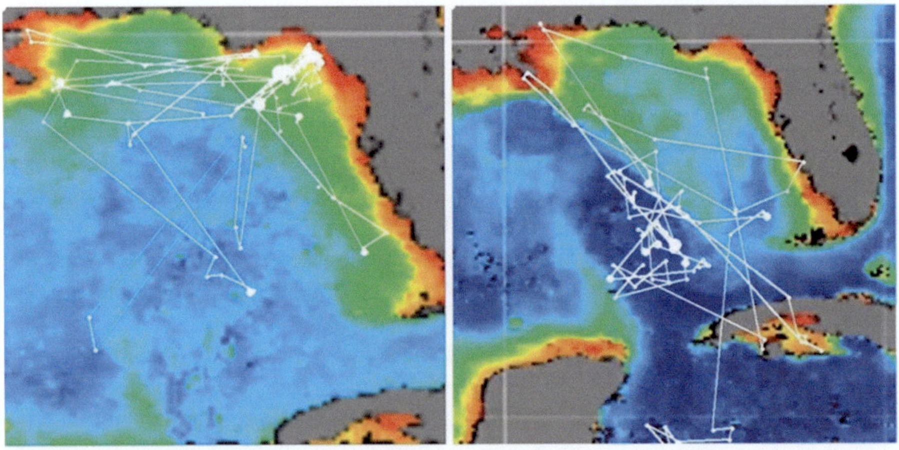

**Fig. 7.12** The 90th frame of the gaze paths of smooth video (*left*) and rough video (*right*)

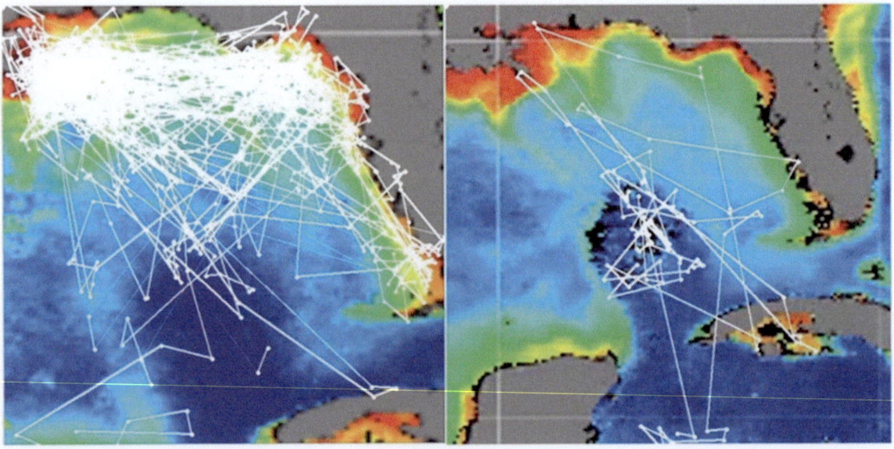

**Fig. 7.13** The final eye gaze paths of smooth video (*left*) and rough video (*right*)

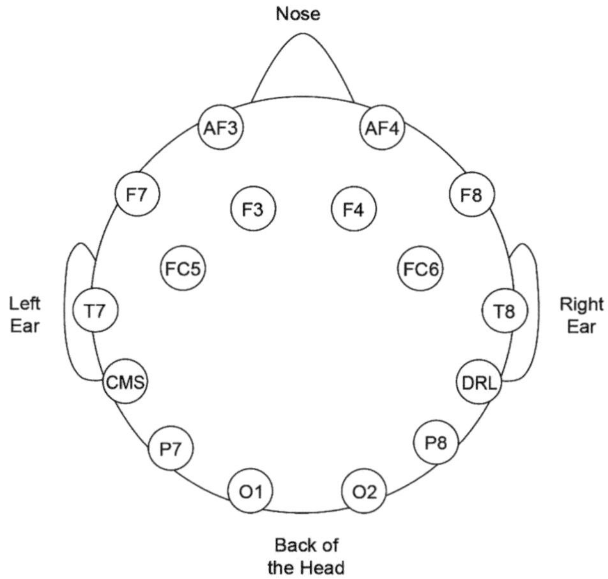


**Fig. 7.14** The locations of the EEG probes AF3, F7, F3, FC5, T7, P7, O1, O2, P8, T8, FC6, F4, F8, AF4 plus two references CMS and DRL

EPOC that has 14 channels plus references CMS and DRL.[15] The channel names are: AF3, F7, F3, FC5, T7, P7, O1, O2, P8, T8, FC6, F4, F8, AF4. See Fig. 7.14. The sampling rate is 128 SPS (2048 internal). The bandwidth is 0.2-45Hz.

We used the open source package EEGLAB to filter the data and we used Automatic Continuous Rejection to reject anomalous sections of the data, e.g. due to blinking. After cleaning up the data, we displayed the EEG signal using a simple script to remove the fixed DC offset voltage and scaled and shifted the plots accordingly to display them. See Figs. 7.15 and 7.16. However, it is rather difficult to compare these two plots directly because the recording duration for the smooth video is about 10 times longer than the time period for the rough video. To get a fair comparison, we aligned the two plots at the same duration scale as that of the rough video with 2800 data points.

To visualize the spatiotemporal patterns of the EEG, we mapped the data to the probe location and used bicubic interpolation to compute the pixels between probes. This helped to create frames for the animated videos of the heat map where the color values represent the signal voltage.

The heat map animations generated while viewing the rough and smooth videos show significant differences. The EEG pattern measured when viewing the rough video shows much more noise and jitteriness. The EEG from the rough video starts

[15] Emotiv EPOC and TestBench™ Specifications (2014). https://emotiv.com/product-specs/Emotiv%20EPOC%20Specifications%202014.pdf

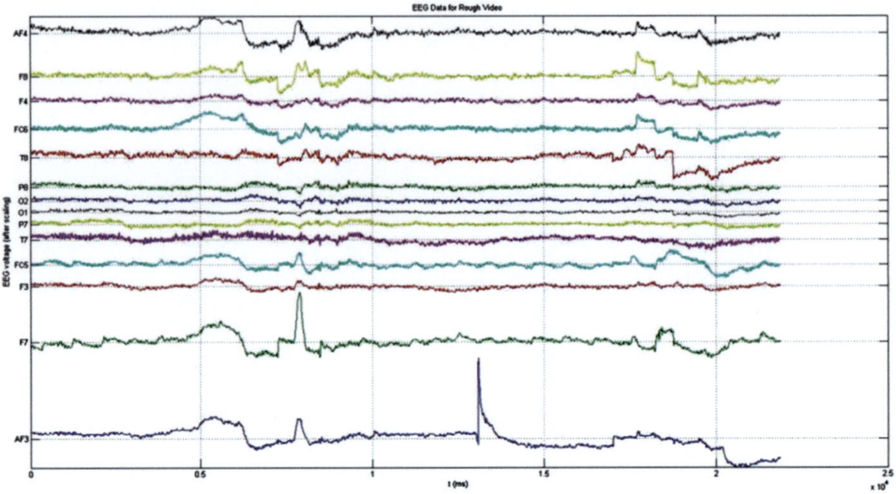


**Fig. 7.15** The 14 channels of signal data from the rough video (before filtering)

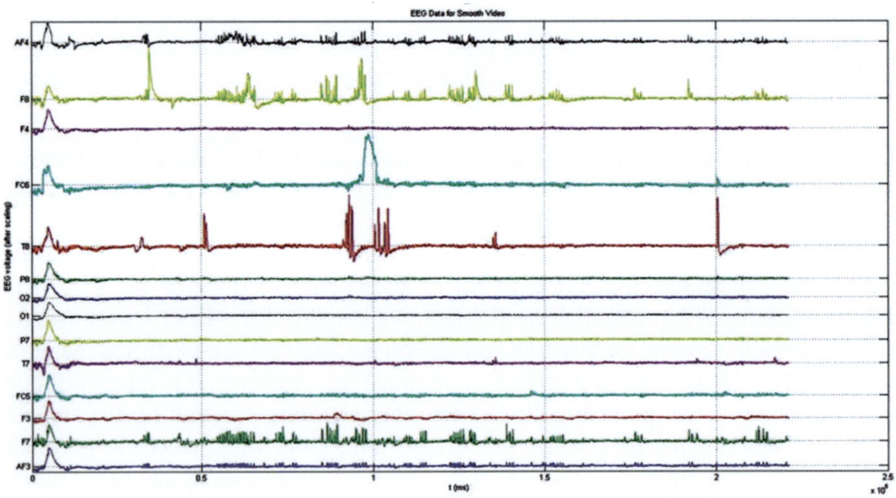


**Fig. 7.16** The 14 channels of signal data from the smooth video (after filtering)

with lower brain activity on both the left and the right sides of the brain. Then, the activity becomes more volatile on both sides. The connectivity between the two sides then increases and the overlapping areas forms a large "kissing" shape. The right brain activity still dominates the EEG signals. However, at one point, the left brain activity dominates the interaction and the right brain activity is almost nonexistent. Then, the balance is reversed again and the right brain dominates the interactions and light kissing between the two sides continues. See Fig. 7.17.

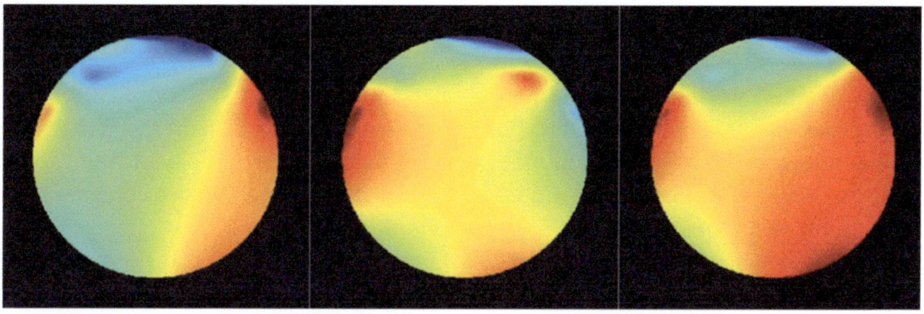

**Fig. 7.17** The EEG heat map of the rough video: low brain activity at the beginning (*left*), the left-brain took over (*middle*), and the right side of the brain interacts with the left-brain

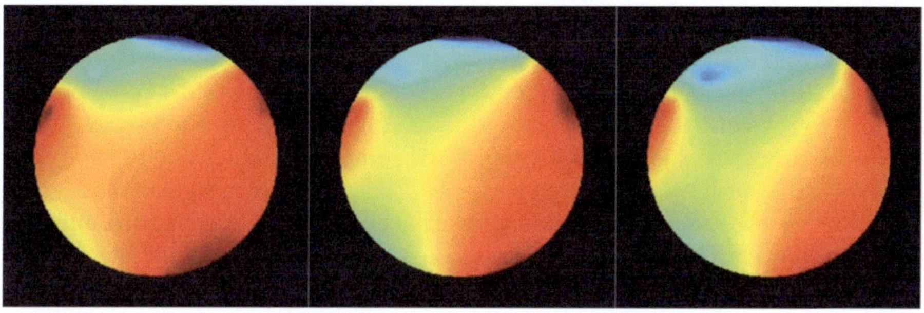

**Fig. 7.18** The EEG heat map of the smooth video: strong right brain activity at the beginning (*left*), the right-brain interacts with the left side of the brain (*middle*), and less connected activities (*right*)

The EEG heat map animation from the smooth video is much calmer than that from the rough video. It starts with higher brain activity on both the left and right sides of the brain. Then, the hot area separates to the left and right, while the right area dominates the interaction most of the time. Both sides occasionally interact through lightly kissing each other. However, the overlapping times are much shorter than those in the rough video. The right brain activity is much stronger than the activity seen in the rough video, indicating that spatial and temporal filtering allows more visual thinking to occur in the right side of the brain, rather than logical reasoning in the left side of the brain. It appears that the rough video causes the EEG signal to vary on a macro level, while the smooth video causes more micro level variations. See Fig. 7.18.

It is worth noting that the EEG is sensitive to muscular movement, especially facial expressions. There are many ways to remove the motion noise. Mathematically, we can use Independent Component Analysis (ICA) to reject anomalous sections of the data. Furthermore, we can monitor facial expression using a frontal viewing camera and remove noise accordingly.

## Summary

In this chapter, we have explored how to map motion data to motion pictures and interactive games, enabling human vision to discover motion patterns. There are four key factors in mapping motion: (1) spatial mapping for individual frames, (2) temporal scaling for visual perception, (3) alignment for object registration, and (4) looping for user experience.

Motion enhances vision. Motion enables measurement, which is the most important function for scientific visualization. Motion enables us to generate an internal three-dimensional model of the world. Motion enables us to have a panoramic view of a large object or scene. In instances where a scene is obscured behind an obstacle, motion would be the only feasible solution for visualization. Motion surveys locations and maps landmarks along the way, as a video odometer,

Gamefication transforms large datasets into virtual worlds where users can actively explore and discover patterns and events in an exciting, ubiquitous, yet affordable way. It can create a first-person view of the virtual world, adding physics engines for object believability, and enabling collaboration over networks. Video games teach us scalable design strategies for computation-intensive tasks, e.g. to streamline the viewing scope, displaying only the data that are in sight of the user. To minimize the event handling workload, a game engine usually defines an effective zone for an event. Randomization is also necessary to spontaneously alter control variables in order to maximize the chances of discovering new motion and event patterns.

Filtering is crucial to decent visualization. It serves two purposes: enhancing patterns and reducing noise. There are many filtering methods, but here, we focused on two essential cognitively-inspired methods: continuity and curvature filters, based on Gestalt principles of continuity of spatial and temporal momentum and minimizing efforts for visual reasoning.

Our eye tracking experiment shows that spatiotemporal filtering helps observers to concentrate on meaningful motion patterns. Furthermore, our EEG brainwave study shows that viewers have more volatile brain activity while viewing noisy, jittery video samples, but more calm and more right brain activity while viewing the filtered, smooth time-lapse images. We also see various interactions between the left-brain and right-brain. The left-brain takes control at a certain point while viewing the jittery video sample. Although our empirical study is preliminary, it points us in a direction toward a more objective study of human cognitive factors in visualization, especially motion pattern and event discoveries.

# CHAPTER 8

# Sensuality

The word *sensuality* comes from the root word sense, which pertains to our five senses: sound, sight, touch, smell, and taste. The Greek word often means a total devotion to the pleasures of the physical senses, sometimes implying outrageous conduct. Sensuality is different from sexuality. Sensuality is a much broader term that goes beyond sexual attraction and eroticism. Sensuality is also subtler and more cognitive than sensation, which is a reflexive feeling or reaction resulting from excitement. In this chapter, we explore how to incorporate sensuality into humanoid robot design, how to analyze the sensuality in human voices, and how to transform a computerized or robotic monotone voice into a sensual voice. We also explore how to detect and analyze sensual body shapes and gaits.

## Humanoid Robosapien

Sensuality is a forbidden fruit and is often listed as one characteristic of impurity in religions. In contrast, sensuality of humanoid robots has been portrayed in popular science fiction movies such as the films "Artificial Intelligence" and "I, Robot." In reality, today's robots and computers can hardly understand or communicate with sensuality, even though they can hear, see, touch, smell, and taste.

Robosapien was the first robot toy based on the science of applied biomorphic robotics that enabled him to act more like a human. His designer was Mark W. Tilden, an established robotics physicist who worked for NASA, DARPA, and JPL on advanced robotic technologies. Robosapien has fully-articulated biomechanical movements and pendulum walking motion. He can respond to a remote controller through an infrared channel to perform up to 67 preprogrammed functions including high-five, throw, dance, and karate moves.[1]

[1] Robosapien manual. http://www.theoldrobots.com/book31/Robosapien-manual.pdf

© Springer-Verlag London 2016

Y. Cai, *Instinctive Computing*, DOI 10.1007/978-1-4471-7278-9_8

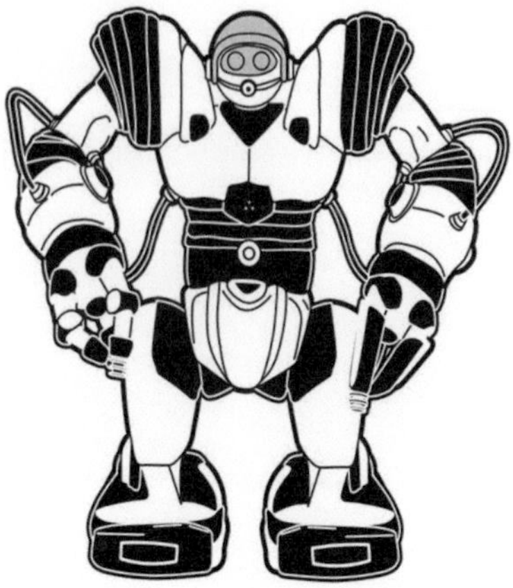

**Fig. 8.1** The humanoid robot Robosapien designed by Mark W. Tilden (Image courtesy of WowWee.com)

However, the most interesting feature of Robosapien lies in a special voice that WowWee called "International Caveman Speech." Robosapien does not simply say "hello" because this is too common. He does not even have a child's voice. Instead, he has an adult's thick and lazy voice that burps, talks back, signs, yells, and even whistles. When the power button is pressed on, Robosapien will awaken with a yawn, stretch and then mutter "uh-huh" like an old cowboy in Western movies. The sensual voice was recorded by the inventor Mark Tilden himself in a tent made of a bed blanket in a hotel room of Hong Kong. To make the best sound effect, he consumed a lot of Thai food and Pepsi drinks. To children, the exaggerated voice, shape, and gait show a novel and sensual personality that stimulates their curiosity and imagination.

The strangest thing about Robosapien is his signoff line "Rosebud," which comes from the 1941 Orson Welles' movie, "Citizen Kane." It was the last word character Charles Foster Kane said before he died in the film. The word "Rosebud" was a key thread of the movie through the research of a newsreel reporter seeking to solve the mystery of Kane's dying word. Tilden explained it: "If you remember, you wait for the entire movie, and you find out that Rosebud was the name of his favorite toy. So just imagine the poetic symmetry – just before Robosapien dies, he has a dream of another toy."[2]

Robosapien possesses contemporary robotic technology such as touch sensors, infrared remote control, programmable gestures, synthesized voices, walking, gripping, and even dancing. Still, his most attractive and unique touches are those designed with sensuality in mind – the manly shape, voice, and gait. These make the robot unique and unforgettable not only to children but to adults as well. This suggests that sensuality does not always translate into sexuality. Rather, sensuality can be implemented in a broader sense for improving the affection and personality of a robot (Fig. 8.1).

[2]Goldman F (2004) A robot for the masses. The New York Times Magazine, November 28, 2004.

## Sensual Voices

It is a well-established phenomenon that the human brain is wired to like female voices. This may go all the way back to our mothers and when we were in the womb. In various studies, fetuses have been shown to respond to their mother's voice but not other female voices. The fetuses showed no distinct reaction to their father's voice. When an infant is born, its vision system is not well developed. Its first sensual experience is simply hearing its mother's voice. This perhaps is the biological reason behind why airport announcers and computer voices are typically female.[3] From a designer's point of view, it is much easier to find a female voice that everyone likes than a male voice that everyone likes. Telephone operators have traditionally been female, making people accustomed to getting assistance from a remote female voice. Similarly, customer service lines feature more female voices than male voices.

Male voices normally convey commanding and threatening signals. To make things worse, many intelligent yet dangerous robots in Hollywood movies were voiced by males, for example, the computer HAL 9000 in the Stanley Kubrick film "2001: A Space Odyssey." HAL 9000 was a synthetic male voice in a faceless monitor and tried to kill the astronaut during the mission in the film. There have also been various recorded robotic voices used for telephone answering machines that have been exaggerated to a slow speed and monotone quality. Because of the bad associations to HAL and similar robotic voices, manufacturers have tried not to use male voices to avoid scaring away consumers. This may explain why in almost all GPS navigation systems on the market, the default voice is female. However, there are some exceptions in selecting voice genders. For example, people normally assume pilots and captains are male. Therefore, some people prefer a male voice for GPS instructions, especially male drivers. Certain men want to change the default female voice to a male voice. In one case, when male voices were not available to a male professor, he asked if a female voice with a Spanish accent or an approximate of his wife's voice could be made available.

When it comes to the consumer applications of computerized voices, the gender of the voice is usually determined by what service or product is using the voice. For example, train stations and airports often use higher-pitched voices because they are easier to hear over large, echoing spaces and the chatter of background noise. As computers and robots have become part of our everyday lives, there is a huge opportunity for the personalization of voices in interactive media. Clifford Nass of Stanford University conducted a psychology experiment to test how 1000 real people responded to virtual people – particularly to their voices.[4] Nass and his team discovered that gender and ethnic stereotyping, often subconscious, is pervasive

---

[3]Griggs B (2011) Why Computer Voices are Mostly Female. http://www.cnn.com/2011/10/21/tech/innovation/female-computer-voices/

[4]Nass C and Brave S (2005) Wired for speech: how voice activates and advances the human-computer relationship. MIT Press

when people encounter voice interfaces. The experimenters also found they could manipulate people's attitudes toward the content of messages by changing the emotional tone of the voice, as well as its physical parameters such as pitch and speed.[5]

## *Voice Actors*

Siri was the first mobile voice-to-voice assistant that applied speech recognition and voice synthesis. In the United States, Siri is a female voice. Apple employed the voice actress Susan Bennett to play "Siri."[6] Her voice was recorded for 4 hours a day for a whole month. Next, her speech patterns were extracted and synthesized. She was over 60 years old when she recorded the voice for Siri. To young iPhone users, Siri sounds like a mother figure with some kind of authority. Apple wanted to make a neutral voice for the US-version of Siri, so they intentionally made her speech sound artificial – half female and half robotic. They could have chosen to make her speech more seamless and human-like, but they wanted her to convey authority over answers as in the voice-over narration in films where male voices usually dominate. Neutral voices are designed to keep a certain distance between the user and the device in order to reduce the risk of abusive behaviors toward the device or the service.

There is no doubt that Siri has more options for voices: male or female, different accents, and various personalities. In the UK, Siri is a male voice. However, we prefer more sensual voices because we often attach to sensual voices.

Marilyn Monroe is an obvious example of sensual voice. Before she became a famous movie star, she was a quiet and shy girl who developed a stutter – a tendency to speak with sudden involuntary pauses or to repeat the initial letters of words. During her years as an aspiring actress, one of the many pieces of advice she received was to lower her tone.[7] Monroe tried seriously. She also studied singing, quickly developing her distinctive, breathy style. She exaggerated the clarity of the "d" and "t" sounds in order to overcome her stutter. Her voice always sounded as if she was just waking up, which has a sensual effect on male listeners.

Iris Lettieri is a Brazilian born newsreader and voiceover artist who is most famous for her airport announcements for Galeão International Airport in Rio de Janeiro. Her sultry, slow, and deep voice, which provided announcements for Galeão for 38 years, has been described as "the most captivating voice in the world" by

[5]Nass C, Steuer J, Tauber E (1994) Computers are social actors. Conference on Computer Human Interaction (CHI), pp. 204–278, 1994

[6]LatestWorldIncidents (2013) Real Voice of Siri is Susan Bennett [HQ]. https://youtu.be/T6boqUfzUD8

[7]Marilyn Facts. http://www.marilynmonroe.ca/camera/about/facts/voice.html

NPR.[8] Her voice, which is much deeper and coarser than most female voices, has a very strong, yet attractive character. A voice announcement is supposed to transmit calmness to passengers, but in this case, it was quite the opposite. A group of Japanese businessmen was reported to have missed their plane home after her siren-like voice kept them lingering in the waiting area. Listeners suggested that Lettieri's voice transmitted an almost mother-like love.

Maria, or Gentle Whispering, is a popular YouTube contributor who has over 430,000 subscribers in 2015. She is most known for her quiet, magnetic, soothing voice that she features in several of her Auto Sensory Meridian Response (ASMR) videos.[9] ASMR is an experimental therapeutic method for generating a perceptible pleasurable tingling sensation in the body in response to visual, auditory, tactile, olfactory, or cognitive stimuli. For auditory stimulation, ASMR speakers normally use high-fidelity stereo microphones when whispering and creating soundscapes in videos. One of Maria's most popular videos features her teaching her audience how to simply fold towels. Her voice is characterized as very soothing, as it has a very soft and breathy quality to it. In the towel folding video, she uses a hushed, whispering voice, hence the name "Gentle Whispering." As a part of the ASMR community, she uses certain tactile and auditory stimuli to trigger a tingling sensation, or autonomous sensory meridian response, which is not meant to be sexual but simply relaxing. It could perhaps even hypnotize video viewers.

## *Voice Transformation*

Voice transformation, or voice morphing, alters the voice of a speaker so that it matches the voice of a target speaker. Voice transformation can be used to obscure a person's identity or to impersonate another individual. In this study, we wanted to transform a "non-sensual" voice into a sensual voice.

The first step was to model the characteristics of voices and identify the differences between the speaker's voice and the target's voice. The human voice is an audible vibration ranging from 20 Hz to 10 kHz. The raw voice data in the time domain can be transformed into the frequency domain using the Short-Term Fourier Transformation (STFT) method, which slides a small sampling window over the raw data to generate a three-dimensional spectrogram of the sound: frequency, volume, and time.[10] Figure 8.2 shows spectrograms of the three sensual voices of Iris Lettieri, Marilyn Monroe, and Maria, and the "non-sensual" voice of Susan Bennett.

---

[8]Julie McCarthy's interview on NPR, March 17, 2007: http://www.npr.org/templates/story/story.php?storyId=8976813

[9]Miller J (2015) Whispering on the Internet is Paying this Women's Rent. http://www.cosmopolitan.com/lifestyle/a40025/gentlewhispering-maria-internets-most-fascinating/

[10]Cai Y (2014) Ambient Diagnostics. CRC Press

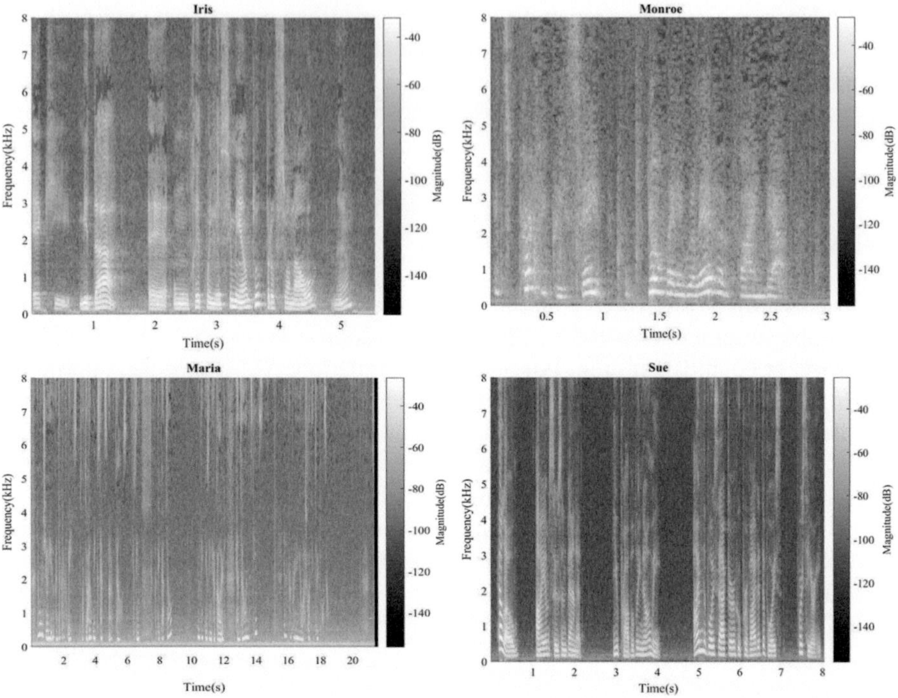


**Fig. 8.2** Spectrograms of the sensual voice of Rio airport announcer Iris Lettieri (*top left*), the voice of Marilyn Monroe (*top right*), the voice of Maria about how to fold a towel (*bottom left*), and the non-sensual voice of the Siri speaker Susan Bennett (*bottom right*)

To further characterize human voices, we use a representation of a short-period power spectrum of sound waves that mimics the behavior of human ears, the Mel-frequency cepstrum coefficients (MFCC). To compute the MFCC, we divide the entire signal into small time windows called frames. The Fast Fourier Transform (FFT) is computed for each frame. Then, the spectrum is segmented into a number of critical bands by means of a *filter bank*.[11] After the MFCCs of the speaker are extracted, they can be quantized to a number of *centroids* using vector quantization algorithms. These centroids constitute a speech feature vector or *codebook* of that speaker in the training phase. The purpose of a codebook is to enable us to compare a number of voices with different data lengths. Instead of comparing voice data, we can compare features defined in the codebooks. In the testing phase, different speakers utter the same words used in training, and the Euclidean distance between the centroids of the speaker in the training phase and the centroids of the speaker in the testing phase are measured. If the Euclidean distance is close to zero or falls within a

[11]Molau S, Pitz M, Schlueter R and Ney H (2001) Computing Mel-Frequency Cepstral Coefficients on the Power Spectrum. Proceedings of International Conference on Acoustic, Speech and Signal Processing, 2001.

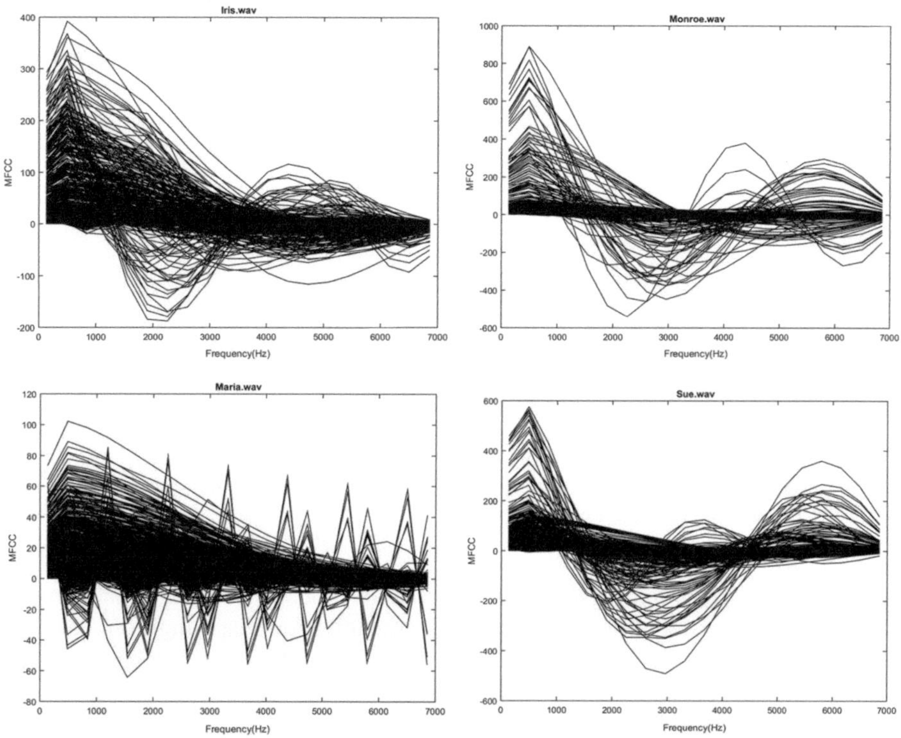


**Fig. 8.3** MFCC spectrogram of the sensual voice of Iris Lettieri (*top left*), the voice of Marilyn Monroe (*top right*), Maria (*bottom left*) and the non-sensual voice of the Siri speaker Susan Bennett (*bottom right*)

predefined threshold, we conclude that it is the same individual.[12] Figure 8.3 shows the MFCC frequency plots of the three sensual voices from Iris Lettieri, Marilyn Monroe, and Maria, as well as the "non-sensual" voice from Susan Bennett.

Intuitively, if we can model and tune the MFCC of the speaker's voice such that the Euclidean distances between the centroid feature vectors of the speaker and the target are within an acceptable threshold, then the reconstructed voice should be very close to the target's voice. Figure 8.4 outlines this process.

The text-dependent voice transformation system consists of two modules. First, the target's speech signal, Marilyn Monroe's voice, is subjected to pre-processing for artifact removal, and is divided into 25 millisecond frames with 10 millisecond overlaps. MFCC speech features are extracted frame by frame from the filtered signal. After we have the frame-wise MFCC, we perform vector quantization to generate a codebook. Vector Quantization (VQ) is a classical quantization technique

[12]Tiwari V (2010) MFCC and its applications in speaker recognition, International Journal on Emerging Technologies, Vol. 1, No.1: 19–22, 2010

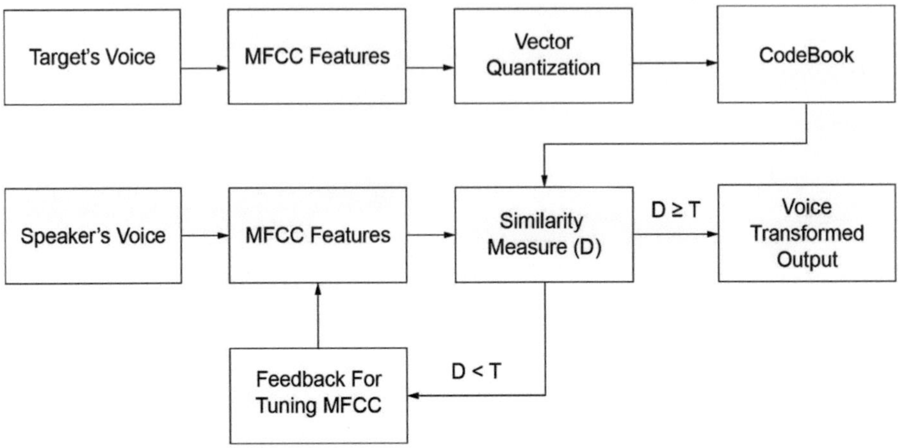


**Fig. 8.4** Overview of the MFCC-based text dependent voice transformation process, where D is the distance and T is the threshold

in signal processing which allows the modeling of probability density functions by the distribution of prototype vectors. Vector quantization maps high dimensional vectors from a large set into clusters of approximately the same number of points closest to a cluster centroid. These cluster centroids are called code words, and the collection of these centroids constitutes a codebook. Figure 8.5 shows a sample of vector space and codebook formulation for Marilyn Monroe's voice. The method commonly used to generate codebooks is the Linde-Buzo-Gray (LGB) algorithm.[13] The pseudo code is as follows:

```
1.  Design a one-dimensional vector codebook
2.  Double the size of the codebook by splitting each current codebook
3.  Find the new centroids for the split codebook
4.  Iterate step 2 and 3 until all the centroids are saturated
```

Now, the speaker's speech signal is subjected to pre-processing and the frame-wise MFCC is extracted, similar to the process followed for the target's speech signal. The middle plot in Fig. 8.5 shows the MFCC for the test subject's voice. The Euclidian distance from a vector instance (MFCC in our case) to the closest code word in a codebook is called VQ-distortion. Our goal is to minimize the overall VQ-distortion. If the overall VQ-distortion of the speaker is within a predefined threshold, then we say the speaker's voice is close to the target voice. Otherwise, we keep tuning the MFCC until we have the VQ-distortion within the threshold.

After the voice is transformed, we may perform an inverse MFCC to reconstruct the speech signal. A rudimentary approach is to recover the short-time magnitude

[13]Kekre HB and Kulkarni V (2010) Performance comparison of speaker recognition using vector quantization by LBG and KFCG, International Journal of Computer Applications, Vol. 3, No. 10, July 2010, pp. 975–8887.

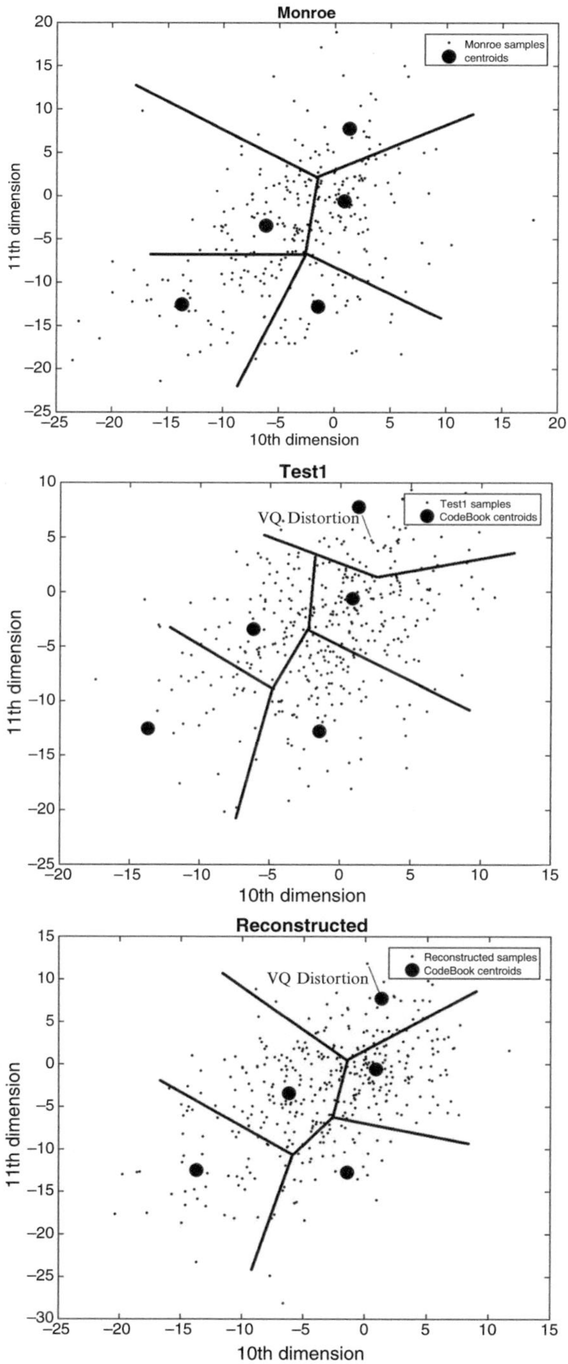


**Fig. 8.5** Sample of the vector space and codebook of Marilyn Monroe's voice (*top*), the test subject's voice (*middle*), and the reconstructed voice (*bottom*)

spectrum given by the MFCCs, and then impose it on white noise.[14] The bottom plot of Fig. 8.5 shows the tuned MFCC of the reconstructed speech signal. Figure 8.6 shows the MFCC-frequency plots of the test subject's voice, Marilyn Monroe's voice, and the reconstructed voice. Figure 8.7 shows the spectrograms of the three voices.

Voice transformation can be further fine-tuned using post-processing methods, such as adjusting the sampling frequency and shifting the pitch (fundamental frequency) of the voice. Pitch-shifting is able to change the pitch of a voice without changing the signal length.

The example above is a rudimentary illustration of the voice transformation. For over a decade, many more sophisticated algorithms have been developed for voice transformation or voice morphing. In addition to mapping codebooks, which is similar to the process discussed earlier, more physiological and acoustic parameters are incorporated into voice modeling. For example, the algorithm of Oavi, et al., transforms the acoustic characteristics of the vocal tract and glottal excitation of the source speaker's voice into that of the target's voice.[15] The algorithm detects the pitch, and separates the glottal excitation and vocal tract spectral features. The algorithm of Cano, et al. models a singer's pitch, timbre, vibrato, and articulation, and then modifies those parameter values to resemble a given famous singer's voice.[16]

The emerging machine-learning algorithms improve the accuracy of voice morphing. Gaussian Mixture Models (GMM) are commonly used for mapping the speech feature coefficients from source to destination.[17] The modified GMM mapping method uses glottal waveform separation and the prediction of excitations to overcome the smoothing problem associated with GMM-based methods. Radial Basis Function (RBF) neural networks are used for voice morphing along with wavelet-based models. The algorithm uses wavelets for extracting speech features, followed by the RBF neural network for modeling the voice.[18] With a better

---

[14]Ellis D (2012) PLP and RASTA (and MFCC, and inversion) in Matlab using melfcc.m and invmelfcc.m. http://labrosa.ee.columbia.edu/matlab/rastamat/

[15]Qavi A, Khan SA and Basir K (2015) Voice morphing based on spectral features and prosodic modification. International Journal of Signal Processing Systems, Vol. 3, No.2, December, 2015.

[16]Cano P, Loscos A, Bonada J, de Boer M, Serra X (2000) Voice morphing system for impersonating in karaoke application. Proceedings of the ICMC 2000.

[17]Chadha A, Savardekar B and Padhya J (2012) Analysis of a modern voice morphing approach using Gaussian Mixture models for Laryngectomees. International Journal of Computer Applications. Vol. 49, No.21, July 20, 2012.

[18]Orphanidou C, Moroz I, and Roberts S (2004) Wavelet-based voice morphing, WSEAS Journal on Systems, Vol.10, No.3, 3297–3302, 2004.

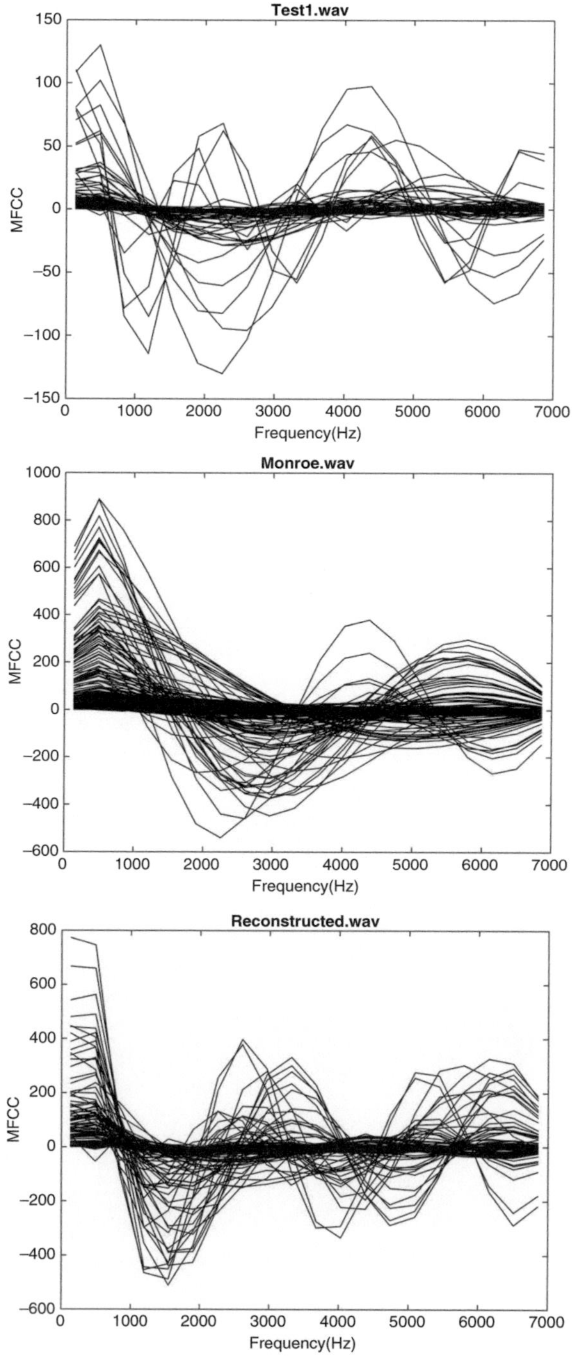


**Fig. 8.6** MFCC transformation from a non-sensual voice to Marilyn Monroe's voice: the *top* is the subject's MFCC-Frequency plot; the middle is the MFCC-frequency plot of all frames of Monroe's voice; and the *bottom* is the plot of all frames of the reconstructed voice

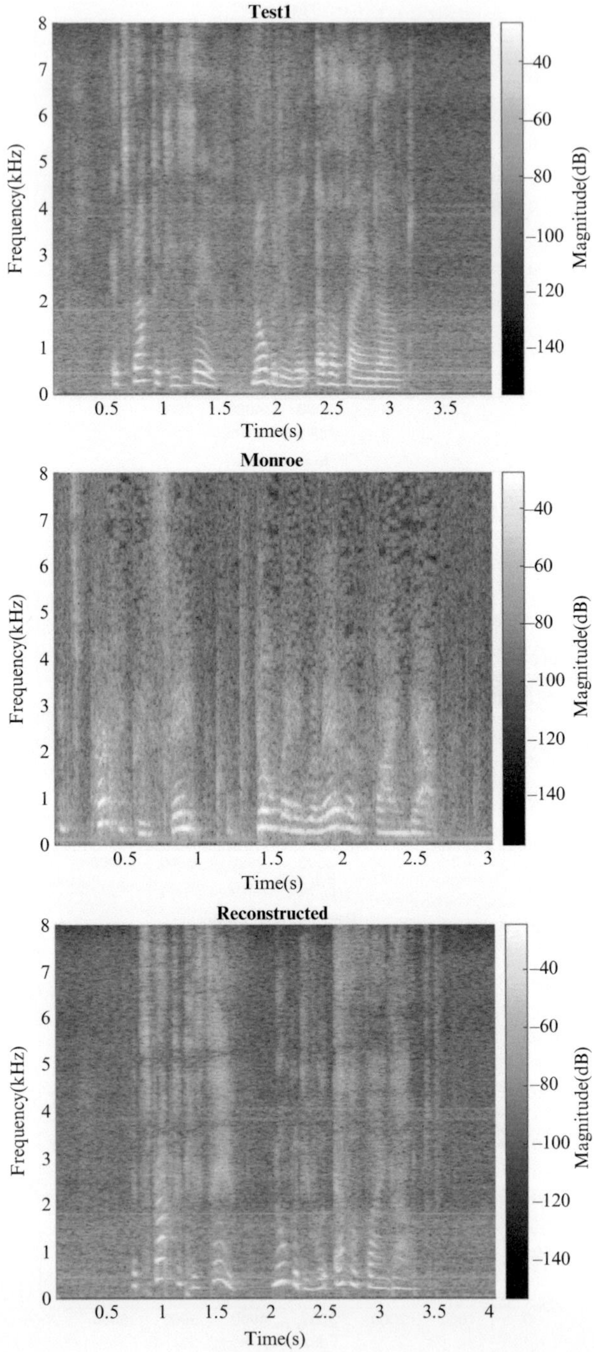


**Fig. 8.7** Spectrograms of a man's original voice and after being transformed to Marilyn Monroe's voice: the *top* is the spectrogram for Test subject 1's voice signal; the middle is the spectrogram for Marilyn Monroe's voice signal; and the *bottom* is the spectrogram for the reconstructed voice signal

understanding of acoustic and physiological parameters, superior machine learning models, and growing amounts of voice samples, voice morphing is moving closer to reality.

## Sensual Shape

Visual sensuality is often related to sexuality, which has a great impact on the way we look and move.[19] So far, most computers and robots are asexual or gender neutral. However, the scope changes gradually as computers and robots are involved with images of the human body. In 1996, a milestone computer vision algorithm was developed for detecting nude figures in a picture, and was published on a peer-reviewed journal.[20] The detection is based on skin color and body contours. The algorithm can be used for image search, classification, and annotation.

The sensuality of the human body has had an indelible imprint on health and culture. Across all eras and countries, three body parts – the breasts, waist, and thighs – are more often referred to as more sensual than other parts of the body. Psychologists from the UK have shown that men focus primarily on a woman's waist-hip ratio (WHR), which is calculated by dividing the circumference of the waist by that of the hips. Their study[21] analyzed thousands of examples from British literature from the sixteenth to the eighteenth century to demonstrate that what men find attractive today was also true hundreds of years ago: men prefer women with an hourglass shape. Modern science reveals that an hourglass shape in women is associated with relatively high levels of the hormone estrogen. Since estrogen levels influence fertility, men seeking to pass their genes on to the next generation would do well to pick an hourglass-shaped woman as a mate. In a similar vein, a sizeable belly on a woman is reliably linked to decreased estrogen, reduced fertility, and an increased risk for major diseases according to research conducted over the past decade.

3D body scanning techniques such as microwave imaging and near-infrared imaging systems can detect 3D human body shapes. Microwave imaging is privacy-invasive because it can penetrate clothes. We may use such body imaging data for health informatics, such as the body mass index (BMI) calculated from the waist-hip ratio.[22] The concern with body scanning is that the detection of sensual body parts is inevitable, and infringes upon human privacy.

[19] King BM (1991) Human sexuality today, 3rd edition, Prentice-Hall International

[20] Forsyth DA, Fleck MM (1996) Identifying nude pictures, *Proceeding. Third IEEE Workshop on Applications of Computer Vision.* 103–108, 1996.

[21] Singh D, Renn P and Singh A (2007) Did the perils of abdominal obesity affect depiction of feminine beauty in the sixteenth to eighteenth century British literature? Exploring the health and beauty link, Proceedings of the Royal Society B: Biological Sciences, Vol. 274, No. 1611, March 22, 2007

[22] Cai Y (2014) Ambient Diagnostics, CRC Press

One possible privacy algorithm would be to detect the breast area from 3D body imaging data and to blur the area or make it transparent. If we cut the 3D scan data horizontally from top to bottom, then we have a stack of profiles of a frontal surface of the body. Assume we have a typical profile of a female chest as a template. See Fig. 8.8. We can use a template matching algorithm to locate the breast area vertically. The matching of the two surfaces is driven by a similarity function. We need to solve two problems. First, a suitable template must be created. Second, a similarity function must be selected so that a minimization algorithm can align the template onto the region of interest. For each plane of the scan data, the back of the body contour can be removed. By assigning the $X$-axis between the two points with the greatest distance, we can obtain the front part of the body contour. This aligns the subject to our template such that the matching is never attempted on a twisted or backward-facing body. See Fig. 8.8.

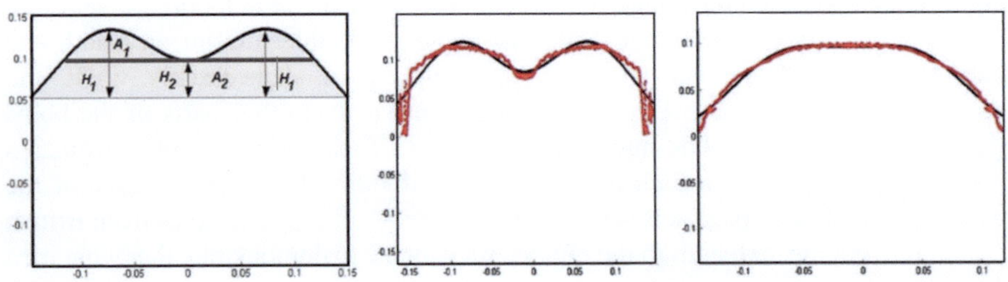


**Fig. 8.8** Variable definitions for the breast template (*left*), matching results for the female samples (*middle*) and male samples (*right*). The *solid black curves* are the template contours. The *dots* are the 3D scan data

We then use three radial basis functions to configure the template for a female breast pattern.[23]

$$y = \sum_{i=1}^{3} a_i e^{-(x-s_i)^2} \tag{8.1}$$

where, $a = a_1 = a_2, b = a_3, s = s_1 = s_2$, and $s_3 = 0$. We can use non-linear regression to compute the variables $a$, $b$, and $s_i$ to match the template with the scan data.

Most shape descriptions depend on particular coordinate systems and viewpoints, meaning that the algorithm can only work within the same space as the training data. Coordinate-invariant measurements aim to compute shape properties from a ratio, rather than absolute values. This reduces the dependency of particular poses, so there is no need to create a unique algorithm for each pose.

[23]Laws J, Bauernfeind N and Cai Y (2006) Feature hiding in 3D human body scans. Journal of Information Visualization. (2006) 5, pp. 271-278

Template matching not only filters out noise, but it also describes the characteristics of a shape. We define the following invariant similarity functions in the coordinate system: height ratio and area ratio. The height ratio is defined as:

$$H_r = \frac{H_1}{H_2} \tag{8.2}$$

It is necessary to match the template to each slice within the detected area, where only the greatest ratio of curvature is kept and used as the final result. See Fig. 8.8 for the definitions of $A_1$ and $A_2$. The area ratio is defined as the ratio of the area of curvature feature ($A_1$) to the total area ($A_2$) of the model by the following formula:

$$A_r = \frac{A_1}{A_2} \tag{8.3}$$

The author's lab tested this algorithm with a subset of the 3D anthropology database CAESAR, which contains 50 males and 50 females aged 16–65 years, where 50 of them are North American, 24 are Asian, and 26 are from the European survey of Italy and the Netherlands. The goal was to find the breast features from known female and male scan data samples. Figure 8.9 shows the test results. From the plot, we can see that there are two distinguishable groups associated with the particular gender of each subject. The male subjects tend to have no curvature features and lie in the lower left range of the graph, whereas female subjects do demonstrate these curvature features and lie in the upper right range of the graph. There is a "dilemma"

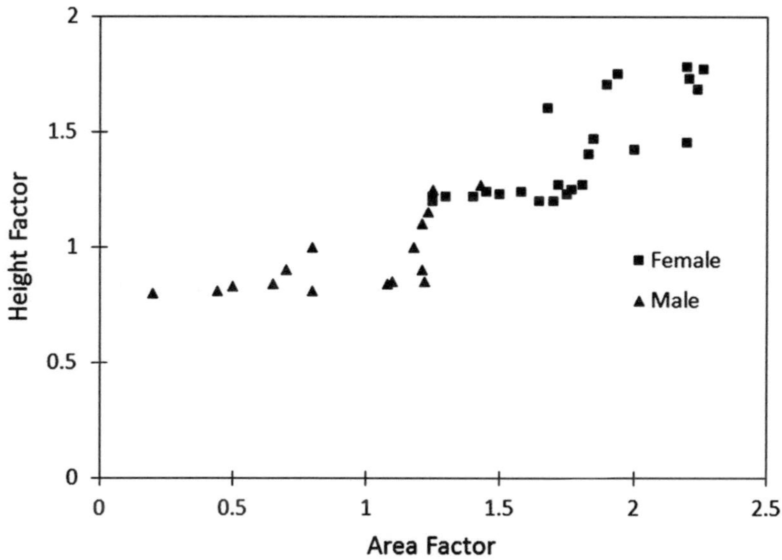


**Fig. 8.9** Classification test results with male and female samples from the CAESAR dataset

zone where some overweight males do possess curvature features. However, the overlapping zone is quite small – less than 8 % of the total one hundred samples. After the area and height factors have been calculated, we can determine the feature area.

## Sensual Gaits

Psychology studies show that we are attracted to others not only by body shape, but also in the way they move. When we encounter another person, we judge the individual's gender, attractiveness, and behavior from his or her body movement or gait. To understand the sensuality of gaits might help us to design future robots that recognize human gaits and even have sensual gaits themselves.

### *Shoulder-Hip Trapezoid and Triangles*

The dynamic visual cues of sensuality can be represented by the movement of joints. For example, the acromion regions on the left and right shoulder bones are two joint points. The greater trochanters near the tops of the left and right femurs are two joint points of the hip. If we connect those four points, we form a shoulder-hip trapezoid, which can be further divided into two triangles when we cross-link all four points. Typically, the upper triangle of a male is bigger than the bottom triangle, while the typical female is just the opposite. See Fig. 8.10.

People exaggerate their walking gestures to display sensuality when they are being watched. In a gait perception study, researchers animated male and female figures with adjustable motion parameters for each joint.[24] The lab experiment involved over 700 individuals who scored three animated representations of people

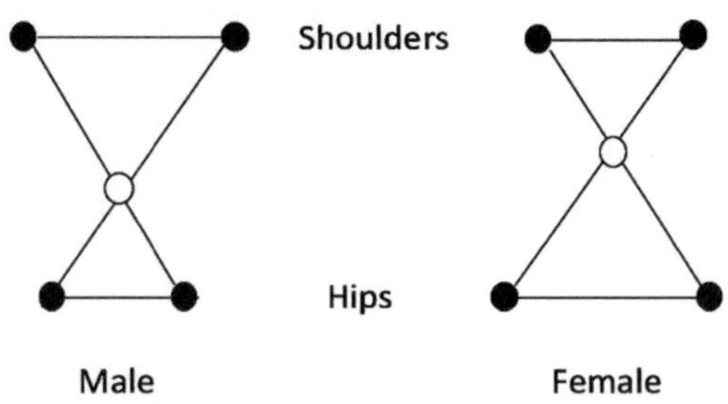


**Fig. 8.10** The shoulder-hip trapezoid and triangles for male and female bodies

[24]Nauert R (2007) Attractiveness is in the way we walk. Psych Central. Retrieved on January 20, 2016 from http://psychcentral.com/news/2007/05/23/attractiveness-is-in-the-way-we-walk/849.html

walking. The perceived attractiveness ratings for women increased by about 50 % when they walked with a hip sway and attractiveness ratings for perceived men more than doubled when they walked with a swagger in their shoulders.

The sensuality of gaits is also influenced by the local culture's definitions of beauty, and of masculinity and femininity. In a psychology experiment that was reported by the Discovery television program,[25] researchers asked two groups of male and female subjects to walk on a treadmill while filming their movements. The researchers gave the subjects an explanation of the purpose of the test. In the first group, the researcher told the subjects that the purpose was to find a comfortable walking speed on the treadmill and that the lab would be analyzing their walking efficiency using a computer algorithm, In the second group, the researcher told the subjects that the system would record walking gait and show it to other people who would then evaluate their sexual attractiveness.

The study showed that the people who thought they were being rated for sexual attractiveness tended to exaggerate their movements, trying to look especially masculine or feminine. This reveals subconscious expectations of sexuality. When we look at a woman walking, we expect her hips to move laterally a little bit more from side to side. When we watch a man walking, we expect his shoulders to swagger a little bit more. A man might also puff up his arms a bit to increase the visible size of his body. We appear to have one mental mechanism that looks for people with sexually attractive walks, which encourages us to walk this way when we know we are being watched. The same urges governed the mating habits of primitive humans in ancient times. A regular walk could have signaled genetic weaknesses and a sign of limited life expectancy. A sexually attractive walk would instead broadcast the signal that you are healthy and likely to survive long enough to raise children. We have evolved to pass our genes into the next generation, thus we are extremely attentive to queues about how healthy a potential mate is and how good their genes are. We have mechanisms that are designed to detect these characteristics in other people.

A group of Belgian researchers used videotaping to study women walking on the street. The exploratory analyses suggested that greater pelvic and vertebral rotation and stride length are characteristics of the gaits of women who have experienced vaginal orgasm. The discerning observer may infer a women's experience of vaginal orgasm from a gait that consists of fluidity, energy, sensuality, freedom, and absence of both flaccid and locked muscles.[26]

Gait patterns are associated with sadness and depression as well. A reduced walking speed, arm swinging, and vertical head movements characterize a depressed gait. Moreover, sad walkers displayed larger lateral swaying movements of the upper body and a more slumped posture. The results indicate that specific gait patterns

[25] Discovery Channel, Science of Sex Appear. https://youtu.be/gwdlq95Tnqc

[26] Nicholas A, Brody S, de Sutter P, de Carufel F (2008) A woman's history of vaginal orgasm is discernible from her walk. J. of Sex Medicine. 2008. Sept. 5(9): 2119–24.

characterize individuals with dysphoric moods.[27] Imagine a computer can analyze gait patterns automatically and diagnose potential psychological and physiological problems. Furthermore, marketing robots may analyze the consumers' behaviors and make personalized commercial messages.

## *Modeling Gaits*

Scientific studies of human and animal gaits have been made possible by evolving imaging technologies. In the 1870s, British photographer Eadweard Muybridge used synchronized cameras to study horse galloping and human walking. His method was able to slow down the motion and compress the runway into a single frame. In the 1970s Swedish experimental psychologist Gunnar Johansson pioneered motion perception studies by attaching light bulbs to the joints of humans and animals and filming the sequences of light motion.[28] Perceiving from as few as half a dozen animated light dots, viewers can recognize biological motion, gender, mood, speed, and sensuality. This sparked the formation of marker-based motion capture technology that we still use today. Instead of light bulbs attached to joints, we now use infrared light reflecting balls or patches.

Human gestures involve three-dimensional (3D) spatiotemporal dynamics. Motion capture data can be represented by a vector of 3D points, skeletons, or polygons over time. The raw motion capture data consists of a set of joints, and each has its $x$, $y$, and $z$ coordinates and time stamps. The raw data can be converted to the standard format for gait tracking, or biovision hierarchical data (BVH). Clearly, the size of a motion capture dataset is large. New affordable structured-light sensors such as the Microsoft Kinect have their own application programming interfaces (APIs) to handle the captured data. Kinect is able to track up to two people's three-dimensional gaits without any markers. The system captures a three-dimensional point cloud of the subject and registers the skeletal joints within. Figure 8.11 is a sample of joint data including the center of shoulder joints, hip joints and the centroid of the body.

In order to describe the joint movements, we can use joint angles as key variables, for example, left hip joint, right hip joint, left shoulder joint, right shoulder joint, etc. Joint angles vary over time. If we project the 3D joint angle to a 2D plane, we can plot the angle-time dynamics. This is a visual and qualitative way to describe rhythmic joint coordination among the joints and differences between male and female movement patterns.

Joint movements in the locomotion of humans and non-human primates are often cyclical, even symmetrical. Figures 8.12 and 8.13 shows the movements of a male

---

[27]Michalak J et al. (2009) Embodiment of sadness and depress – gait patterns associated with Dysphoric mood. Psychosomatic Medicine. Vol. 71, pp.580–587, 2009

[28]Johansson G (1973) Visual perception of biological motion and a model for its analysis. *Perception and Psychophysics* 14, 201–211, 1973.

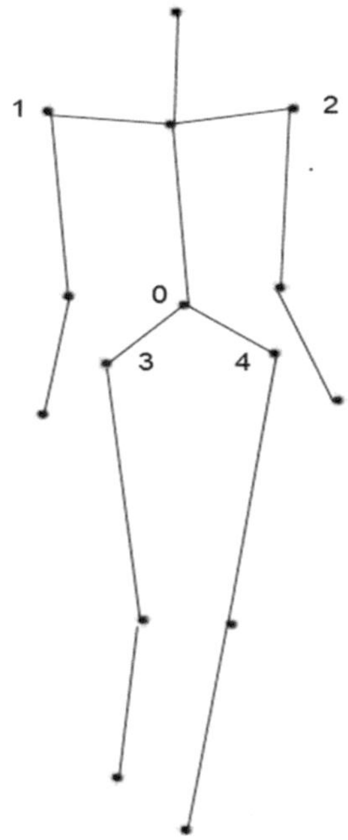


**Fig. 8.11** Motion capture data contain spatiotemporal features, including (1) left shoulder joint, (2) right shoulder joint, (0) centroid, (3) left hip joint, and (4) right hip joint

and female's shoulder joints and hip joints in the motion capture video in which the subjects walked on a treadmill, facing the camera. The two plots contain the angular movements of the shoulder and hip joints around the centroid of body. You can see the traces are nearly periodic; and the man swaggered his shoulders while the woman moved her hips laterally from side to side.

For a quantitative way to describe joint movements we may use the Fourier series, which was discovered by the French mathematician Jean-Baptist-Joseph Fourier in 1822. According to Fourier, any cyclic pattern, including sound, heat, and vibration can be expressed as the sum of a series of sines and cosines of different wavelengths or frequencies. The Fourier Transformation is used to compute the coefficients for each term in the series. Typically, the first 3 coefficients for the sines and the first 3 coefficients for cosines would provide over 96 % of the information contained within the gait data for humans and non-human primates.[29] A gait dynamic pattern can be represented by the 6 coefficients and compared numerically. This is helpful in quantitatively comparing human gait patterns, for example, comparing sensually-exaggerated walking versus normal walking, male walking versus female walking, etc.

[29]Webb D and Sparrow WA (2007) Description of joint movements in human and non-human primate locomotion using Fourier analysis. Primate. 48:277-292. 2007

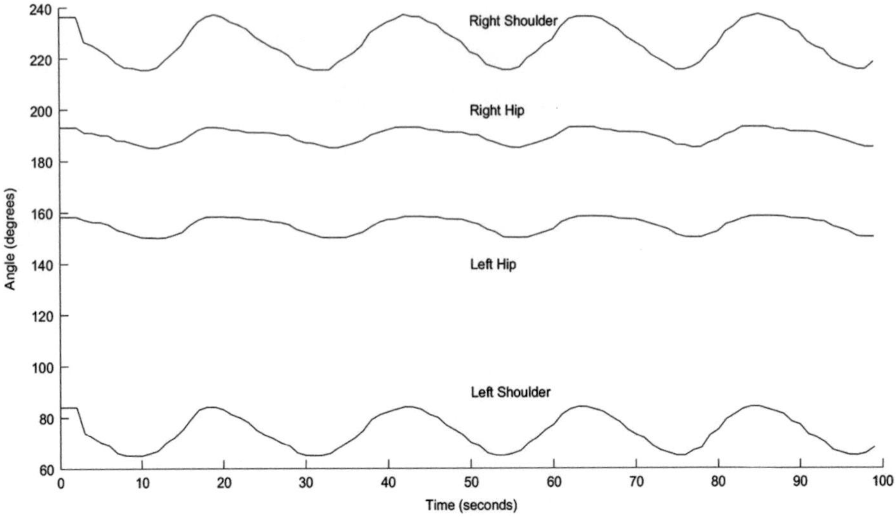


**Fig. 8.12** The angular movements of a man's shoulder and hip joints around the centroid

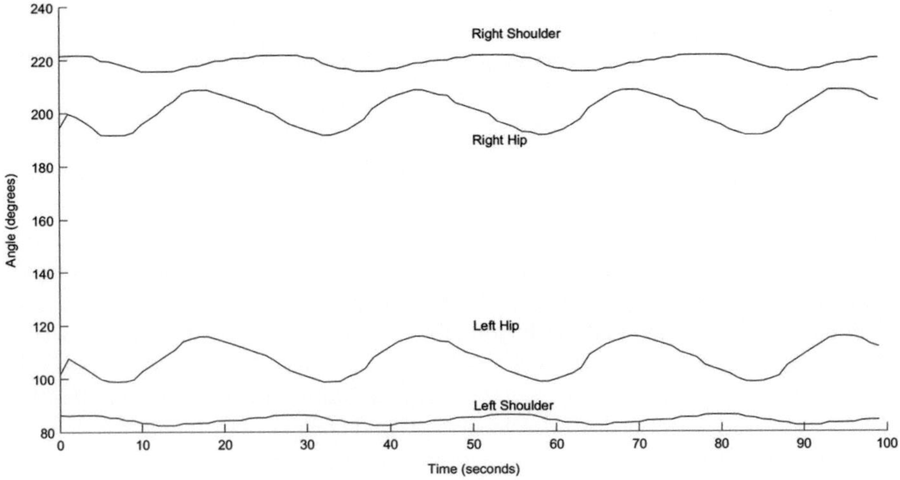


**Fig. 8.13** The angular movements of a woman's shoulder and hip joints around the centroid

The Fourier series is a powerful analytic tool used to study joint movements, specifically joint angle dynamics. However, it is limited by the number of joints and the degrees-of-freedom of those joints. There is a more general model for comparing two gaits – Dynamic Time Warping (DTW).[30] Given two time sequences of gaits

[30]Dynamic programming example (dynamic time warping). http://www.phon.ox.ac.uk/jcoleman/old_SLP/Lecture_5/DTW_explanation.html

(e.g. angles or positions of joints), we want to evaluate the feature distance between the two. For example, sensual walking gaits versus non-sensual walking gaits. We ask the question, how similar are the two and how do we measure their alignment? Traditional measurements such as Euclidean distance will produce a poor similarity score. A more elastic and intuitive alignment measurement is needed. This would allow similar shapes to be matched even if they are out of phase in the time axis. Dynamic Time Warping can also be used to find the best alignment between time sequences. In many practical cases, we only need to use a portion of all joint positions. In our case, if we simply want to track the movement of the shoulders and hips, then we only need to use the left and right shoulder joints and left and right hip joints.

When comparing the performance of DTW to other algorithms such as Hidden Markov Models (HMMs), studies show that DTW yields a higher performance than HMMs.[31] Comparing the performance of the Kinect versus color camera-based gesture recognition, results show that Kinect-based gesture recognition is far more robust than other methods. In addition, we may use the open code software CodePlex for Dynamic Time Warping for gait recognition.[32] CodePlex includes a gesture recorder, gesture recognizer, and sample gestures. Users can save their gait data for later use.

## Summary

Sensuality currently presents a gap between humans and robots. This gap is narrowing as we gain more understanding about human sensuality in voices, body shapes, gaits, and other dynamic senses. In this chapter, we explored how children respond to robots with cartoonish, sensual voices and the existing stereotypes of computer voices in modern devices, such as GPS. Sensual voices often include mother-like features that convey love and caring. We also explored how to transform a non-sensual voice into a sensual one such as the voice of Marilyn Monroe. This can be implemented by altering the speaker's Mel-frequency cepstrum coefficients (MFCC) to match the MFCC of the target's voice. To make matching more efficient, vector quantization can be used to reduce the number of feature dimensions. For fine-tuning the voices, we can adjust sampling frequency and pitches. Adding acoustic parameters can further improve the fidelity of the target voice, along with machine-learning algorithms for automated tuning and optimization.

---

[31] Carmona JM, Climent J (2012) A Performance Evaluation of HMM and DTW for Gesture Recognition. Progress in Pattern Recognition, Image Analysis, Computer Vision, and Applications. Lecture Notes in Computer Science Volume 7441, 2012, pp 236–243. http://link.springer.com/chapter/10.1007%2F978-3-642-33275-3_29

[32] Rymix (2011) Kinect SDK Dynamic Time Warping (DTW) Gesture Recognition. http://kinectdtw.codeplex.com/

Sensual body shape detection of human horizontal and vertical profiles can be used for designing more sensual humanoid robots, adaptive tailoring systems, and visual privacy algorithms. Shoulder-to-hip trapezoids and triangles are key elements for analyzing sensuality in human gaits. These multi-dimensional dynamics can be modeled using Fourier series and Dynamic Time Warping methods. Such techniques can potentially be applied to medical diagnoses, marketing, gesture control, and human identification.

# CHAPTER 9

# Euphoria Dynamics

Why do we get hungry and then feel satiated when we eat? How do these biological compulsions arise? Our thoughts, feelings, and bodily sensations correlate with the activation of specific parts of our neural circuitry. Such a dynamic process happens by means of neurochemical changes in our brain that keep us alive and reproducing. In this chapter, we explore a mathematical and electronic model of euphoria dynamics, and the potential for a computer to understand, or even to *have* such a behavior.

## Orgasmatron

Euphoria is an addictive feeling that the brain seeks to maximize. It is the pleasurable feelings, or "reward" produced by the chemical agent *dopamine*.[1] Austrian-American psychiatrist Wilhelm Reich once said, "The pleasure of living and the pleasure of the orgasm are identical. Extreme orgasm anxiety forms the basis of the general fear of life." Most drugs of abuse directly or indirectly target the brain's reward system by flooding the circuit with dopamine. What drives euphoria is the intense orgasmic feeling of pleasure derived from the surge of dopamine. Orgasm is a special state of euphoria and it is an involuntary excitement that arises from a series of continuously-charging stimuli.

Can someone have orgasmic euphoria without any feeling or stimulation? In the 1973 futuristic comedy movie *Sleeper*, Woody Allen created a fictional electromechanical device called the "*orgasmatron*." It was a cylindrical closet that instantly triggered orgasm. As we often see in cases where science fiction precedes real science, doctors stumbled on a real-world orgasmatron in a routine

[1]Marnia, Your brain on sex. On Reuniting healing with sexual relationship. June 24, 2005. Retrieved on Feb. 29, 2016: http://www.reuniting.info/science/sex_in_the_brain

© Springer-Verlag London 2016

Y. Cai, *Instinctive Computing*, DOI 10.1007/978-1-4471-7278-9_9

neurosurgical procedure. By applying bionic implants rather than climbing into a closet, a real-life orgasmatron gave the user even more spontaneous pleasure. American medical doctor Stuart Meloy implanted an electrode into a woman's spinal cord to treat her chronic pain, but accidentally triggered the patient's orgasmic response.

The spinal cord is an extension of the brain and transmits sensational signals. These signals stimulate a part of the brain called the *orbitofrontal cortex*, which is associated with the pleasurable feelings. Neurosurgeons routinely implant an electronic device in the orbitofrontal cortex of Parkinson's patients' brains to treat the disease by creating tiny electrical shocks in that area. They soon discovered that the device could be used to stimulate orgasms as well. However, the current orgasm chip is too crude and intrusive to be used widely. In reality, no one could sustain a burst of orgasms every second. It should be subtler, allowing for more control over the power so that the user may turn the chip on or off as needed. To help understand the subtlety and dynamics of orgasm, let us look at a simulation model.

## The Burst Firing Effect

Orgasm is a complex phenomenon. As with all emotions, it is the anticipation of a given event which intensifies feelings. This anticipation may be instinctual (hardwired) or learned. So far, we do not fully understand the underlying electrical neurochemistry in this process. We do know that an orgasm has roughly three basic stages: *build-up*, *plateau*, and *discharge*. At the build-up stage, a reward value increases quickly in response to repeated stimuli at a high frequency. As the reward value reaches a certain threshold, further gain of the reward value slows down and converges to a static value, which is the plateau stage. Finally, at the discharge stage, the accumulated energy is released rapidly. The orgasm disappears at the peak of the reward. The system may experience a sudden short-circuit phenomenon, such as a current jump, flash of light, or even a black out. This is also referred to *le petite mort*, or "the little death."

The *burst firing effect* in a boost-discharge circuit occurs when the stimuli pulses are very close in time. Given a series of high frequency input, the output may build up drastically. Similar to when striking a bell repeatedly, when you strike the ball faster, the sound becomes louder, because the sound waves are superimposed over each other and their sum is greater than the individual strikes within a long interval (Fig. 9.1). The burst firing effect occurring in a non-linear circuit involves energy storage, switches, and stimulus intervals.

To build up, sustain, and discharge energy requires a special biophysical mechanism in the brain: it must operate at high speeds; it must perform a variety of linear and non-linear computational operations; and it must interface with items in the physical world such as sensors and actuators. The membrane potential is an ideal physical variable within the nervous system that meets these three requirements: it can do rapid computations within a fraction of a second and

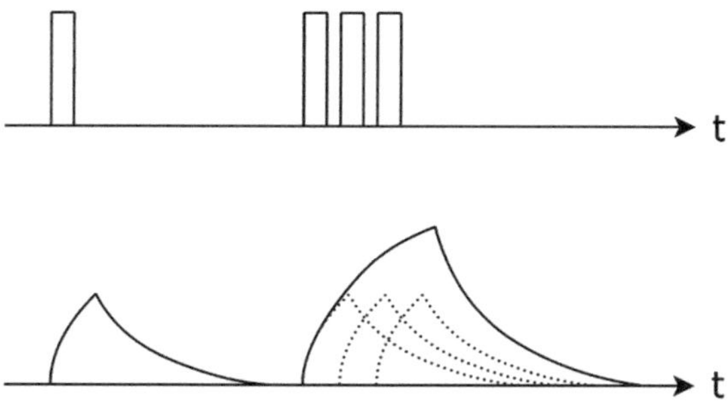


**Fig. 9.1** The Burst Firing Effect. When the intervals between stimuli are close, the response will be stronger than when the stimuli are isolated

propagate large distances in the brain, and it controls a vast number of switches such as ionic channels for implementing sophisticated nonlinear operations. These channels transduce visual, tactile, auditory, and olfactory stimuli into changes in the membrane potential, and such voltage changes relate back into the release of neurotransmitters or the contraction of muscles.[2]

A biological membrane is comprised of two layers of molecules separated by a very thin insulating layer. From an electrical point of view, it will act as a *capacitor* (C), which can store energy. The capacitance of the membrane allows for the build up a charge on both sides of the membrane, where the charges are proportional to the product of the specific membrane capacitance and the membrane potential. Measurements also indicate the membrane functions as a rectifier. Furthermore, biologists have found membrane potential oscillations within squid axons indicating the presence of inductive elements within the nerve. This neurobiological phenomenon is similar to the behavior of coils in an electric circuit across which the voltage is proportional to the current change. Biological membranes with certain types of voltage- and time-dependent conductance can behave as if they contained inductors (L), or so-called *phenomenological inductance*.[3] In a linear inductance circuit, the voltage is proportional to the product of the inductance and the change rate of the electrical current in the opposite direction. Finally, protein molecules inside the membrane act as channels, gates, or switches in the lipid barrier through which particular information can be transferred from one side to the other.

Now, we have channels, gates, and two types of energy storage: capacitance and inductance. Both are voltage- and time-dependent, which creates a rich repertoire of computational primitives. Each of these primitives has the ability to implement a variety of linear and non-linear, high-gain operations.

---

[2]Koch C (1999) Biophysics of Computation: Information Processing in Single Neurons. Oxford University Press. 1999

[3]Cole KS (1972) Membranes, Ions and Impulses. University of California Press: Berkley, CA, 1972

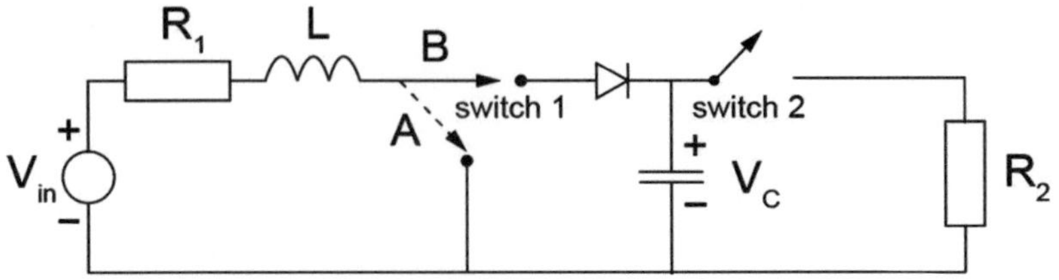


**Fig. 9.2** The boost-discharge circuit

Assume the membrane resistance (R) can be ignored. We can configure a boost-discharge circuit as shown in Fig. 9.2. This is similar to widely-used circuits for power voltage regulation, and camera flash charging and discharge. The underlying principle of the camera flash is a charge pump circuit built using a capacitor and a boost circuit discharging into a bulb filled with xenon gas. For a very bright flash, xenon atoms must be excited to high energy levels. This requires the electrons to be imparted with very high energies. To accomplish this, the terminal voltage across the tube must be raised to high values.

Let us now look at our simulation diagram. This circuit works in two states: switch 1 is at position A or B. $v_{in}$ is a small excitatory synaptic input. $v_c$ is the output voltage across the capacitor. When the switch is turned to the A position, current flows through the inductor and energy is stored in its magnetic field. When the switch is turned to the B position, the energy from the excitatory voltage input and the inductor (*L*) is transferred to the electric field of the capacitor (*C*). A voltage appears across the inductor which is proportional to the change in the current passing through it. The rectifier prevents any reversal of the current. Continuously switching the switch between position A and B makes the circuit behave as a charge pump. The result is to boost the voltage across the capacitor to a level much higher than the excitatory input voltage.

Although this circuit resembles a classic RLC circuit, we cannot simply use RLC equations to model it because the switch reroutes the circuit, and the diode prevents the reversal of current. Switching can be described by two parameters: the frequency ($\omega$) and duty cycle (*D*). These parameters can be translated into the time the circuit spends in each cycle in state A ($dt_A = 100\text{D}/\omega$) and state B ($dt_B = 100(1-D)/\omega$). With that, we can isolate it into two circuits and analyze each one independently, with the end result of the components of one state feeding into the next. See Fig. 9.3.

When Switch 1 is at the A position and the circuit is in its initialization state, the battery charges the inductor L. We have,

$$v_{in} = L\frac{di}{dt} + R_1 \cdot i \tag{9.1}$$

When Switch 1 is at the B position, the resistor, inductor, diode, and capacitor are all in series, thus, Kirchhoff's voltage law applies:

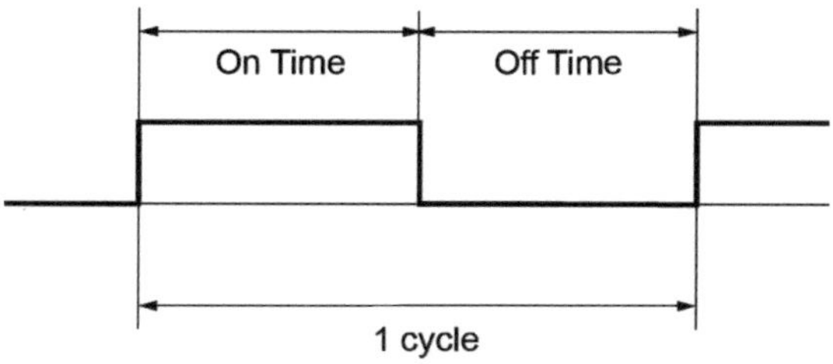


**Fig. 9.3** The input electrical pulses of a voltage

$$v_{in} = v_{R_1} + v_L + v_C + v_D \tag{9.2}$$

where $v_{in}$ as the constant voltage of the DC power, $v_{R_1}$ the voltage drop across the resistor, $v_L$, of the diode, $v_D$ of the diode, and $v_C$ of the capacitor. Here, let us assume the voltage drop on diode $v_D$ is ignorable and can be set to zero. The rest are variable, varying with respect to time. Applying Ohm's law to calculate the voltage across $R_1$, L and C, we have:

$$v_{in} = i \cdot R_1 + L\frac{i}{dt} + v_C \tag{9.3}$$

According to Kirchhoff's current law, the current flowing through all components is exactly the same, and because inductors resist change in current, the current through the components (which are all in series) is forced to be the instantaneous current of the inductor. Rearranging the equation above, we have:

$$v_C = v_{in} - L\frac{i}{dt} - i \cdot R_1 \tag{9.4}$$

$$\frac{dv_C}{dt} = \frac{i}{C} \tag{9.5}$$

Combining 9.4 and 9.5, we have the second order differential equation of the RLC circuit:

$$v_C = v_{in} - LC\frac{dv_C{}^2}{dt^2} - R_1\frac{dv_C}{dt} \tag{9.6}$$

When Switch 2 is closed, it discharges the current through the resistor load $R_2$. In this discharge mode, it is a simple RC circuit. Thus, the discharge equation is:

$$\frac{dv_C}{dt} = -\frac{i}{C} = -\frac{v_C}{R_2C} \tag{9.7}$$

The equations from 9.6 through 9.7 can be solved by physical circuit simulation or numerical computation.

## Physical Circuit Simulation

We can simulate boost firing with an electronic circuit, but a few adjustments are necessary. The physical Switch 1 can be replaced with a transistor for fast pulse generation. When the transistor is in cutoff mode and no current flows through it, then all the components are in series, which is the situation when Switch 1 is at the B position in Fig. 9.2. When the transistor is in saturation mode, then it becomes a short circuit. After flowing through the resistor and inductor, the current is shorted straight to ground, bypassing the components to the right of the transistor. This is the same situation when Switch 1 is at the A position.

This experimental physical circuit can be built with parts commonly available at electronic hobby stores such as RadioShack in the US, including an 82 mH inductor, an 82uF capacitor, a 0.3 V Schottky diode, and a standard NPN transistor. A resistor is not necessary because the inductor itself has a decent amount of internal resistance (around 300 Ω) (Fig. 9.4).

This physical circuit simulation is different from many prevailing neuroscience models because it includes the duty cycle in the simulation. Here, duty cycle refers to the percentage of time that an input voltage is on versus the percentage of time it is off. A 20 % duty cycle would indicate that the switch is on 1/5 of the time and off for 4/5 of the time. Figure 9.5 shows the output voltage is a function of the frequency for a fixed duty cycle. The higher the frequency is, the higher the output. Figure 9.6 shows the impact of duty cycle on the burst firing effect. Initially, lower duty cycles are better, but given more time, higher duty cycles prevail.

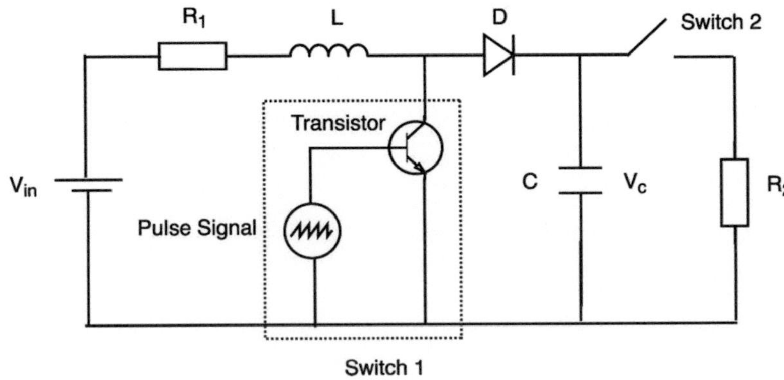


**Fig. 9.4** Diagram of the physical circuit. The switching is accomplished using a transistor, which can switch rapidly between cutoff and saturation mode

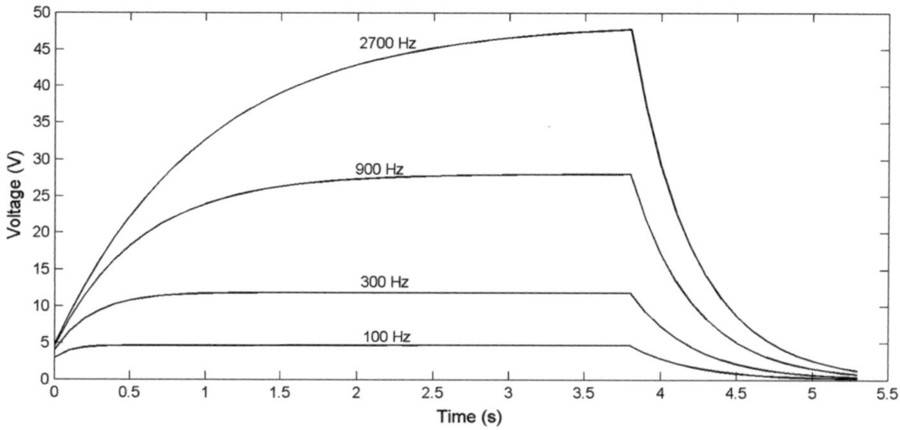


**Fig. 9.5** The effect of frequency on the boosting curve (duty cycle held constant). Increasing frequency always leads to faster boosting. L = 82 mH, R = 100Ω, C = 82 uF, Vs = 5v, $V_d = 0.1$v, $V_s = 90$v

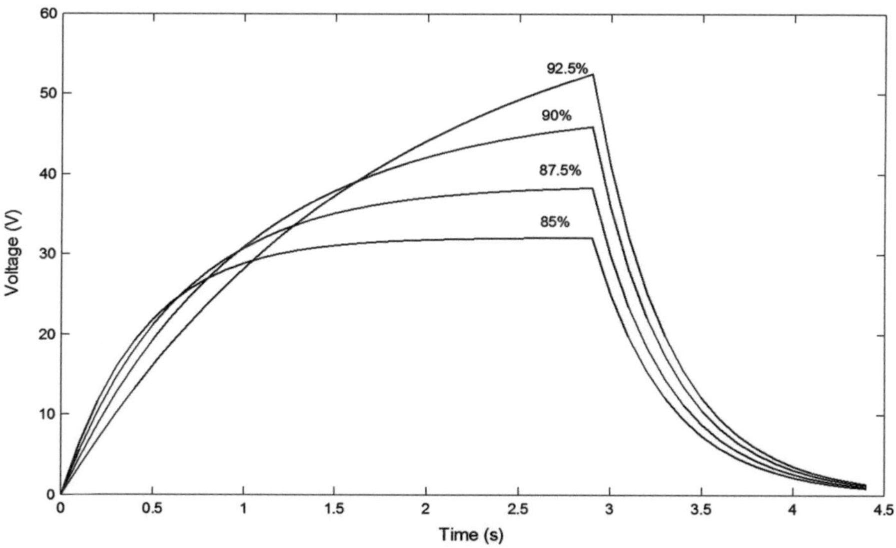


**Fig. 9.6** The effect of duty cycle on the boosting curve (frequency held constant). Initially, lower duty cycles are better, but given more time, higher duty cycles prevail. L = 82 mH, $R_l = 100\Omega$, C = 82 uF, $V_s = 5$, $V_d = 0.1$, $\omega = 2.7$ kHz

## Optimal Input

There are two independent parameters of Switch 1 that affect the output of the burst firing effect: *frequency* and *duty cycle*. As shown in Figs. 9.5 and 9.6, with everything else held constant, operating for longer lengths of time before discharging will always result in higher voltages, since the diode prevents the reversal of current. Amazingly, the relationship between frequency and boosting is also positive throughout, unlike a LRC circuit where the output voltage peaks at the resonant frequency.

To maximize the output of the circuit, we need to calibrate the duty cycle for a given time and frequency. The duty cycle is the only parameter with an upper limit (i.e. 100 %), and is also the only parameter where a higher value does not always translate into a higher output. A very clear and simple counterexample is a duty cycle of exactly 100 %; the circuit never resembles the case of when Switch 1 is at the B position, and therefore the capacitor is never charged. In Fig. 9.5, we see the plot of voltage versus time for the boost firing circuit with varying duty cycles but with frequency held constant. It can be seen that the curves for different duty cycles are shaped differently, and different duty cycles are ideal at differing points in time. Lower duty cycles are more logarithmically-shaped, rapidly increasing at first, but then rapidly flattening out, meaning that they are ideal for short periods of time. In contrast, higher duty cycles are more linear, losing out to low duty cycles initially, but surpassing them long after the low duty cycles flatten out.

The reason for these behaviors can be explained by analyzing the inductor and capacitor. For low duty cycles, more time is devoted to charging the capacitor when Switch 1 is at the B position. As a result, the capacitor charges to a higher voltage. However, it will plateau quickly, since the combined voltage of the input voltage and the voltage across the inductor is insufficient to charge the capacitor further. To generate and maintain high voltages across the inductor, switching has to occur at a higher frequency. In the end, higher duty cycles and higher frequencies win out given more time. As seen in Fig. 9.6, the 85 % duty cycle starts out the highest, then followed by the 87.5 % duty cycle, then 90 %, and so on. The ideal input would be a frequency that is as high as possible coupled with a dynamically adjusted duty cycle that increases with the voltage of the capacitor.

## Simulation of Dopamine Dynamics

As we mentioned earlier, dopamine provides the feelings of pleasure that lead to addictive orgasm. Dopamine is one of the chemical signals in the brain that passes information from one neuron to another in the tiny spaces between them. A spike in dopamine appears to be the "motivation" or "pleasure", but in fact, dopamine is simply signaling feedback for predicted rewards. For example, if a boy has learned to associate a cue, such as a video game controller, with a fun video game, he might start experiencing increases in dopamine levels in response to the sight of the video game controller as his brain predicts the reward. Conversely, if the boy does not get to play the video game, then dopamine levels can decrease, and that is not a good feeling.

Dopaminergic neurons are a class of neurons that produce the neurotransmitter dopamine. In rats they normally fire at a low frequency of 5 Hz, but these neurons also exhibit the burst firing patterns at frequencies of 20 Hz or more, which can be observed *in-vivo*, using fast scan cyclic voltammetry (FSCV).[4] This burst firing leads to rapid changes in dopamine concentration to the point of overflowing. Burst firing can be simulated with high-frequency stimulation of dopaminergic neurons. The time course of this evoked overflow arises from the balance of uptake and release. The overflow of dopamine [DA] is governed by the Michaelis-Menten equations before the stimulation is terminated ($t < t_s$) and after the stimulation is terminated ($t \geq t_s$).[5]

$$\frac{d\,[DA]}{dt} = f \cdot [DA]_p - \frac{V_{max}}{K_m/\,[DA]+1};\ t < t_s \tag{9.8}$$

$$\frac{d\,[DA]}{dt} = -\frac{V_{max}}{K_m/\,[DA]+1};\ t \geq t_s \tag{9.9}$$

where $[DA]_p$ is the dopamine released per pulse; $V_{max}$ and $K_m$ are properties of the system kinetics and were measured independently. $V_{max}$ is the maximum rate of dopamine transport; $K_m$ is the substrate concentration at which the reaction rate is half of $V_{max}$.

Equation 9.8 represents the balance of the injection of dopamine and dopamine removal before the stimulus pulse is terminated. When the concentration of dopamine is much greater than $K_m$, the injection stage is just a straight line, like a spike. With $t$ as time and $f$ as frequency, the increase in dopamine is $t \times f \times [DA_p]$, which is the amount of dopamine injected during each pulse. This is essentially a simplified model of the charging part of our RLC model. When L and C are very large, then the time constant is large enough that the charging curve is almost linear.

Equation 9.9 represents the rate of dopamine loss after the stimulus pulse is terminated. Again, when the concentration of dopamine is much greater than $K_m$, then the discharge stage is close to linear. This trend corresponds to the discharging of the capacitor in our model (Eq. 9.7).

The study from Wightman's lab showed that repeated administrations of chemicals such as cocaine might lead to an increase of dopamine overflow because of changes in the brain.[6] When neurons are stimulated with a constant frequency, the model predicts that extracellular dopamine concentration will increase for 7 pulses at 10 Hz. With each pulse, an increase in dopamine concentration is

---

[4]Wightman RM, May LJ, Michael AC (1988) Detection of dopamine dynamics in the brain. Analytical Chemistry, vol. 60, No. 13, July 1, 1988.

[5]Wightman RM, Amatore C, Engstrom RC, Hale PD (1988) Real-time characterization of dopamine overflow and uptake in the rat striatum. Neuroscience, vol. 25, No. 2, pp. 513–523, 1988.

[6]Addy NA, Daberkow DP, Ford JN, Garris PA, Wightman RM (2010) Sensitization of rapid dopamine signaling in the neucleus accumbens core and shell after repeated cocaine in rats. Journal of Neurophysiology. 104: 922–931, 2010.

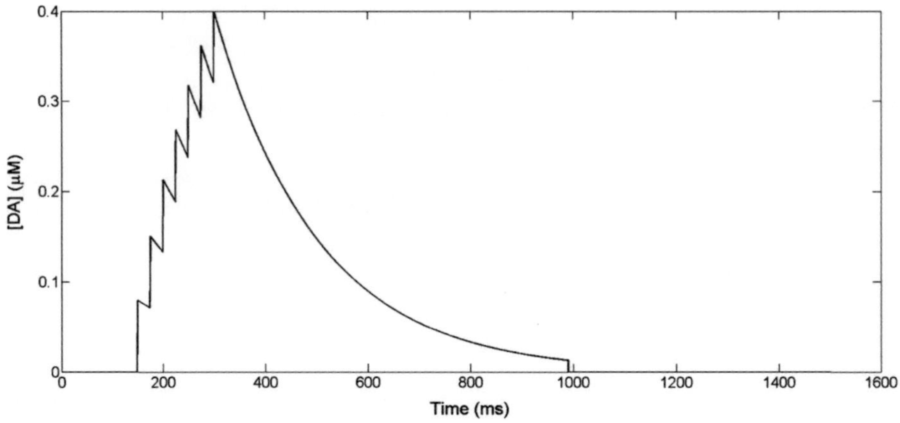


**Fig. 9.7** Simulation of stimulated-dopamine release at 40 Hz. See more details in reference (Wightman RM, Amatore C, Engstrom RC, Hale PD (1988) Real-time characterization of dopamine overflow and uptake in the rat striatum. Neuroscience, vol. 25, No. 2, pp. 513–523, 1988)

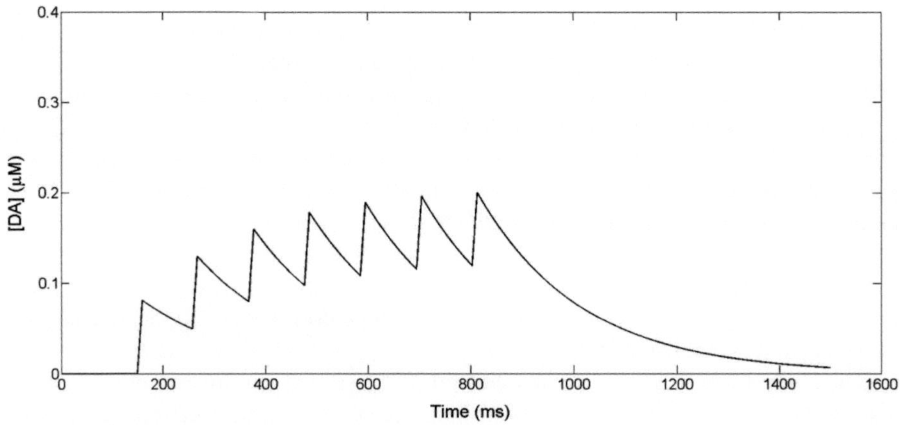


**Fig. 9.8** Simulation of stimulated-dopamine release at 10 Hz in the same case study above

associated by the amount designated as $[DA]_p$. The rate of uptake of dopamine during and following the stimulation is generated by our RLC model, rather than the Michaelis-Menten equation.[7] Figures 9.7, 9.8 and 9.9 show the results from the RLC model.

[7]Garris PA, Ciokowski EL, Pastore P, Wightman RM (1994) Efflux of dopamine from the synaptic cleft in the nucleus accumbens of the rat brain. The Journal of Neuroscience, Oct, 1994, 14(10): 6084–6093.

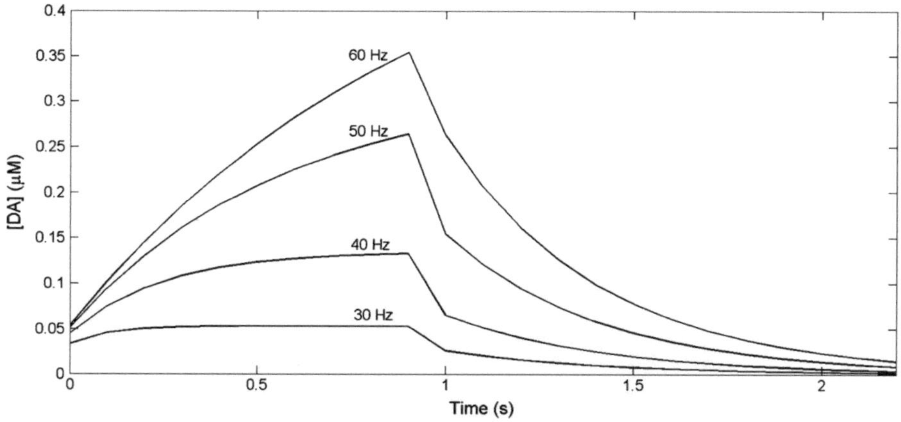


**Fig. 9.9** Simulation of frequency response. The overflow of dopamine during the 120-pulse stimulus at 60 Hz, 50 Hz, 40 Hz, and 30 Hz

## Understanding Music Patterns

Music imitates orgasm. Climax is a basic rule in classical composition, especially in the Western world, where music is understood as a form of gesture and drama. A musical climax often comes as a Fibonacci row, or comes at a time adhering to the Golden ratio, at approximately the 2/3 mark of the musical piece. According to composer Jakub Polaczyk, over 90 % of classical music has this kind of structure because it is consistent with the natural feeling of proportions. For some composers, a retro climax is used instead. Mykietyn in Poland used mirroring for the climaxes in his music, often at the 1/3 mark of the piece.

Here, we try to explore how to recognize a climax in music. Let us start with French composer Maurice Ravel's famous classical piece "Bolero," which has only one melody that is developed until the end. The music imitates the dynamics of orgasm, from buildup, to plateau, to the climax at the very end. Along a linear path, the repetitive melody becomes stronger and stronger towards the end of the piece. The comedy film *10*, written and directed by Blake Edwards, used Ravel's "Bolero" for the musical backdrop in a romantic scene in Mexico. The comic effect comes from a sense of false alarm: the highly anticipated climax did not occur as expected, due to a stuck phonographic needle.[8]

How do we visualize the movement of "Bolero" from build up to climax? We can use a spectrogram in which the sound file is visually represented on a three-dimensional display where one axis represents time, a second axis represents frequency, and the third axis represents sound volume. Figure 9.10 shows a spectrogram of Ravel's "Bolero," where we can see the single melody repeating

[8]Film "10" from WikiPedia. http://en.wikipedia.org/wiki/10_(film)

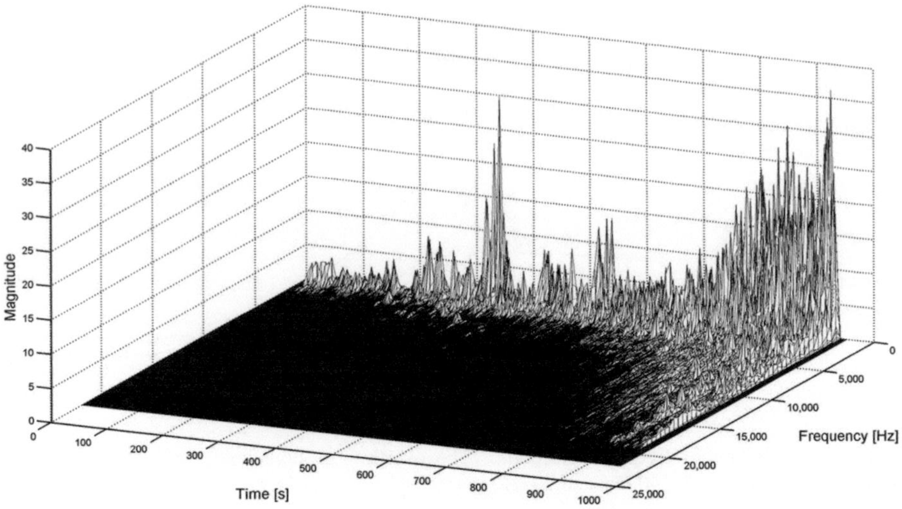


**Fig. 9.10** Spectrogram of Maurice Ravel's "Bolero," in which one axis represents time, the second axis represents frequency, and the third axis represents sound volume. From the spectrogram, we can see the single melody repeating many times each with increasing volume and frequency until it reaches the climax at the end

many times, each time with increasing volume and frequency until it reaches a climax. Interestingly enough, "Bolero" incorporates the climax at the end rather than the Golden Ratio spot near the 2/3 mark of the piece, much like the edge of a cliff. The audience anticipation and variation of the music create an orgasmic tension.

If we want a computer to understand the structure of an artwork, the algorithm must learn to appreciate melody and rhythm, which are essential temporal elements of art, especially in music. A simple but rare form found in music is to repeat the same pattern several times in the form of AAA. For example, most disco music varies only subtly over the course of many repetitions, each time growing in intensity. To avoid a potentially monotonic repetition, more variants are added as seen in forms AABA, AABABA, ABAB and ABABA. The most common and accessible pop song form to evolve over the last 50 years is the form ABABCBB, specifically, *intro/verse1/chorus1/verse2/chorus2/bridge/chorus3/chorus out.* This musical framework contains simply a verse and a chorus, with variety being created by the sounds of parts that would otherwise become boring. Despite having so many variations in musical forms and patterns, musical dynamics always contain built up, plateau, and discharge stages. Between the plateau and discharge stage, there is normally a distinguishable variety called the "bridge" or "middle 8" to hint at the approaching climax or ending.

Melody can be represented using symbols and musical notations that visualize aurally-perceived musical patterns. The modern five-line staff notation system can be traced to its roots in medieval Europe. Although it is the most accurate and popular notation system used today, it is not easy to index for the purpose of searching for particular music. On the online folk song search web site Folk-TuneFinder, users need to key in the melody of the song they are searching for using a simulated keyboard.[9] In 1975, Denys Parsons developed the Parsons Code for Melodic Contours to simplify the notation used to identify a piece of music through melodic motion – the motion of pitch up or pitch down. Rather than representing the absolute level of a pitch, Parsons Code simply represents the slope of pitch between consecutive notes. Parsons Code actually converts a figure into a string of words, similar to a DNA strand, but using only three letters:

u = up, if the note is higher than the previous note
d = down, if the note is lower than the previous note
r = repeat, if the note is the same pitch as the previous note

For example, "Twinkle, Twinkle, Little Star" can be represented as: "rururddrdr-drd urdrdrdurdrdrd drururddrdrdrd." As you can see, Parsons Code compresses the melody into a manageable three-letter string, which is convenient for computers to search for, compare, and compose.

In addition to melodic patterns, rhythmic patterns can also be used to search for music. Across all cultures, people move their bodies to the rhythms of music by clapping, tapping, drumming, humming, dancing, or rocking. For many people who are not musically trained, tapping to express a rhythm is surprisingly easy. While pitch-based music search such as humming or whistling are potentially more descriptive in conveying song melodies, they are not always as intuitive as tapping, nor are they as accurate for the average musically-untrained ears.[4] For popular songs, the rhythm of a song's instrumentals is often equivalent to the rhythm of the words in the lyrics. Each syllable or two of a word might represent one beat. Even if the average user does not remember the specific words of a song, he or she can often recall and reproduce the rhythm or even the melodic tone, which would be enough information for retrieving a song from an online database.

Tapping-based music search querying prompts users to tap the rhythm of a song on the space bar of their computer keyboard to retrieve a list of songs containing a similar rhythm. The algorithm of the tapping-based music search contains two parts: Rhythmic Coding and String Matching.

Rhythmic Coding converts a monophonic tapping sequence into a string of letters. A computer can record the tapping sequence over time. Each tap contains a key depressing and key releasing cycle. To simplify the computation, the algorithm only considers the timing of the user's keyboard striking for each tap. An entire tapping sequence consists of a number of beat durations. In order to allow for global tempo independence, we need to normalize the duration of each beat. The average duration of a beat is calculated, then each beat's duration is divided by the value of the average duration. Similar to Parsons Coding that converts melodies to a string, Rhythmic Coding converts the durations of beats into a string.

[9]Folk Tune Finder. http://www.folktunefinder.com/

s = same, if the similarity threshold is less than a value
u = up, if the beat duration is longer than the last one
d = down, if the beat duration is shorter than the last one

For example, the folk song "Are You Sleeping" with the lyrics "Are you sleeping? Are you sleeping? Brother John, Brother John?" can be translated to the string: "sssssssssu dsu." Rhythmic Coding compresses the musical notation into a simple, three-letter string. It is relatively more compact than Parsons Coding. To avoid estimating the base duration of beats, Doraisamy and Ruger proposed using ratios between note onset times by comparing the differences between consecutive note onset times.

String Matching finds approximate matches between the input letters and candidate strings of letters. Every song entered in the database is preprocessed using Rhythmic Coding. When a query is received from the user, the input rhythm is analyzed and the letter string is also generated. Since they are all strings, it is instinctive to apply classic string matching algorithms. Instead of trying to find an exact match, we can use an approximate string-matching algorithm to find the closest matching songs, considering the input string contains some errors within a tolerable range. Sun and Manber's fast approximate string matching algorithm is the most popular one used today. In this algorithm, the difference between two strings is defined by the Levenshtein Distance, which measures any spelling errors. For example, the Levenshtein Distance is 1 in the following cases:

sud → sd = omission of a letter
sud → suud = insertion of a letter
sud → uud = substitution of a letter

The distances between the input string and each string in the database are sorted into ascending order. The song's string in the database with the smallest Levenshtein Distance to the input rhythm string is the most probable match.

According to a study by Peter et al., the rhythmic coding algorithm has a decent error tolerance. For example, users can tap at their own pace without compromising the accuracy of the result. Users may also tap anywhere within the song and errors in the input are tolerated.[10]

## Euphoric Dynamics of Social Media

American essayist H. L. Mencken once said "Government under democracy is thus government by orgy, almost by orgasm." This is true of most social media today, largely run by "democracy" such as Twitter, Facebook, Google+, Flickr, and Instagram. It is worth noting that each social media platform carries its own resonance frequency, response time, and strength.

[10]Peter G, Cukierman D, Anthony C, Schwartz M (2006) Online music search by tapping. In: Lecture notes in computer science (LNAI), vol 3864. Springer, Heidelberg, pp 178–197

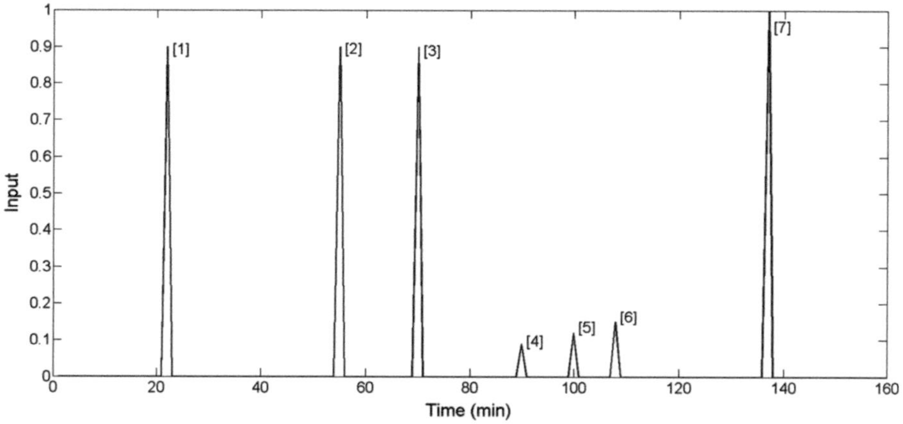


**Fig. 9.11** The World Cup events as stimuli of pulses with various intervals and duty cycles in terms of weight, where: (**1**) match started, weight=90 %; (**2**) Omar González (USA) shown *yellow card*, weight = 90 %; (**3**) Halftime (0 USA and 0 Germany), weight = 90 %; (**4**) game resumed, weight = 9 %; (**5**) Goal! Thomas Müller (GER) scores – USA 0, Germany 1, weight = 12 %; (**6**) Kyle Beckerman (USA) shown *yellow card*, weight = 15 %; (**7**) match ends (USA 0, Germany 1), weight=100 %

Many social media posts are triggered by events that have *event resonance frequencies*. These are usually noteworthy, widely observed events, such as sports games and presidential elections. The Twitter Analysis of the World Cup Match: USA versus Germany on June 28, 2014 by Michael Troung and Tristan Mikofeit shows how Twitter posts rose and fell during the match. Figure 9.11 shows the visualization of a 5 % collection sample of Twitter posts containing the hashtags #USMNT, #USAvsGermany, and #USAvsGER during the duration of the match. As you can see, Twitter activity during the game was very dynamic. The posts reached their first climax at the halftime mark and the ultimate climax at the end of the match.[11]

Events like the World Cup match have a natural "frequency," which consists of important points in the event. Each point can be assigned a weight, reflecting the importance of that point and how much it affects Tweets. This is analogous to a duty cycle. The points in the match that we considered, along with their relative weights, are shown in Fig. 9.11. Time is shown in minutes, according to the standard notation in FIFA matches.

Different events, due to differing importance, naturally vary in their impact on social media networks. The end of the game has the highest impact, not only because it is the conclusion of the match, but also because people have more time to tweet after the game with undivided attention. The match start, halftime, and first yellow card have roughly the same weights, closely following the value of the importance

[11]Troung M and Mikoleit T (2014) Twitter Analysis of the World Cup Match: USA vs Germany, June 26, 2014. http://public.tableausoftware.com/profile/michael.truong#!/vizhome/WorldCupUSAvsGermanyTweets/WorldCupUSA-GERTweets

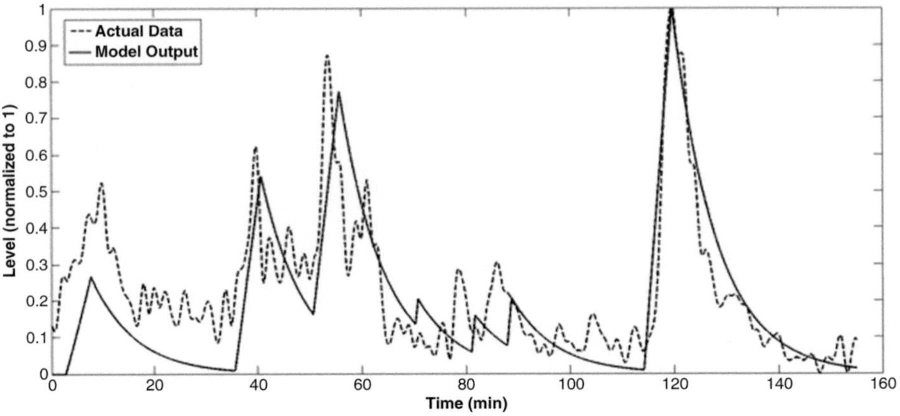


**Fig. 9.12** The Twitter analysis of the World Cup match: USA versus Germany on June 28, 2014. The *dash line* is the actual data and the solid line is the output from our model (L = 670,000, $R_1 = 100$, C = 2, $V_s = 10$, $V_d = 0$, $R_2 = 250$)

of the end of the match. Next up is the second yellow card. There are two plausible reasons that it has a substantially lower impact; (1) that it has less shock value than the first violation, and (2) because it was less controversial than the first violation. Germany's goal, the next on the list, is even lower. Clearly, Americans were not very enthusiastic about their opponent scoring a goal. If the language of the keywords taken in this survey was German, the results may have been quite different. At the bottom of the list is when the game resumes after halftime, which is not a significant event, but is included anyway. For this purpose, the weights were standardized to 100, with 100 being the highest, and everything else was relative to 100. Figure 9.12 shows the simulation result from the RLC model.

A presidential election in any democratic country is also a big event. For example, according to Pytolab's analysis, there were over 8 million tweets related to the 17 French presidential candidates from September 1, 2011 to June 1, 2012, including the first round of the presidential election on April 22, 2012 and the second round on May 6' 2012. The events we used are shown below. Figure 9.13 shows the normalized post count per candidate. As you can see, Twitter activity becomes more and more volatile towards the final election and reaches its climax at the closing moment. Here, we see that the second round of elections has the largest weight by a wide margin. In France, only the second runoff election determines the final winner, so people naturally pay the most attention to that event. In contrast, the first round only determines who gets to be in the runoff, which was widely expected to be the candidates from the two main parties, Hollande's Socialist Party, and Sarkozy's UMP (Union pour un mouvement populaire). The first round included many third party candidates such as Marine Le Pen of the far right National Front Party. For the primary elections earlier, Hollande was widely expected to win the Socialist primary, so there was not much enthusiasm for that event[12] (Fig. 9.14).

[12]Laurent Luce blogs. http://www.laurentluce.com/!

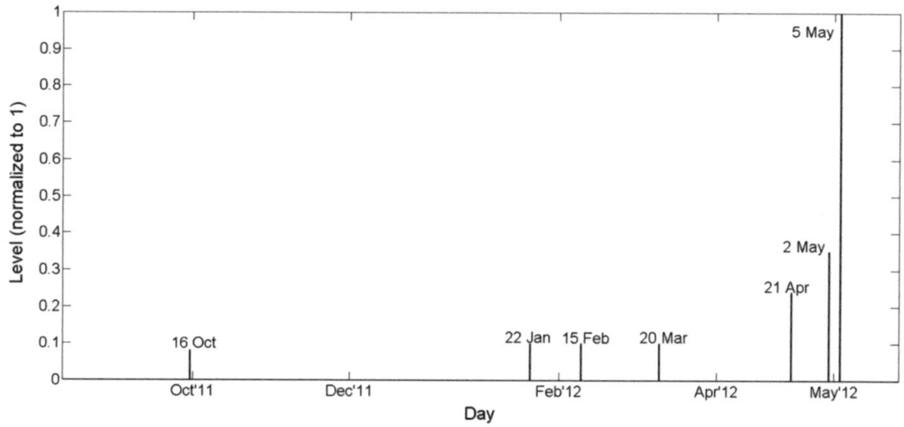


**Fig. 9.13** Twitter posts during the French Presidential Race in 2011–2012: (**a**) Oct 16, 2011, François Hollande wins the Socialist Party primary, weight = 8 %; (**b**) Jan. 22, 2012, François Hollande officially announces his presidential campaign, weight = 10 %; (**c**) Feb. 15, 2012, Nicolas Sarkozy announces his bid for a second presidential term, weight = 10 %; (**d**) March 20, 2012, official campaign begins, weight=10 %; April 21–22, 2012, first round of Presidential election, weight = 24 %; (**f**) May 2, 2012, first and only televised debate between Hollande and Sarkozy, weight = 35 %; (**e**) May 5–6 2012, second round of the Presidential election. Hollande won, weight = 100 %

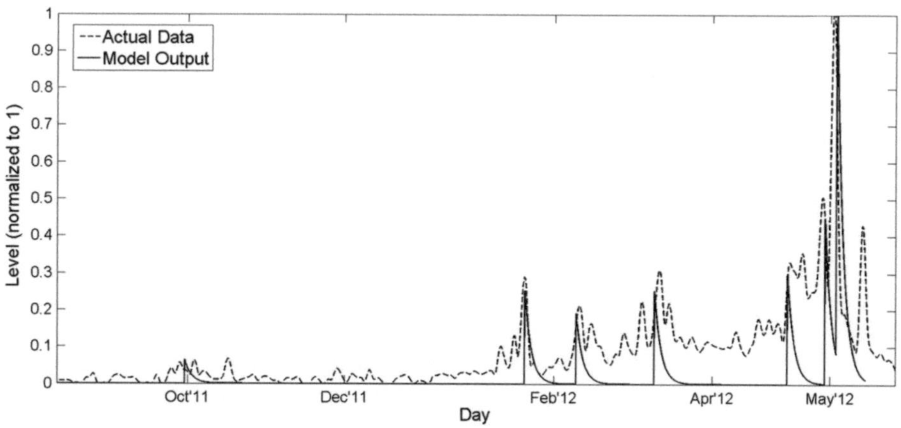


**Fig. 9.14** The simulation of the Twitter posts during the French Presidential Race in 2011–2012 (L = 10,000,000, $R_1$ = 2000, C = 3 F, $V_s$ = 10 V, $V_d$ = 0, $R_2$ = 50,000)

Twitter trends can be simulated using the same boosting model, and even the horizontal axis (time) can be analogous. However, there is no direct way to correlate voltage with the number of tweets, which are the respective y axes. For the first set of data about the World Cup, the magnitude was around 3 hours, or 10,000 seconds. For the French election, the magnitude was around 6 months, or 15,000,000 seconds. For such long periods of time, values of L, R, and C had to be adjusted accordingly.

Twitter posts can be used for sentiment analysis. With natural language processing tools such as DocuScope and NLTK (Natural Language Toolkit), a computer is able to automatically classify a tweet as a positive or negative tweet, based on sentiment. DocuScope has a built-in positive and negative phrase library that acts like an expert system. It simply performs string matching for tweets and visualizes their moods as positive or negative. On the other hand, the Natural Language Toolkit is a classifier that must be trained with a list of manually-classified tweets, including both positive and negative tweets. For example, "I like the game" is a positive tweet. "I am disappointed about the game" is a negative tweet. Then, we must extract a list of word features from the tweets, summarizing every distinct word ordered by the frequency of its appearance. To create a classifier, we need to determine which features are relevant. The Naïve Bayes classifier uses the prior probability of each label as the frequency of each label in the training set, and the contribution from each feature. In our case, the frequency of each label is the same for the 'positive' and 'negative' categories. For example, the word 'like' appears once in every 5 of the positive training tweets and in none of the negative tweets. This means the likelihood of a 'positive' label can be multiplied by 0.2 when the word 'like' is included in the input. After training with word features and sample tweets, the classifier is able to determine the sentiment of input tweets. According to the case studies of tweets made during World Cup match in 2014 and the French presidential election in 2012, the sentiment dynamics correlate to the event resonance frequencies. For example, both positive and negative tweets reached climax at the end of the World Cup match, and declined rapidly the next day when people went back to work.

Twitter is mainly an event-driven social media, which can be traced back to its origin. The founder of Twitter was fascinated by urban emergency response work and the brief updating messages sent among response teams and their coordinators. That is one reason why each Twitter message is limited to 130 words or less because that is just enough for broadcasting news, status, or location along with photos or videos. Since there are not many original news stories on Twitter, most tweets are links of news from other sites rather than original posts.

Conversely, Facebook has no word limit for text posts and allows more space for visual posts, facilitating more diverse interactions. However, a majority of users are infrequent contributors of original content. It is often the case that a very small number of savvy users have massive groups of friends or fans and post frequently, but the rest of the network members have much fewer friends or very few posts. Many users simply read posts but do not leave comments or post their own content, engaging in one-way interactions. We call center-of-attention sites or savvy users "hubs" or "gateways." Hubs are essential in setting the *tribal resonance frequency*, which keeps a group of social media alive. Individuals with high numbers of contacts can be represented by the Power Law, or Zipf's law, which usually refers to the "size" of an occurrence of an event relative to its rank. It was named after George

Kingsley Zipf, a Harvard linguistics professor, who discovered that a very small group of words makes up the most frequently-used words in the English language. The majority of all words are not used very often at all.

For some social media groups, the tribal resonance frequency could be as high as daily, as seen in college student groups or within a family unit. However, some social media platforms have intrinsically low tribal resonance frequencies, such as LinkedIn and ResearchGate. The anticipated posting frequencies are much lower than in other social media avenues. For instance, LinkedIn is designed for sharing job information. Since most users do not search for or change jobs every month, the information traffic is quite low. Users might grow frustrated with LinkedIn when it tries to artificially increase its traffic, such as by sending multiple generic messages every day saying, "I'd like to connect with you on LinkedIn," or "XYZ has a work anniversary today." Obviously, the senders are not really interested in making a connection but are simply building up the contacts they have for future spam messages.[13] Similarly, ResearchGate aims to create a sharing community among research authors. It does not have much traffic either because researchers do not publish or read papers every day. Adding more followers to the research network does not necessarily create a natural tribal resonance. Without enough true interactions, such social media platforms will eventually die out. As our model shows in Figs. 9.5 and 9.6, events occurring too slower than the intrinsic resonance frequency will significantly impair the system output.

## Discharge in Social Media

Social media avenues often have their own build-up, plateau, climax, and discharge cycles. A recent study by Facebook Data Scientist Carlos Diuk reveals that relationships start with a period of courtship. On Facebook, messages are exchanged, profiles are visited, and posts are shared on each other's timelines (e.g. the homepage or wall of the user). Diuk's team studied a group of people who changed their status from "Single" to "In a Relationship," and also identified an anniversary date as the start of their relationship. During the 100 days before a relationship starts, there is a slow but steady increase in the number of timeline posts shared between the future couple. When the relationship starts on "Day 0," the number of posts begins to decrease. Around 12 days before the relationship begins, a peak of 1.67 posts per day is observed. After 85 days into the relationship, the posts on the timelines drop to the lowest point of 1.53 posts per day. It is assumed couples decide to spend more time together in person. Then, the online courtship

[13] Why LinkedIn is losing value to me? http://sales20network.com/blog/?cat=94

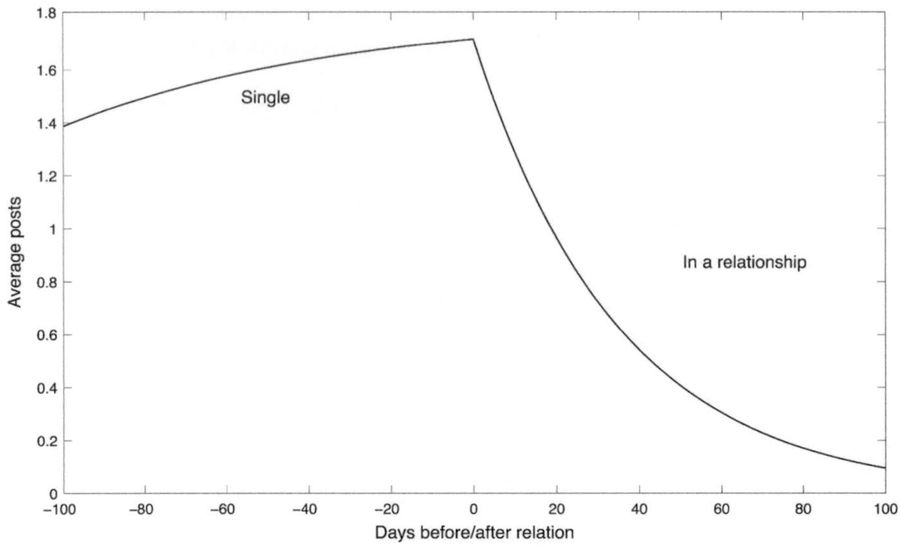


**Fig. 9.15** Days before and after relationship ($L = 1000{,}000$ H, $R_1 = 2$ kΩ, $C = 3$ kF, $R_2 = 1$ kΩ)

is off, and interactions in the virtual world give way to more interactions in the physical world.[14] Figure 9.15 shows the simulation of this phenomenon with the RC discharge model.

Diuk's study also reveals a significant increase in the number of positive emotional posts on the timelines of the subjects. This was found by counting the proportion of words expressing positive emotions such as "love," "nice," "happy," etc. minus the proportion of words expressing negative feelings such as "hate," "hurt," "bad," etc. The proportion of positive over negative emotion level being expressed in timeline posts increases dramatically even between Day 0 and Day 1 of a Facebook relationship.

The discharge phenomenon may be used for predicting a relationship. However, it may over-simplify the problem because a decrease in timeline posts could be caused by changing jobs, extended travel, etc. It is worth noting that the study is based on a strong assumption: romantic couples update their status and posts on their timelines. In addition, the average timeline posts dropping from 1.65 to 1.53 may not be significant enough to indicate a major event in a relationship. Nevertheless, the discharge phenomenon can be viewed as a metaphor for a broader range of applications, including trust computing.

[14]Diuk C (2014) The formation of love. Facebook Data Science. Feb 14, 2014. Retrieved on Feb 29, 2016:https://www.facebook.com/notes/facebook-data-science/the-formation-of-love/10152064609253859

## Discharge in Trust Systems

Trust is a belief system that can be analogous to orgasmic dynamics. Trust is built-up over time by a series of small actions, similar to the boost process observed in orgasm. The discovery of a single lie is enough to create contagious doubt over every other truth expressed. In this case, truth quickly diminishes down to zero, much like throwing a discharge switch. Cognitively, trust is an anticipatory system in which a person anticipates something will happen at a certain resonance frequency. When the stimulus frequency matches the anticipated frequency, a resonance can be reached. Therefore, trust can be built-up gradually and reach a plateau. In the financial world, one can borrow a small amount of money and pay it off in time to establish credit history. This system works best when the transaction frequency is within an anticipated transaction frequency range.

## Summary

In this Chapter, we built a mathematical and electronic model to represent orgasmic dynamics, which is important for a computer to recognize orgasmic behaviors, or even to *have* such a behavior.

The key elements of the model include: (1) energy storage, (2) a stimulus frequency, and (3) the duty cycle of the stimulus pulses. The physical energy storage components include an inductor and a capacitor. With the inclusion of a switch to determine the directional path of the current, the output voltage can be "pumped" up or boosted to a value that is higher than the input voltage. Increasing the switching frequency always leads to faster boosting. The duty cycle also has an effect on the output dynamics. For a given stimulus frequency, lower duty cycles boost the output faster than those having higher duty cycles. But larger duty cycles prevail to generate higher voltage outputs. Euphoria dynamics can be viewed as a key metaphor in human interaction, such as the generation and recognition of music patterns. Furthermore, the model can be used to simulate the dopamine overflow process and the dynamic patterns of posts on social media such as Twitter and Facebook, including the withdrawal behavior in social media and trust systems.

# CHAPTER 10

# Describing Faces

Over the last several decades, facial recognition technology has become increasingly accurate. However, despite advances in technology, these systems are not yet as good as what was envisioned in crime scene investigation (CSI) films. For example, facial recognition technology failed in the Boston Marathon bombing manhunt in 2013.[1] The two bombers, Dzhokhar and Tamerlan Tsarnaev, were both in the facial image database. There were photos of the suspects, but the system could not find a match, or at least could not come up with one before the suspects were identified by humans. Under the best circumstances, facial recognition can be extremely accurate, returning the right person as a potential match within ideal conditions, e.g. front-view faces where all photos are shot from the same angle and with the same lighting. To reach that level of accuracy in real-world footage, which is often blurry, with different poses and lighting, computers almost always require a degree of skilled human guidance. According to the NIST report on the evaluation of 2D still-image facial recognition algorithms, the facial recognition accuracy rate decreases linearly with the logarithm of the population size of the image database. In all cases, human adjudication is ultimately necessary for verification.[2]

Today's computers still miss the important human instinct of facial recognition, in spite of computers' superb computational speed and massive memory. The purpose of this chapter is not to introduce another new facial recognition algorithm. Instead, we will explore how humans describe facial elements; how do we see groups of faces? how do we combine computerized facial reconstruction tools with "crowd-sourced" facial descriptions to produce an accurate composite face?

[1]Gallagher S (2013) Why facial recognition tech failed in the Boston bombing manhunt. http://arstechnica.com/information-technology/2013/05/why-facial-recognition-tech-failed-in-the-boston-bombing-manhunt/

[2]Grother P, Quinn GW, and Philips PJ (2011) Report on the evaluation of 2D still-image face recognition algorithms. NIST Interagency Report 7709, August 24, 2011.

© Springer-Verlag London 2016

Y. Cai, *Instinctive Computing*, DOI 10.1007/978-1-4471-7278-9_10

## How Do We See a Face?

What are the essential elements of a face?[3] In the Neolithic and Iron Ages, crafting tools were very primitive. Ancient people managed to carve faces on rocks. Facemasks in China and North America provide perhaps the earliest representations of faces on record. These prehistoric faces were made of circles, including the facial contour, eyes, and mouth. The eyes, nose, or mouth, and the facial contour comprise the essential elements for a face. Relatively speaking, the mouth conveys more expression than the nose. On the other hand, the nose is more stationary than the mouth, which helps in representing more biometric features. In forensic science, a skull reveals the location of the eyes and the tip of the nose bone. Thus, the *facial triangle* is essential in representing a face, starting with two eyes, then either the center of the upper lip of the mouth, or the tip of the nose (or the tip of the nose bone from a skull).

In the 1960s, Russian psychologist Alfred Yarbus studied human eye movements and vision using a high-precision eye gaze tracker and recorded gaze trajectories on film.[4] The traces reveal eye fixation paths and durations. Yarbus' studies of face viewing typically involved the presentation of isolated faces. He asked the subjects to view a photographic portrait for 3 minutes with no instructions. The results showed cyclical fixation behavior while viewing the faces: the gaze traced the eyes, nose, mouth, and face contours iteratively. These eye movements were not simply mechanical tracings, but rather, they reflected the viewers' attention and inquiries. There is a strong tendency to look at the eyes and mouth of a face in a portrait.

Yarbus' portrait studies also revealed that scanning patterns were varied and not always cyclical. They depended on the context of lighting, pose, and objects outside of the area of the face. For example, when an eye was in relative shadow on the face, it received only a small fraction of the overall allocation of gaze time. Furthermore, Yarbus tested how the eyes moved during a high-level task when certain instructions were given prior to viewing a face. For example, when subjects were prompted to estimate age, remember everything in the picture, remember characteristics of the clothing, etc., the plotting of eye movements pointed to differences in the allocation of gaze across various instruction conditions. After 43 years, Tatler et al. revisited Yarbus' data, summarizing it with heat maps and statistics, as well as performing validation experiments.[5] The team confirmed that the eyes and mouth still received the greatest fraction of gaze allocation. However, the cyclical scanning between eyes and mouth reported by Yarbus was less evident when the face was viewed within

---

[3]Cai Y (2003) How Many Pixels Do We Need to See Things? Lecture Notes in Computer Science (LNCS), ICCS Proceedings, 2003.

[4]Yarbus A (1967) Eye movements and vision. Plenum, 1967

[5]Tatler BW et al. (2010) Yarbus, eye movements, and vision. I-Perception. 2010: 1(1): 7-27. http://www.ncbi.nlm.nih.gov/pmc/articles/PMC3563050/#R41

the context of a body, when given instruction, or placed within a complex scene. To humans, viewing facial features is a selective process driven by economy of attention and context.

In 2016, the author conducted an experiment with an eye tracker to test how people view a face. The lab experiment included 40 subjects, half male and half female graduate students ages between 21 and 27. They were given a set of face photos and asked to view them without any cue questions for 60 seconds. The eye gaze plots indeed reveal that the eyes and mouth on a face are frequently scanned. However, the results do not show much tracing around the contour of a face as was shown in Yarbus' experimental results in the 60 seconds. See Fig. 10.1. Perhaps, people have changed their face viewing behavior after six decades? With overwhelming social media and electronic devices, are we changing our visual cognition behavior? We must find out with more rigorous experiments.

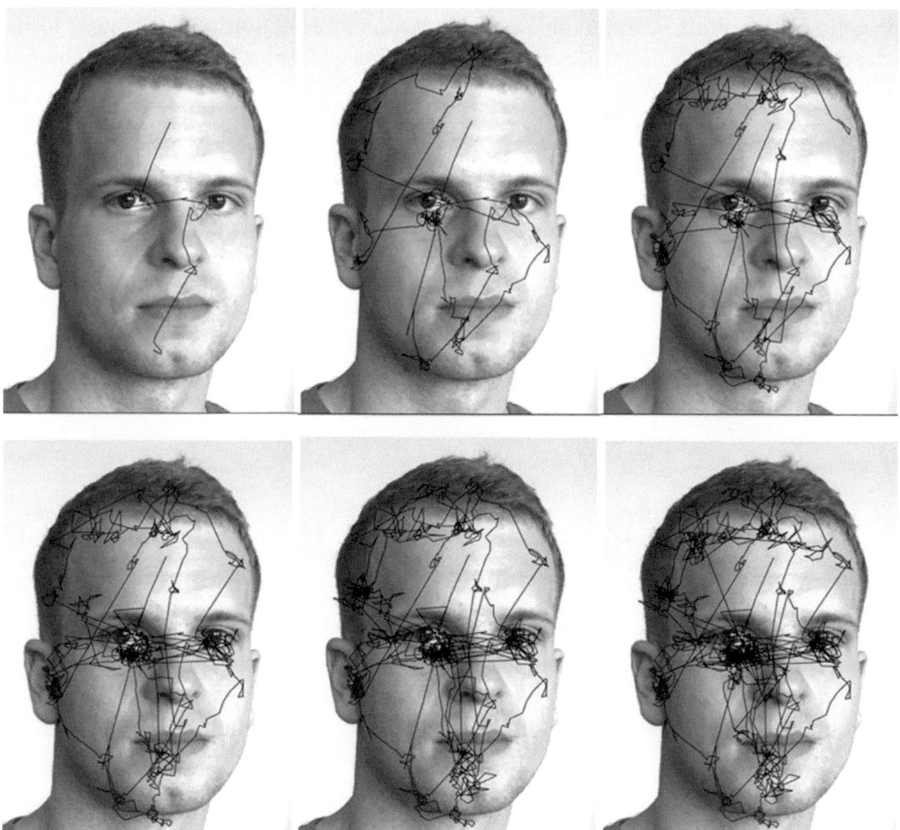

**Fig. 10.1** The time-lapse screenshots of a subject's gaze traces on a portrait at 2, 10, 20, 30, 40 and 50 seconds (from the *top left* to the *bottom right*). Our eye tracking lab experiments reveal the gaze fixations at the facial triangle area formed by eyes and mouth (Image courtesy of Viswa Tej Koganti)

Modern fMRI studies reveal that facial cognition is a holistic process involving several parts of the brain. Researchers use blood oxygen level dependent (BOLD) contrast fMRI to determine which areas of the brain are activated when looking at images of cars and faces. They found the entire region of the fusiform face area (FFA) and occipital face area (OFA) links to form a network that acts to distinguish faces. The processing of faces in the human brain is known as a "sum of parts" perception.[6] However, the individual parts of a face must be processed first in order to ultimately put all of the pieces together. At the early stage of this process, the occipital face area (OFA) detects the eyes, nose, and mouth as individual pieces. On the other hand, the fusiform face area (FFA) is responsible for overall structural information like the arrangement of facial parts (e.g. the eyes are above the nose and the nose is above the mouth). Facial cognition involves both component and structural processing. The fMRI discoveries are in fact consistent with Yarbus' eye tracking data in which gaze fixations moved from one facial component to another in order to iteratively map the structure of the face. Successful computerized face detection algorithms such as the Viola-Jones object detection model appear to be acceptable models of the processes at the component-level (OFA) first, and then at the structural level (FFA).

Facial cognition involves many neural networks. These networks also include visual and emotional processing systems. For example, looking at a face with a fearful expression, compared to neutral faces, increases activity in visual recognition in the right fusiform gyrus, which correlates with increased activity in the amygdala (the almond-shaped brain area responsible for emotional reactions) in the same situations.[7] "Face blindness" occurs when the connection is broken between facial recognition and the emotive parts in the brain. In this case, a patient sees his mother's face as an object, but not as his mother. In light of this, all of today's computer face recognition algorithms are actually "face blind" because there is no emotion attached to the facial recognition process itself.

## Facial Geometries

Portrait drawings use figurative abstraction to represent faces. Ancient Chinese ink drawings use lines to portray abstract human faces. Western paintings use light and shadow to render faces, creating photorealistic three-dimensional illusions. In many rural areas of the world, people are still not accustomed to seeing their faces in photographs where faces are cast in shadow, which disturbs their facial landmarks.

---

[6] Gold JM, Mundy PJ, Tjan BS (2012) The perception of a face is no more than the sum of its parts. Psychological Science 23 (4): 427–434.

[7] Vuilleumier P, Pourtois G (2007) Distributed and interactive brain mechanisms during emotion face perception: Evidence from functional neuroimaging. Neuropsychologia 45: 174–194.

Facial geometry has been used in forensic sketches to reconstruct faces based on biometric features. In order to represent scale- and rotation-invariant biometrics, proportions or ratios are commonly used.[8] For example, the Nasal Triangle, or Naso-Orbital Triangle, encodes the three points connecting the outward corners of the eyes and the end of the nose. The Facial Triangle, or Labio-Orbital Triangle, encodes the three points connecting the outward corners of the eyes and the center of the lower lip line. Both triangles are primary abstraction forms of overall facial structure. An infant's nasal triangle is normally equilateral, representing a so-called "baby face." An adult's nasal triangle is normally isosceles. Sometimes, we see a scalene shape in nasal or facial triangles, which represents injury or illness. The nasal triangle is also a handle for 3D transformations of scaling, rotation, and translation. With the three points, one can register one facial image to another. In addition to facial triangles, we can use ratios to represent biometric features. For example, the Mandibulo-Facial Index is the ratio of the width of the jaw relative to the width of the face. The Iridio-Chelial Index is the ratio of the width between the eyes relative to the width of nose. The Intercanthal Index is the ratio of the distance between the eyes relative to the ocular width.

Facial geometry contains 20–40 measurements. Are all of these measurements equally important to facial recognition? It would be ideal to select only the most important features as a subset for use in facial reconstruction or recognition. This reduction would make facial geometry easier for humans to process, shortening training time, and enhancing generalization by reducing the over-fitting data. Here, we start with the open source face data MSRA-CFW – a dataset of celebrity faces on the web from Microsoft.[9] The MSRA-CFW dataset contains 202,792 faces of 1583 people. About 10 % of the images are too small or too blurry. There are approximately 180,000 faces of 1300 people that are considered good-quality images. We can use 80 % of the final dataset as the training dataset and the remaining images as the testing dataset.

Facial landmark extraction is the first step in facial geometry calculations. To extract facial features, we can use Active Shape Model (ASM) tools such as STASM library for locating up to 77 facial landmarks.[10] Then, we can define a set of facial geometric measurements for analysis. Here, we consider 20 facial geometric measurements that can be derived from facial landmarks. For example, the length of lips (labial) is the distance between the horizontal coordinates of the two corners of the lips. A facial geometry is a vector of facial geometric measurement values. In this case, we have a 20 dimensional vector.

We can rank geometric features by their importance. The more important a feature is, the more likely it should be selected. Learning Vector Quantization (LVQ) is a supervised classification algorithm. The key is choosing an appropriate measure

[8]George R (2007) Facial Geometry. Charles C Thomas Publisher, Ltd.

[9]Microsoft MSRA-CFW dataset: http://research.microsoft.com/en-us/projects/msra-cfw/

[10]Milborrow S (2013) Stasm 4 User Manual. http://www.milbo.org/stasm-files/stasm4.pdf

of distance or similarity to the ground truth for training and classification.[11] We can also use an automatic method for geometric feature selection such as Recursive Feature Elimination or RFE.[12] It is a process of selecting a subset of relevant features by recursively eliminating less important features. The Random Forest algorithm can be used to explore all of the possible subsets of the geometric features. Fortunately, the statistics algorithm package Caret R provides the relevant tools for ranking importance (LVQ) and recursive feature elimination (RFE).[13]

According to the LVQ calculation results, we have six most important facial features: the Facial Triangle, or Labio-Orbital Triangle, the Nasal Triangle, or Naso-Orbital Triangle, Endocanthal-Alar Index, Facial Length, Mandibulo-Facial Angle, and Nasal Width. These features can reach more than 85 % accuracy.[14] From the result of the RFE automatic feature selection algorithm, four features are significant enough to represent a face: Nasal Width, Nasal Triangle, or Naso-Orbital Triangle, Facial Triangle, or Labio-Orbital Triangle, and Endocanthal-Alar. If we combine results from these two algorithms, the top six facial geometries are: Facial Triangle, or Labio-Orbital Triangle, Nasal Triangle, or Naso-Orbital Triangle, Endocanthal-Alar, Facial Length, the Mandibulo-Facial Index, and the ratio of Nasal Width to Nasal Length. Within these top six features, three of them are related to the nose and two are related to the eyes. That makes intuitive sense because the nose and the eyes play important roles in facial structure. According to a survey of the American Society of Plastic Surgeons, nose and eye reshaping are listed as the most popular plastic surgery procedures. The overall facial shape is also important. All the six features are structure-related.[15] Figure 10.2 shows the top six facial geometries.

## Semantic Representations

Human faces provide the richest vocabularies of visual imagery in any modern language. Creative literature is a well-known source of such descriptions, where human features are often described in vivid detail. In addition, reference collections in the English language that focus on visual imagery, such as textual and pictorial dictionaries, never fail to have major sections devoted to descriptions of the human face. These sections are typically devoted to anatomical features rather than social

[11]Schneider P, Hammer B, and Biehl M (2009) Adaptive relevance matrices in learning vector quantization. Neural Computation 21: 3532–3561.

[12]James G, Witten D, Hastie T, and Tibshirani R (2013) An introduction to statistical learning. Springer. P.204. 2013.

[13]Brownlee J (2014) Feature selection with the Caret R Package. Sept. 22, 2014. Machine Learning Mastery. http://machinelearningmastery.com/feature-selection-with-the-caret-r-package/

[14]Yan Y and Wang X (2015) Facial recognition based on facial geometrics. Final Report of Cognitive Video, ECE, Carnegie Mellon University, December 2015

[15]Ten most popular plastic surgery procedures, CBS News, 2012. http://www.cbsnews.com/pictures/10-most-popular-plastic-surgery-procedures/

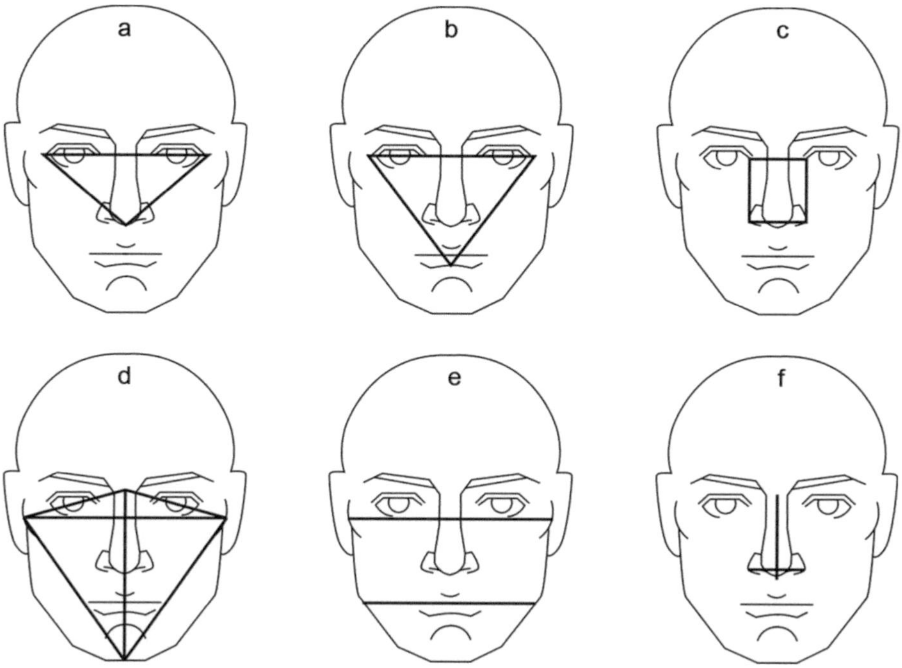


**Fig. 10.2** The top six facial geometries: (**a**) Nasal Triangle, or Naso-Orbital Triangle, (**b**) the Facial Triangle, or Labio-Orbital Triangle, (**c**) the Endocanthal-Alar Index, (**d**) Facial Length, (**e**) the Mandibulo-Facial Index, and (**f**) Nasal Width to Nasal Length

and cultural descriptions of faces. The mapping between images and faces that we have been exploring are built more upon stereotypical and analogical associations.

Here, we briefly present an overview of a variety of semantic visual description methods, including multiple resolution, semantic differentiation, symbol-number, and analogy. Then, we introduce the computational implementation of these human descriptions in visual and verbal forms and the interaction between the two.

## *Multiple Resolution Descriptions*

Human descriptions are classifiers for shape, color, texture, proportion, size, and dynamics in *multiple resolutions*. For example, one may start to describe a person's torso shape, then her hairstyle, face, eyes, nose, and mouth. Human feature descriptions have a common hierarchical structure, such as figure, head, face, eyes, etc. Like painting a portrait, verbal descriptions can be built up in multiple resolutions. The words may start with a coarse global description and then "zoom" into sub-components and details, or vice versa. Let us first look at the following examples, where the underlined sections represent the global levels of a description, the **bolded** section show component-based descriptions, and the *italicized* sections are the details:

> For a lean face, *pitted and scarred*, very **thick black eyebrows** and **carbon-black eyes** with *deep grainy circles of black* under them. A *heavy five o'clock shadow*. But the **skin under all was pale** and unhealthy-looking.[16]

> Otto has a face like a very ripe peach. His hair is fair and thick, growing low on his forehead. He has **small sparkling eyes**, *full of naughtiness, and a wide, disarming* **grin** which is too innocent to be true. When he grins, **two large dimples** appear in his **peach blossom cheeks.**[17]

> Webb is the oldest man of their regular foursome, fifty and then some – a lean, thoughtful gentleman in roofing and siding contracting and supply with a calming, gravel voice, his long face broken into *longitudinal strips by creases* and his **hazel eyes** almost lost under an **amber tangle of eyebrows**.[18]

Creative writers often mimic our eye movements much like the eye gaze trajectories that Yarbus plotted on film. Fixations seamlessly move from facial structural information to facial components, and then back to structural information, stimulating brain activity in the fusiform face area (FFA) and the occipital face area (OFA) subconsciously.

## *Semantic Differential Representation*

The Semantic Differential method measures perceptual and cognitive states in numbers or words aligned with a scale. For example, the feeling of pain can be expressed with adjectives, ranging from weakest to strongest. Figure 10.3 shows a chart of visual, numerical and verbal expressions of pain in hospitals: No Hurt (0), Hurts Little Bit (2), Hurts Little More (4), Hurts Even More (6), Hurts a

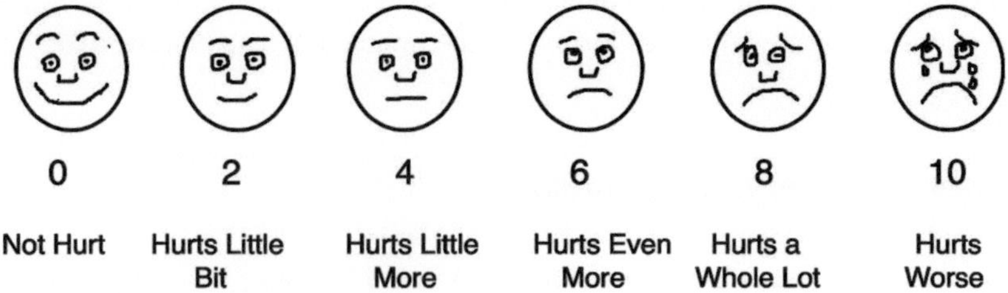


**Fig. 10.3** Expressions of pain in pictures, numbers, and words

[16] Doctorow EL (1980) Loon Lake. New York: Random House

[17] Isherwood C (1952) Goodbye to Berlin. Signet.

[18] Updike J (1996) The Rabbit is Rich. Ballantine Books.

Whole Lot (8) and Hurts Worst (10). This pictorial representation is very useful in patient communication where descriptions of pain type (e.g., pounding, burning) and intensity (e.g., a little, a lot) lack a robust and differentiated vocabulary.

Perceptual feelings can be quantified with a mathematical model. When the change of a stimulus ($I$) is very small, we will not detect a change. The minimal difference ($\Delta I$) that is just noticeable enough is called perceptual threshold, and it depends on the initial stimulus strength $I$. At a broad range, the normalized perceptual threshold is a constant, $\Delta I/I = K$. This is Weber's Law.[19]

Given the perceptual strength $E$, as the stimulus $I$ changes by $\Delta I$, and the change of $E$ is $\Delta E$, we have the relationship:

$$\Delta E = K \frac{\Delta I}{I} \tag{10.1}$$

Let $\Delta I$ be $dI$ and $\Delta E$ be $dE$, thus we have Weber-Fechner's Law:

$$E = K \cdot \ln(I) + C \tag{10.2}$$

Where, $C$ is constant and $K$ is the Weber Ratio, $I$ is the stimulus strength and $E$ is the perceptual strength. The Weber-Fechner Law states that the relationship between our perceptual strength and stimulus strength is a logarithmic function.

The Weber-Fechner Law enables us to describe responses to signal strength in a focused point perceptual basis, such as the feeling of pain, color, pixel intensity, and so on. However, it becomes significantly more complicated when used to describe facial features. First of all, facial features contain one-dimensional (e.g. nose length), two-dimensional (e.g. mouth), and three-dimensional measurements (e.g. chin). Second, facial features are organized in geometric structures. We not only perceive individual components, but also the relationships between features. For example, a long nose may be associated with a long face. All measurements are relative. Third, we have different sensitivity thresholds towards different facial features. Figures 10.4 and 10.5 are examples of lab experiments with variable facial features using a computer to generate a synthetic 3D face. The empirical experimental results show that people are more sensitive to the changes in mouth size than changes in nose length. The human vision system has a horizontal field of view of about 120° for binocular vision without rotating the eyes, but only a 55° vertical field of view. Perhaps this is a reason why we make more accurate horizontal measurements than vertical. Furthermore, people pay more attention to mouths than noses because a mouth presents information about words, emotions, and sensuality.

Here, we define the frame distance as the distance between two frames. For example, the distance between N1 and N2 is 1. The distance between N1 and N4 is 3. An experiment was designed for 10 subjects to identify any differences between pairs of images in Fig. 10.4 within 5 seconds. They were presented with pairs

[19]WikiPedia: Web-Fechner's Law. https://en.wikipedia.org/wiki/Weber-Fechner_law

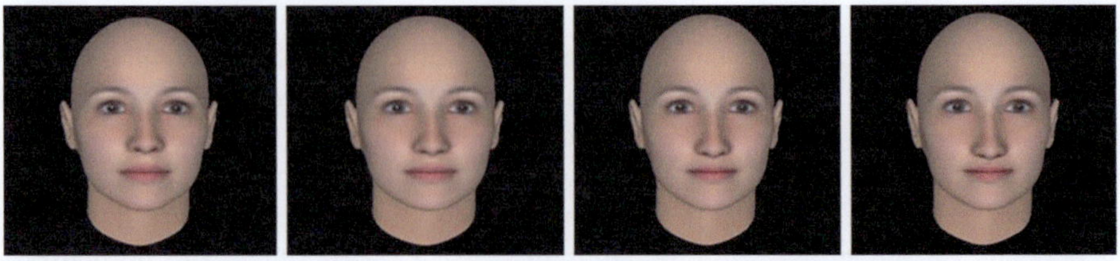

**Fig. 10.4** Variable nose lengths on a synthetic three-dimensional face: from left to right: N1, N2, N3 and N4

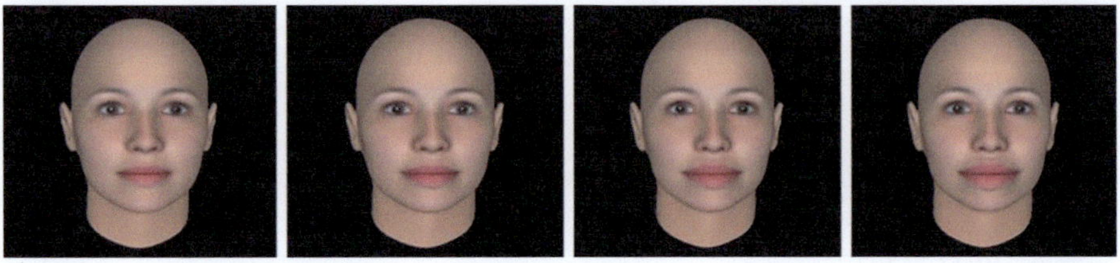

**Fig. 10.5** Variable mouth sizes on a synthetic three-dimensional face: from left to right: M1, M2, M3, and M4

starting from a frame distance of 1 moving upwards, and were judged on whether they could correctly differentiate the difference in nose size. The experiment was then repeated with Fig. 10.5 containing images of varying lip size. For the images of varying nose sizes, 100 % of subjects correctly recognized the nose difference when the frame distance was 3; the recognition rate dropped to 80 % when the frame distance was 2; the recognition rate dropped further to only 13.3 % when the frame distance was 1. For the images of varying mouth sizes in Fig. 10.5, 100 % of subjects correctly recognized mouth size differences when the frame distance was 2 or 3; the recognition rate dropped to 66.6 % when the frame distance was 1. Overall, more subjects recognized differences in lip size than in nose size. Furthermore, more subjects were able to perceive subtle changes in mouth size than in nose size.

The empirical results also show that people can easily recognize extreme length or size of facial components (e.g. the long nose N4 in Fig. 10.4 and large mouth M4 in Fig. 10.5). People make fewer errors and need less response time in estimating more extreme lengths or sizes. In contrast, people make more errors and need more response time in estimating length or sizes close to the norm. This is because *every time we view a face, our brain is searching for differences with respect to the norm.*

## Symbol-Number Descriptions

In many cases, numbers can be added to provide even greater granularity of the facial descriptions. For example, the FBI's Facial Identification Handbook comes with a class name such as bulging eyes, and then a number to give specific levels and types.[20] The FBI has already created a manual for witnesses, victims, or other suspect observers to use in identifying possible suspect features. The catalog presents several images per page under a category such as "bulging eyes." Each image in such a category has bulging eyes as a feature, and the respondent is asked to identify which image has bulging eyes most closely resembling the suspect. The handbook is an extremely efficient and effective tool for both forensic sketch artists and police detectives. It is most commonly used as a tool in helping a witness or victim convey the features of a suspect to the sketch artist in order to render an accurate composite sketch.

## Analogical Descriptions

From the multi-resolution point of view, analogies are coarse descriptors compared to symbolic-number descriptions. Instead of describing features directly, people often find it easier to refer a feature to a stereotype, for example, a movie star's face. Analogical mapping includes structural mapping (e.g. face to face), or component mapping (e.g. Lincoln's ear and Washington's nose). Children often use familiar things to describe a person's characteristics, for example using the word "cookie" to describe a round face.

Analogies are culture-based. In the Western world, several nose stereotypes are named according to historical figures. Many analogies are from animal noses or plants. Figure 10.6 illustrates examples of the nose profiles as described above. We use simple line drawings to render the visual presentation.

Analogies trigger experiences, which involve not only images but also dynamics. The far right image in Fig. 10.6 shows a 'volcano nose', which triggers a reader's physical experience of pain, eruption, and explosion. In this case, readers not only experience it, but also predict the consequence (e.g. acne pimples popping). It is an analogy with a novel physical process.

Given a verbal description of a nose, how do we visually reconstruct the nose profile using only minimal elements? In this study, we use a set of 5–9 "control

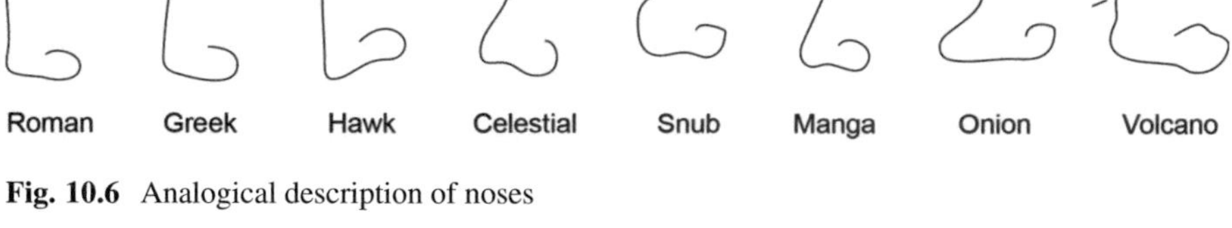


**Fig. 10.6** Analogical description of noses

[20]FBI Facial Identification Catalog (1988), published in Nov. 1988

points" to draw a profile. By adjusting the relative positions of the control points, we can reconstruct various stereotypical profiles and many others in between. To smooth the profile contour, we can apply the spline curve-fitting model.[21]

## Shape Recognition in Words

Mapping shapes to words is a process of pattern recognition. Here, we use the point distribution model (PDM) to describe a new shape from familiar shapes. The point distribution model is the visual abstraction of a shape that converts pixels into points on the contour of a shape.[22] We can represent a shape as a parameterized deformable template.[23] The template can be an average shape of the object, for example, an average face. A face can be reconstructed by deforming the average face with the control points. In this case, to memorize a face is to store the vector of control points; to recognize a face is to find the vector of control points that is closest to that of the input face.

For simple shapes such as face profiles, we use heuristic rules to classify the shapes. For example, in Fig. 10.7, the labeled squares represent points that the user can move. The control points *a* and *g* are end-points that are not allowed to move. They are anchored to the appropriate place on the picture of the face. The control points *b* through *f* are movable.

We may use heuristics to classify noses. For example: *Very Straight Nose:* If the slopes of line *a-b* and line *b-c* are the same as line *c-d*, then it is a very straight nose; *Convex Nose:* If the slope of line *b-c* is less than the slope of line *c-d*, or the slope

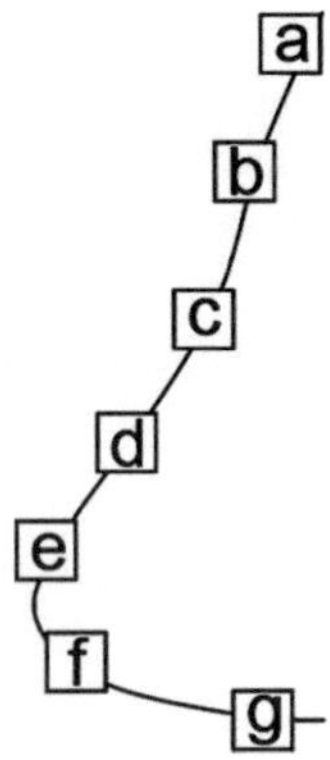


**Fig. 10.7** Representing the profile of a nose with control points

[21] Web site: 2007. http://en.wikipedia.org/wiki/Spline_(mathematics)

[22] Cootes TF, Taylor CJ, and Graham J (1992) Training models of shape from sets of examples. In D.C. Hogg and R.D. Boyle, editors, Processings of the Bristish Machine Vision Conference, Leeds, UK, pages 9–18, Springer Verlag, London, 1992

[23] Black A and Isard M (1998) Active Contours. Springer

of line *c-d* is less than the slope of line *d-e*, it is a convex nose; *Concave Nose:* If the slope of line *b-c* is greater than the slope of line *c-d*, or the slope of line *c-d* is greater than the slope of line *d-e*, it is a concave nose; *Snub Nose:* If point *f* is above and to the left of point *g*, and the angle between line *f-g* and the horizontal is greater than 15°, it is a snub nose; *Hawk Nose:* If point *f* is below and left of point *g*, it is a hawk nose.

We also need to check the constraints for the element. For example, if a nose does not meet these criteria, it is certainly unrealistic. For example, *Orientation:* points *b-f* must be to the left of points *a* and *g*; points *b-e* must be to the left of the point which comes before them alphabetically. Point *f* may be to the left or right of point e; *Ordering:* points *b-f* must be below the point that comes before them alphabetically. Note that it is valid for point *g* to be above point *f*; this represents a hawk nose; *Nose Tip:* The angle between line *f-g* and the horizontal must be less than or equal to 30°, whether it has a positive or negative slope.

## Pareidolia Phenomenon

*Pareidolia* is a phenomenon that allows us to "see" faces and other objects in completely unrelated objects and settings. Carl Sagan theorized that this provided an evolutionary advantage, as it allowed humans to better recognize each other, even under poor visual circumstances. Even though these images may seem not significant nor life changing, they are a really fun figment. Figure 10.8 shows pareidolia in a coffee cup, where two bubbles form eyes and the oval shape below appears to be an open mouth.

In 1976, the Viking satellite photographed a rock on Mars that looked like a human face. Its appearance depended on the exact angle and position of the sun. In 2001, Mars Global Surveyor provided another image of the same spot and found it was simply a rock with no face.[24] This example is a clear case of pareidolia, our tendency to see faces or other recognizable shapes in random formations. Now, computer vision algorithms also exhibit symptoms of pareidolia. Figure 10.9 shows the faces detected by the popular algorithm in OpenCV.

How do we extract facial features from photos at the pixel level? The Viola-Jones algorithm views the light intensity of a face in an image, which are then decomposed into a bunch of binary strips, or so-called "Haar-like" features. See Fig. 10.10. For example, the eyebrows are typically darker than the cheeks, and the bridge of the nose is typically brighter than the two sides of the face around it. A feature is essentially all of the pixels in one area of the image added together, minus the pixels in another area added together. The low-level classifier, or so-called "weak classifier," takes the value and classifies it as "feature" or "non-feature" based on a

[24]Luntz S (2014) Is there a Man on the Moon? http://www.iflscience.com/space/there-man-moon#Rckl0FsjIRhyy0Gl.99

**Fig. 10.8** Pareidolia effect: seeing faces everywhere

**Fig. 10.9** Computer pareidolia cases: faces detected in images of Mars and trees

certain threshold and polarity. For example, if the threshold is 50 and the polarity is "greater than," the weak classifier will classify an image as a feature if the sum of the intensity in one area minus the sum of the intensity in another is "greater than 50."

The Viola-Jones algorithm aims to find multiple rectangular features in order to detect an object in an image. It uses a cascade architecture to combine the weak feature classifiers and their weighted errors to make a strong classifier. The cascade architecture has interesting implications for the performance of the individual classifiers, because the activation of each classifier depends entirely on the behavior of its predecessor. The Viola-Jones algorithm has a high detection speed. However, it also has a few weaknesses. For example, it is very sensitive to lighting conditions and works best when portraits are well-lit, such as in a passport photo. It is also sensitive to training objects, such as skin color, different poses, and so on. The algorithm is also sensitive to scaling, rotation, and shearing. While detection speed can reach real time, the algorithm's training process is very slow, and can take an order of days or weeks. This is essentially a brute-force approach for learning all the possible combinations of features, polarities, and thresholds.

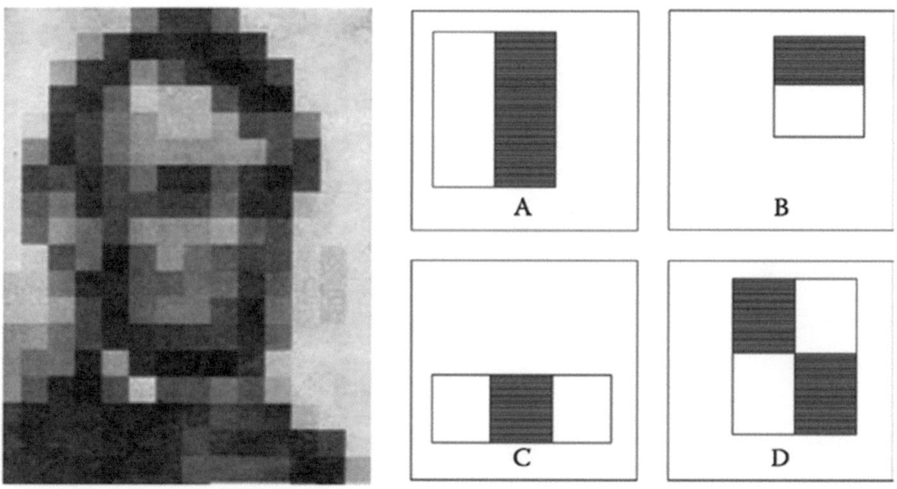


**Fig. 10.10** Haar-like wavelet-based features

## The Mean Faces

In our lives, we have seen many faces. How we store all of these faces in the brain is still a mystery. Let us assume all faces are superimposed on top of each other to create a "mean face." This is equivalent to the *pixel-based average*. As you can see, the mean face is blurred because the landmarks are not aligned. To align the facial landmarks, we need to annotate the key feature points as landmarks, such as the corners of the eyes, nose, mouth, and ears. Depending on the person, some landmarks are not as obvious, such as the corner of the chin. Figure 10.11 shows the

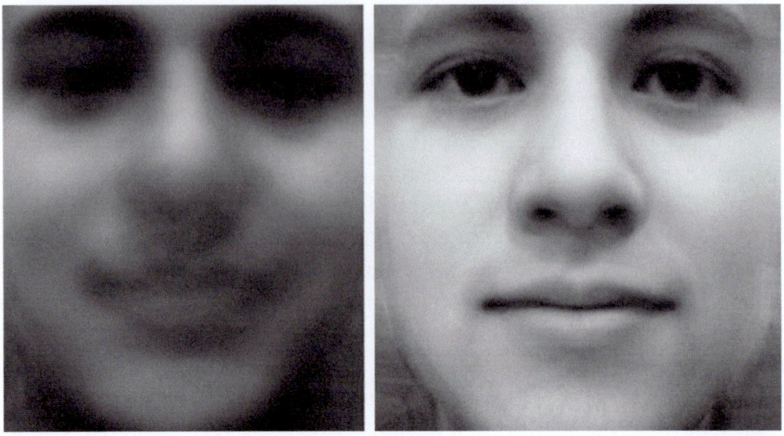

**Fig. 10.11** Mean face based on pixels (*left*) and based on landmarks (*right*)

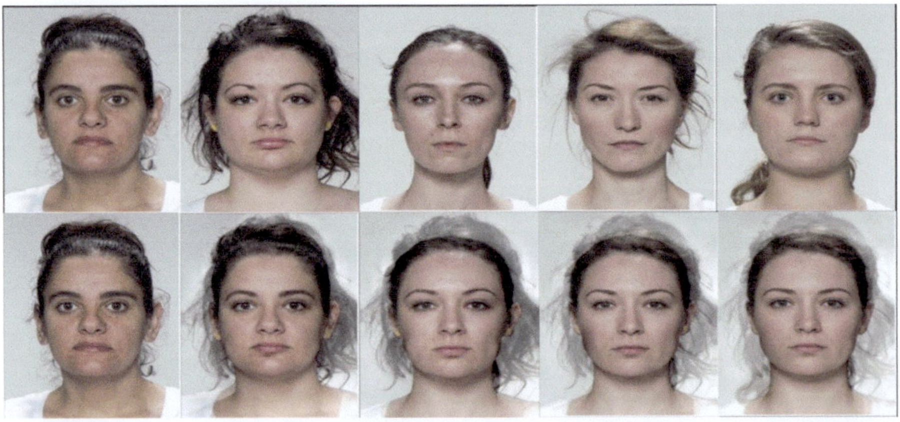

**Fig. 10.12** The mean face becomes prettier as more faces are added to the mix

mean faces from both the pixel-average and landmark-average methods. It appears that humans prefer the landmark-based average face because it contains a better definition of the features.

Now, let us select some actual faces from the database and average them. The more faces we add, the more symmetrical the mean face becomes. All of the distinguishing features disappear until the most generic face emerges. By most standards, the mean face appears to be prettier! An average face represents beauty. An average face is symmetrical and signifies a healthy body. It is our instinct that drives us to be attracted to healthy-looking faces. See Fig. 10.12.

On the other hand, our instinct also drives us to remember and recognize individual facial features. In many cases, nuances such as scars, dimples, even hairs make a big difference in facial recognition.

## Crowdsourcing for Facial Reconstruction

Facial recognition is instinctual to humans. We describe faces in words, gestures, even sketches. Being a forensic sketch artist is a unique career that requires the translation of verbal descriptions or forensic evidence of a suspect into a visual portrait or a sculpture. This reconstruction process is usually interactive and incremental. The forensic sketch artist works with witnesses multiple times to add more details in the facial features.[25] With rapidly-growing social media, crowdsourcing forensic footage online is becoming increasingly possible. Three days after the Boston Marathon bombing, the FBI published video footage and still-images of the two suspects. The New York City police department also posted surveillance footage on their website to enable citizens to identify the suspects.[26] Crowdsourcing and image tagging also help to harvest human knowledge about faces and to narrow the gap between human-like semantic representation and computational facial feature models.

Nearly half of all violent crimes go unsolved in the US. The composite sketch from eyewitnesses is one of the only tools detectives have in the absence of other evidence. In Stratford, Connecticut, 2015, a serial burglar broke into ten homes in the region. One of the eyewitnesses, an 11-year old girl named Rebecca Depietro came forward to draw a stick figure of a suspect. Even though the stick figure was rudimentary (containing just a few outlines of hairs, a beard, eyes, a nose, and eyebrows) police took the 2-min doodle seriously. They matched the sketch to the suspect who eventually confessed to the serial burglaries, including the break-in at the Depietro's home.[27]

Most people have no difficulty recognizing a face, but trying to describe it from memory is a challenge. Professional sketch artists interview victims and eyewitnesses about a suspect's face, and their drawings go out to the public with the goal of soliciting tips for law enforcement. Here, we wanted to discover how accurate these sketches are, and if there might be a better way to put these memories onto paper or build a computer model.[28] There is not really any good data on how often a sketch helps police catch a criminal on the loose, but more often than not, these sketches do not achieve a very strong likeness. There are certainly some notable exceptions, such as the sketches made of the Oklahoma City bomber

---

[25]Dove US, Youtube video, retrieved on Jan. 23, 2016. Dove real beauty sketches. https://www.youtube.com/watch?v=XpaOjMXyJGk

[26]NY City Crime Stop web site: http://a056-crimestoppers.nyc.gov/crimestoppers/public/wanted.html

[27]NBC News, 11-years-old's sketch helps police catch serial burglar. Oct. 16, 2015. http://www.9news.com/story/news/weird/2015/10/16/stick-figure-burglar-arrested/74039792/

[28]Vox Program: Police sketeches may become a thing of the past, Observatory #5, April 22, 2015. https://www.youtube.com/watch?v=h81SuD2pltM

Timothy McVeigh, Swedish serial rapist Kurt Niklas Lindgren, and Baton Rouge serial killer Derrick Todd Lee, or in cases where the suspect has a very distinctive look. Oftentimes, composite sketches are not very informative to the investigation. A sketch is arguably better than nothing, but there is always a risk when releasing a bad sketch to the public.

Psychologists have long known that we process human faces holistically, rather than on the level of individual features. For example, people generally recognize a person's age, race, and gender from the overall facial structure instead of individual parts. Our brains are not well-suited to describe individual facial features. However, when creating a hand-drawn sketch, that is usually what victims and eyewitnesses are asked to do.

There have been many attempts to improve upon the hand-drawn sketch. In the 1960s, police tried using mechanical kits of interchangeable features to piece together a suspect's face, much like a puzzle. Then, computer programs became more customizable, but still relied on feature-by-feature reconstruction, for example, the two-dimensional facial composite software FACES.[29] The latest generation of computer programs takes a more holistic approach to facial structure composition. FaceGen and EvoFIT are two prime examples. FaceGen is a three-dimensional facial composite software program that was originally designed for film character animation.[30] It has a set of parameters to control the structure and features of a face. It allows photos to be input to generate texture-mapped three-dimensional models that can be aligned to any angle. Because FaceGen was designed for professionals with over a hundred parameters, it is rather challenging for a novice user to construct a composite face. EvoFIT,[31] on the other hand, is a two-dimensional software program designed for forensic facial compositing with fewer parameters at each step. With EvoFIT, eyewitnesses and victims select from a screen of complete faces, and, over time, a complete face *evolves*. After the victim or eyewitness specifies the gender, age, and race of the subject, the software presents randomly-generated faces rather than individual features. All the user needs to do is to choose a few faces that most resemble the suspect. Using a genetic algorithm, the selected faces are blended together and a new set of faces appears. The user can also adjust holistic variables, such as attractiveness. When they tested this program in the lab, participants were able to generate celebrity faces from memory. When EvoFIT

[29]FaceID. http://www.facesid.com/products_faces_le.html

[30]FaceGen wib site: http://facegen.com

[31]Frowd C et al. (2012) Catching evn more offenders with EvoFIT facial composites. Third International Conference on Emerging Security Technologies, 2012. http://www.evofit.co.uk/wp-content/uploads/2012/11/Frowd-et-al-2012-Catching-Even-More-Offenders-with-EvoFIT-Facial-Composites-final.pdf

was used in five police departments, it led to an average of 40 % more arrests than traditional methods and up to an average of 60 % more arrests in the most recent trial.[32]

## *Spearman-Brown Prediction Formula*

Rapidly-growing surveillance camera systems and mobile devices provide massive video or still image footage online. In the past, law enforcement agencies such as the FBI and NYPD have posted forensic footage on social media for soliciting tips. In fact, publicly posted forensic footage allows a massive number of online users to be "eyewitnesses." Here, the question is: can the average user with minimal training use a software tool to make facial composites? How do we then merge the facial composites? What would be the reliability of facial composites derived from the masses?

The verbal or visual descriptions employed in facial reconstruction vary dramatically in their reliability. How many witnesses shall we employ? Assume we have a witness who reconstructs a face from memory. Intuition suggests that we would gain reliability by adding the ratings of a second witness because the second witness' random errors tend to cancel out those of the first witness. Intuition suggests that adding more witnesses, all of whom agree with each other to roughly the same degree, should further increase our effective reliability. This intuition is supported by a very old and well-known result discovered independently by Charles Spearman and William Brown in 1910.[33,34,35] The Spearman-Brown prediction formula is helpful in understanding the non-linear relationship between the effective reliability ($R$), the mean reliability of individual test ($r$), and the *test* length ($n$).

$$R = \frac{nr}{1 + (n-1)\,r} \tag{10.3}$$

This formula predicts the effective reliability of a new *test* composed by $n$ individual tests in parallel. The formula can also be rearranged to predict the number of *tests* required to achieve a degree of effective reliability $R$.

$$n = \frac{R\,(1-r)}{r\,(1-R)} \tag{10.4}$$

---

[32]EvoFIT Case Studies. http://www.evofit.co.uk/case-studies/Retrieved on Jan. 31, 2016

[33]Spearman CC (1910) Correlation calculated from faulty data. British Journal of Psychologu,3, 271–295.

[34]Brown W (1910) Some experimental results in the correlation of mental abilities. British Journal of Psychology, 3, 296-322.

[35]Rosenthal R (2005) Conducting judgment studies. In J. A. Harrigan, R. Rosenthal & K. R. Scherer (Eds.), Handbook of nonverbal behavior research methods in the affective sciences (pp. 199–236). NY: Oxford

In this study, the *test* is a set of facial descriptions from an individual witness. Given $n$ witnesses with the mean of individual reliability $r$, we might predict the overall effective reliability $R$ by combining all $n$ facial description inputs. Furthermore, the mean of individual reliability and effective reliability can be represented by the measurement of similarity between the ground truth data and the composite faces. Figure 10.13 shows examples of the effective reliability values, given mean reliability values of 0.2, 0.4, 0.6, and 0.8, while the number of witnesses increases from 1 to 25.

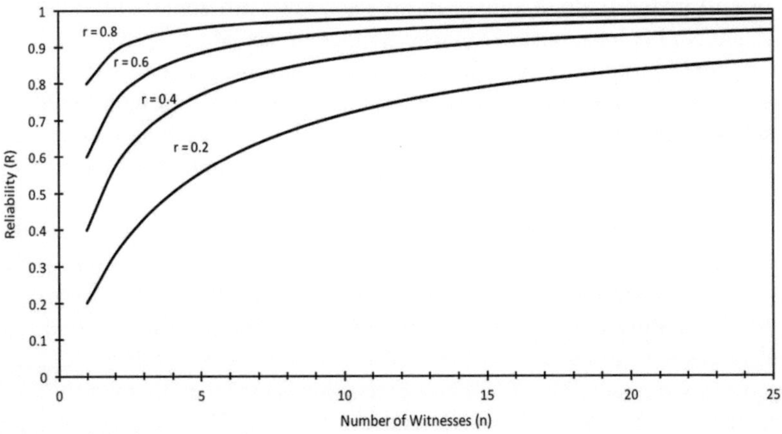


**Fig. 10.13** The effective reliability (R) of the composite results from witnesses (n) with mean of individual reliability (r), predicted by the Spearman-Brown formula

The curves show that the effective reliability will increase as the number of witnesses increase. The higher the mean reliability, the faster the effective reliability increases. In our case, the hypothesis is that the more reconstructed faces we have, the more similar to the actual face the composite face becomes. The formula assumes that the skills, knowledge, and accuracy of the facial descriptions from individuals are consistent at a certain level. For example, if a highly reliable description test was lengthened by adding many poor-quality inputs, then the effective reliability will likely be much lower than that predicted by this formula.[36]

## *Facial Composite Experiment*

Does the Spearman-Brown prediction formula work with crowdsourcing facial composites? We conducted a lab experiment to test the theory. A surveillance

[36]WikiPedia: Spearman-Brown prediction formula. Retrieved on Feb. 10, 2016. https://en.wikipedia.org/wiki/Spearman-Brown_prediction_formula

camera was set up above a door of a room to take video footage in which a man walked toward the door. The video captured his face from various angles, but never at a perfectly frontal pose. The resolution of the video was 1080p HD at 30 frames per second, which is higher than the typical resolutions of surveillance cameras. Still, facial details were not as clear as in most portrait photos.

Twenty-five teams were recruited to build a composite face from the video footage. Each team consisted of two people, ages between 21 and 25, half male and female. The teams were allowed to browse the footage freely, using play, pause, and replay in order to observe the face thoroughly. Each team was given up to 30 min of training to work with the facial composite software FACES. When the teams worked with FACES, they put the video footage on one screen and the FACES composite display on another screen. They composed the face from its features, such as eyes, nose, mouth, hair, and shape of the head, adjusting the length or size of each part. After the teams completed the facial composite, they took a screenshot of the 2D frontal face image.

In order to evaluate the accuracy of these facial composites, high-resolution images were collected as the "ground truth", including photographs of the actor's face from the front view, the left side, and the right side. Biometric measurements were extracted from the frontal facial image in Fig. 10.14. For this, we used STASM, an active shape model algorithm that can automatically extract 77 facial landmarks. Here, we used only 29 facial landmarks and 15 facial geometric measurements. See Fig. 10.15.

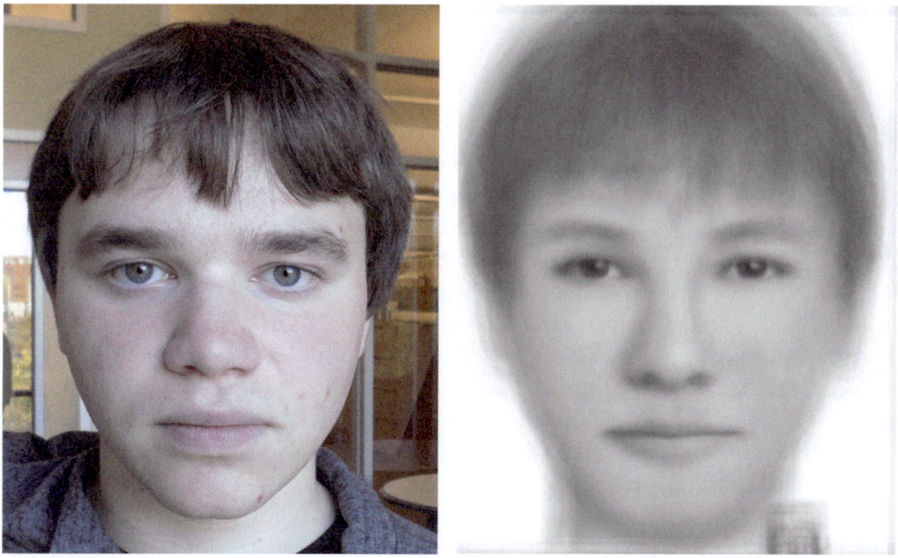

**Fig. 10.14** The high-resolution photo of the frontal face (*left*) and the composite result from 25 inputs (*right*)

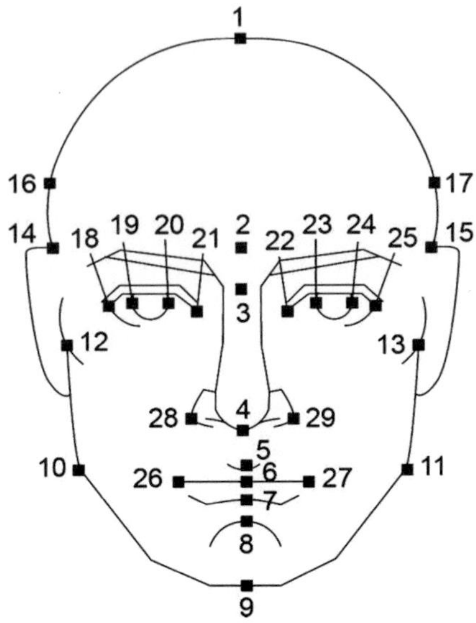


**Fig. 10.15** The biometric landmarks on the face of the ground truth: *1*. top of the head, *2*. between the eyebrows, *3*. top of the nose, *4*. bottom of the nose, *5*. top of the lip, *6*. center of the lip, *7*. bottom of the lip, *8*. chin cleft, *9*. bottom of the chin, *10*. right jawline, *11*. left jawline, *12*. right cheekbone, *13*. left cheekbone, *14*. right ear top, *15*.left ear top, *16*. right side of head, *17*. left side of head, *18*. right corner of right eye, *19*. right corner of right iris, *20*. left corner of right iris, *21*. left corner of right eye, *22*. right corner of left eye, *23*. right corner of left iris, *24*. left corner of left iris, *25*. left corner of left eye, *26*. right corner of mouth, *27*. left corner of mouth, *28*. right nostril, *29*. left nostril

This study included two experiments to test how humans visually evaluate the crowd-sourced facial composites based on the frontal 2D images generated from the 2D facial reconstruction tool. Figure 10.16 is an overview of the data flow. Given over 25 facial composite inputs, we used the face averaging software tool PsychoMorph to linearly combine a number of randomly selected faces. To make things fair, we conducted at least three trials. Figure 10.16 shows the "mean face" using 2, 4, 6, 8, and 10 composite 2D faces from FACES. The 2D final results of the 25 faces are shown in Fig. 10.14. We asked over 20 subjects to score the similarity between the facial composites and the ground truth. The facial composites were randomly presented to the subject. Each facial composite was given a score between 0 (not similar at all) and 1 (almost identical).

We then identified the similarities between the ground truth and the composites based on computerized facial geometries. We used STASM to locate the 29 facial landmarks and then derived 15 facial geometric measurements, stored in a vector, for

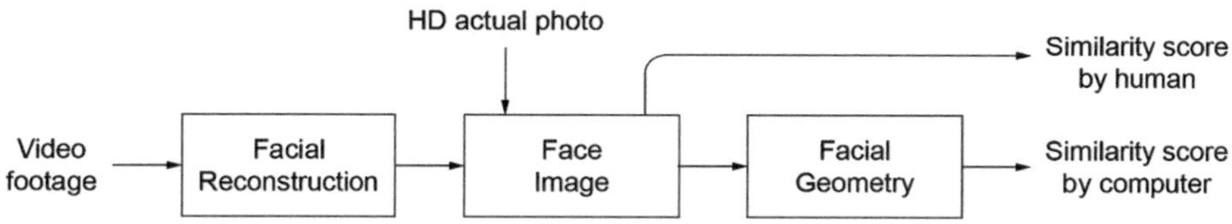


**Fig. 10.16** The workflow of the five experiments for facial composition and similarity evaluations by humans and computers

all of the composite faces from merged 2, 4, 6, 8, 10, . . . , 25 faces. See Fig. 10.17. We then calculated the Cosine Angle Similarity Coefficient between the merged composite faces and the facial geometric vector of the ground truth.

The results are plotted in Figs. 10.18 and 10.19. According to human subject evaluation, the similarity between the 2D composite face and the ground truth increases as the number of averaged composite faces increases. The slope steeply increases between one face and six faces, and the degree of similarity reaches its peak at around six faces. After six faces, the gain of similarity becomes small and reaches saturation after ten faces. According to the Cosine Angle Similarity Coefficient study, the facial geometric similarities of composite faces increase as the number of faces increases. Generally, the gain of similarities slows down after ten faces.

In conclusion, the crowdsourcing of facial composites follows the trend of the Spearman-Brown prediction formula even though an individual's composite reliability is very limited. The similarities typically reach a peak at around six composite faces and the gain of similarity slows down after about ten composite faces.

The measurement of facial geometries is not perfect in this study. It contains 15 facial geometric features. Most of them are ratio-based, enabling the measurement to be scale-invariant. However, it is not necessary that these features represent the most discriminatory of facial features, such as the height and profile of the nose.

There are also many other methods for "averaging" composite faces aside from the geometric averaging of landmarks. Mode averaging enables majority values to "vote" for the "average" value, while eliminating minority values. This alternative method would have an impact on the outcome of the composite faces. Some outstanding features might be lost in the voting process.

Perhaps the most critical question is whether composite faces can preserve the most discriminatory facial features as the number of faces increases. From Figs. 10.18 and 10.19, we found that the average faces grow more and more symmetrical and 'featureless" in the reconstructed images. The similarities of the composite faces may have a 'saturation point,' for example, around 6 or 10 faces. After the saturation point, adding more composite faces may add very little to helping with identification, or, in some cases, make it even worse.

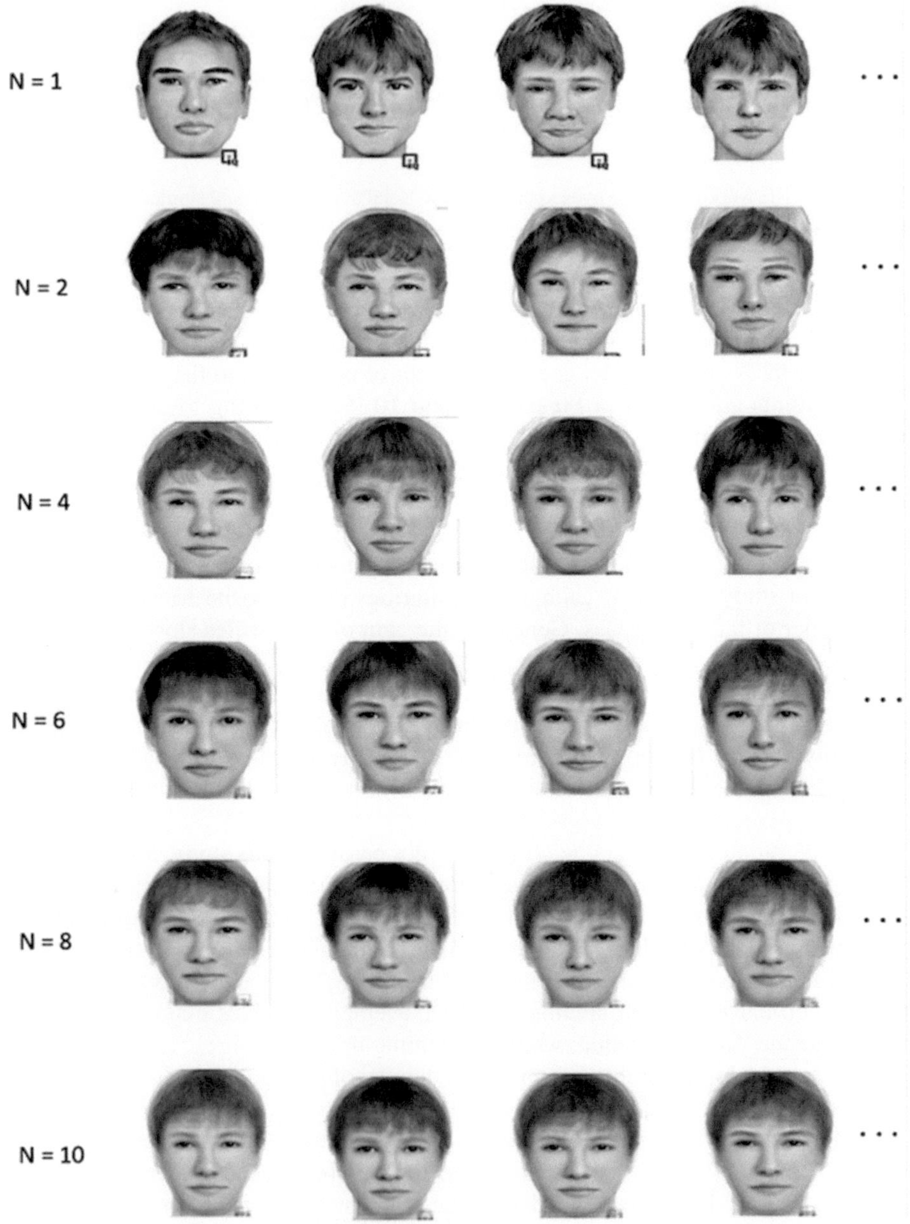


**Fig. 10.17** The "average face" from a combination of samples from 25 composite faces, where N is the number of samples randomly selected from the composite faces

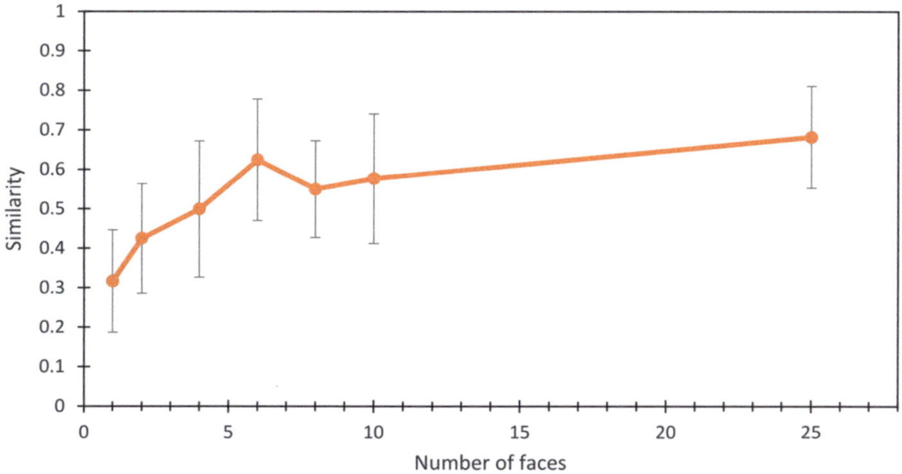


**Fig. 10.18** Human evaluation of the similarity between the ground truth and the reconstructed faces

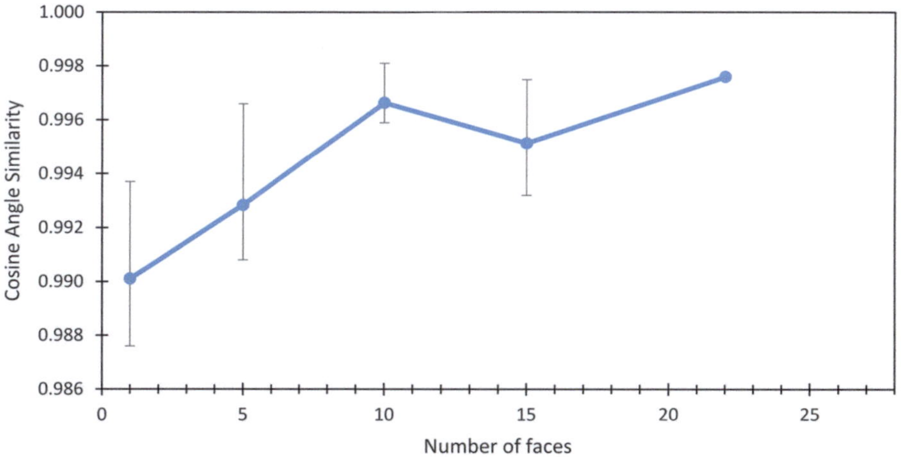


**Fig. 10.19** Computer evaluation of the cosine similarity coefficient between the ground truth and the reconstructed faces

## Summary

In this chapter, we discussed the most representative features of a face: the facial triangle with the eyes and nose or mouth. The facial triangle is validated by psychological studies of eye tracking data from the 1960s. The fMRI studies also show that facial recognition is a holistic process – a sum of individual features.

Faces can also be represented verbally, or using symbol-numbers systems, analogy, Haar-like wavelets-based features, stereotypes, and facial geometry. Based upon optimal feature selection, the top six facial geometries are: the Facial Triangle, or Labio-Orbital Triangle, the Nasal Triangle, or Naso-Orbital Triangle, the Endocanthal-Alar, Facial Length, the Mandibulo-Facial Index, and Nasal Width.

*Paeidolia,* the illusion of faces in unrelated objects, is found in both humans and computers, and is instinctual for survival. The mean face demonstrates that an average face is a symmetrical and healthy face, therefore being beautiful by universal standards. The mean face also tells us that we recognize faces not only by facial structures, but also by nuances such as hair and scars.

In order to recognize a face, one must mentally reconstruct a face. Rapidly-growing surveillance camera systems and mobile devices provide massive amounts of video and still image footage online. Online forensic footage allows a large pool of online users to serve as "eyewitnesses." Can the average user with minimal training use a software tool to create facial composites? How do we then merge the facial composites? How reliable would the facial composites be when derived from the masses? Our lab experiments suggest that crowdsourcing facial composites follow the trend of the Spearman-Brown prediction formula, even though an individual's composite reliability is very limited. The similarities between composites and the high-resolution reference images ("ground truth") normally reaches its peak around six composite faces and the gain of similarity slows down after ten composite faces.

Averaging composite faces may not preserve the most discriminatory facial features as the number of composite faces increases. We found that the average face becomes more and more symmetrical and "featureless" in the composite process. This suggests that the similarities of the composite faces might have a "saturation point," for example, around 6 or 10 faces. After the saturation point is reached, adding more composite faces does little to help the identification process. This experiment was just a preliminary study of the potential of crowdsourcing facial composites. There is much room for improvement, including better strategies for measuring facial geometries, more intuitive and holistic composition tools, and more user-friendly and robust facial modeling technologies.

# CHAPTER 11

# Figurative Thinking

Charles Darwin believed that language was half art, half instinct. Languages originated from sharing survival skills in ancient times. In the Stone Age, people learned to make tools through observation, mimicry, and gestures, eventually developing verbal languages. What makes human languages unique are the written languages that have been passed down over time. Figurative thinking is our instinct. It follows the principle of least effort as George Zipf predicted for words. Future intelligent systems are moving toward more figurative thinking with more figurative contents and interactions. In this Chapter, we explore the origin of figurative languages, digital encoding methods, and applications in passcode, programming, and social media.

## Figurative Abstraction

American writer and filmmaker Susan Sontag once said, "Art is a form of consciousness." Art requires media to carry it out. If we looked at ancient carvings, wood blocks, bamboo, and rice paper, all the way to computers and mobile phones, we would find that media and writing technologies played an important role in language development in terms of form and capacity.[1] About 30,000 years ago, the oldest figurative drawings were carved on the walls of the Chauvet Cave in southern France. Animal figures were drawn with charcoal on the cave walls. Using this primitive media, the figures were rendered in astonishing detail, including eyes, noses, and mouths.

[1]Nayab N (2014) How Has Technology Changed Communication? Retrieved on February 22, 2015, from http://www.brighthubpm.com/methods-strategies/79052-exploring-how-technology-has-changed-communication/

© Springer-Verlag London 2016

Y. Cai, *Instinctive Computing*, DOI 10.1007/978-1-4471-7278-9_11

**Fig. 11.1** Neolithic Rock Art in the Helan Shan Mountains, China

Petroglyphs, on the other hand, were mainly created outdoors using quartz to hit the rock surface to create marks. In Italian, this is called "*graffiti,*" meaning carved pictures. With such a rudimentary tool and a hard rock surface, it is difficult to carve figurative details. Thus, the *figurative abstraction* form of communication was born. Figurative abstraction is the simplified representation of figures using geometric outlines and symbols, similar to the maps, signs, and icons we see today. Figure 11.1 shows examples of Neolithic rock art in the Helan Shan Mountains in Northwest China. This ancient rock artwork represents ancient life, using lines, symbols, and shapes.

The strategies of figurative abstraction include *projection*, *exaggeration,* and *simplification*. The overall purpose of figurative abstraction is to use limited resources to maximize the representational features of an object. *Projection* is the mapping of three-dimensional objects onto two-dimensional spaces, presenting a profile view of objects. There are infinite ways for such projection to occur. The most important question here is, which direction is most representative of an object? In many cases, sideways projection or frontal projection is used. The ancient Egyptians mainly used sideways projection to highlight the profile features of human figures. As shown in Fig. 11.1, the ancient Chinese used sideways projection for animations and ridership to highlight the profile features of animals. However, for human figures, they used frontal projection to show the head, torso, arms, and legs. Simplification was often used in rock art portrayals, ignoring the details inside of objects.

*Exaggeration* emphasizes significant details on projected shapes. For example, from a typical sideways projection, animals standing upright will not show all four legs because the legs furthest away from the viewer are usually obscured. However, early petroglyphs often showed all four legs despite the line of sight. In addition, petroglyphs often exaggerated the size of spears and genitals, and elongated the height of warriors. Exaggeration is a positive feedback process that enhances representational features. Cartoons are a good example of exaggeration. Even though the features and proportions in a cartoon are exaggerated, people can usually still recognize the figure. Neuroscientists have studied the function of exaggeration in our brain circuitry and its applications in computer vision.[2]

However, figurative abstraction itself is not general or stable enough to be its own language. Although petroglyph figures have many similarities, each one has slight variations. For example, on Newspaper Rock Moab, Utah, USA, there are four wheels. Each wheel has different features. Some have two rings and others have only one ring. There are also many different footprints. Some have six toes and others have four or five toes. When we see a photograph, we have a visual experience given to us by the photographer for that particular scene. In order to generalize or distinguish objects in a photograph, we transform our visual experiences into concepts through the process of *simplification*.

Taking a mountain for example, there are many ways to portray a mountain. In Chinese, the character for "mountain" is a form with three vertical lines on top of a horizontal line. These vertical lines not only suggest the peak, but also represent the curvature of the mountaintop because the heights of the three vertical lines are not equal. This graphical representation appears arguably more interesting and realistic than a specific photograph of a mountain (Fig. 11.2).

Figurative abstraction not only provides a bridge between visual thinking and abstract thinking, but it also creates the building blocks of visual composition. In the early days of human civilization, people needed to draw one picture to

[2]Cleese J (2001) The human faces. BBC Program. Retrieved on Feb. 8, 2016. https://en.wikipedia.org/wiki/The_Human_Face

**Fig. 11.2** Evolution of the Chinese character "mountain" in styles of oracle, engrave, zuan and kai

represent one meaning. That's not efficient. Later, people discovered that we could combine basic elements in order to create more sophisticated forms and meanings. This is similar to the function of hidden layers in a neural network, which turn simple linear combinations of input data into a more powerful non-linear model for representing complex and subtle patterns. As a result of this discovery, new pictographic languages emerged.

## Design a Pictographic Language

If pictographs are early versions of written languages, then Chinese language is the only pictographic language that has survived for over 3,600 years. Today, over 955 million people (14.1 % of the world's population in 2007) use Chinese characters. The Chinese language is a media of Han culture as well as an artistic creation, bringing visual, musical, and poetic aesthetics to the rest of the world.

The evolution of Chinese characters followed the advances of writing technologies and materials. There are four major milestones marking its evolution. During the Shan Dynasty (3,600 years ago), ancient people engraved words onto bones (oracles). The characters were pictographic, irregular, and inconsistent in the number of strokes used for each word, given the irregular shapes and sizes of the turtle bones used. Oracle words are highly compressed semantically due to the small surface area of turtle bones. Later, in the Iron and Bronze Ages, people etched words onto metal surfaces ("*jin wen*," or metal text). The characters were still pictographic during this period, but had become more regular in size and were better aligned to vertical, horizontal, or arc orientations. Cast metals provided more space for writing and language development, but were still limited to use only in noble society. Engraved metals appeared on containers, weapons, bells, and seals. For example, the bronze saw of the King's family of Kingdom Wu (530 BC – 527 BC) contained 75 words engraved in two lines. About 2,000 years ago, carving methods were replaced by ink brushes on wood, bamboo, and rice paper. Pictographs gradually transformed to symbolic stroke groups ("*bian pong*") with stylish, long, curvy, thin lines. Each character was contained within a rectangular shape. From there, the calligraphy style "*Xiao Kai*" was born. The invention of woodblock printing technology further standardized characters into stroke groups, sizes, and shapes. The most popular calligraphy style today is "*Kai Shu*," in which each character is constrained within a square shape, and is more symbolic.

In 100 AD, Xu Zhen composed an epic dictionary, "*Shuo Wen Jie Zi*," where he systematically studied the characters in the "*Xiao Kai*" style, and summarized at least four composition principles: (1) pictograph, (2) signage, (3) composition, and (4) pictophonetics. A *pictograph* is a drawing for aiding in memory, or for sharing with others. For example, a mountain is a picture of three peaks with a bottom. *Signage* is the addition of some hint or symbol to a pictograph, creating new meaning to its origin. With the growing information flow, pictographs and signage were not expansive enough for true expression. Thus, ancient people devised more ways to create new words. *Composition* combines two or more stroke parts to create one new word. In the early forms of characters, as seen with ancient Chinese oracles, pictographs were easily mixed up. *Pictophonetics* was a way to add a sign or sound note to an existing character to make a new word with a distinguishing phonic feature. According to the dictionary "*Shu Wen Jia Zi*", which comprises a total of 9,353 characters: there are 364 (4 %) pictographs, 125 (1 %) symbols, 1,167 (12 %) compositions, and 7,697 (80 %) pictophonetic characters. This shows that the majority of Chinese characters have moved from single pictographs and symbols to composed or pictophonetic groups.

## Structural Composition Grammars

Imagine we are going to create a new pictographic language for instinctive computing. We have to design a set of rules or "grammars" for two-dimensional composition with building blocks. Without constraints and structures in place, composition would be a miserable experience. Let's extract the grammars from ancient Chinese characters that might be used to develop future pictographic languages.

Chinese characters are composed in two-dimensions. How did Chinese characters emerge from the early building blocks ("*bian pong*")? In his dictionary "*Shu Wen Jia Zi*," Xu Zhen summarized 540 building blocks that composed 9,353 words. Some of the pictographic building blocks can be stand-alone characters. In this case, they are not building blocks anymore. We shall call this Composite 0.1. Most characters are composed in one of following structures:

- *Vertical* structures, including top-bottom (1.1) and top-middle-bottom (1.2);
- *Horizontal* structures, including left-right (2.1) and left-middle-right (2.2);
- *Enclosed* structures, including full closure (3.1) and semi-closure (3.2), such as left-bottom (3.2.1), left-top (3.2.2), right-top (3.2.3), left-bottom-right (3.2.4), top-left-bottom (3.2.5), and left-top-bottom (3.2.6) enclosures;
- *Hanging* structures in which building blocks are attached to another building block like hanging fruits (4.1–4.4);
- *Pyramid* structures in which three identical building blocks are piled up like a pyramid (5.1) (Fig. 11.3).

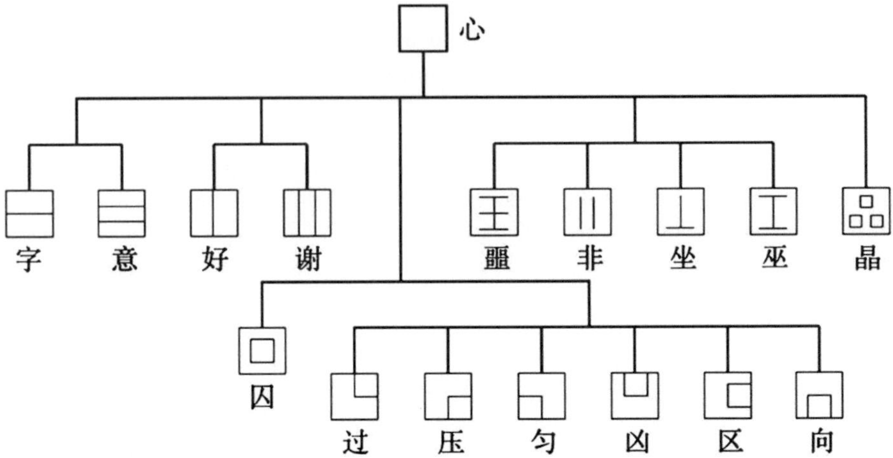


**Fig. 11.3** Typical 2D composition structures of Chinese characters

The pictographic building blocks are network "hubs" that connect words together. Figure 11.4 is a network of word creation from the two building blocks "mouth" and "man." For example, the building block "mouth" is a pictograph that mimics the shape of an open mouth. Adding a dot to the mouth creates a new symbolic word, "sweet." Piling three "mouths" like a pyramid creates a composed word, "taste." Combining "mouth" and "man," creates the pictophonetic character, "*recite*". Figure 11.4 is just a small portion of a larger network, where 540 building blocks create over 9,353 words. Figure 11.4 illustrates the typical combinations. With composition grammar, not only can we analyze existing characters, but we can also generate new ones. This grammar can also be generalized to compose other pictographic languages or logograms.

Composition grammar enables people or computers to generate new characters, and even new languages. Take, for example, the music notation system for the 3,000 year-old Chinese stringed instrument *guqin*. Each character in this notation contains left and right hand positions and actions. The character in Fig. 11.5 has five parts: the upper left part indicates the finger name, the upper right part indicates the bar position, the middle part indicates the first striking action, below that is the string number, and the bottom part indicates the second striking action. The character shown below is saying that the left thumb is at the 9th position and the right index finger swipes down at the 4th string, and then swipes up. *Here, one Chinese character equals 22 words in English.* However, this notation does not include accurate rhythmic timing like the five-line notation system. The tempo is up to the musician's own interpretation. It would be an interesting and challenging project to automatically translate these ancient notations to modern five-line notations.

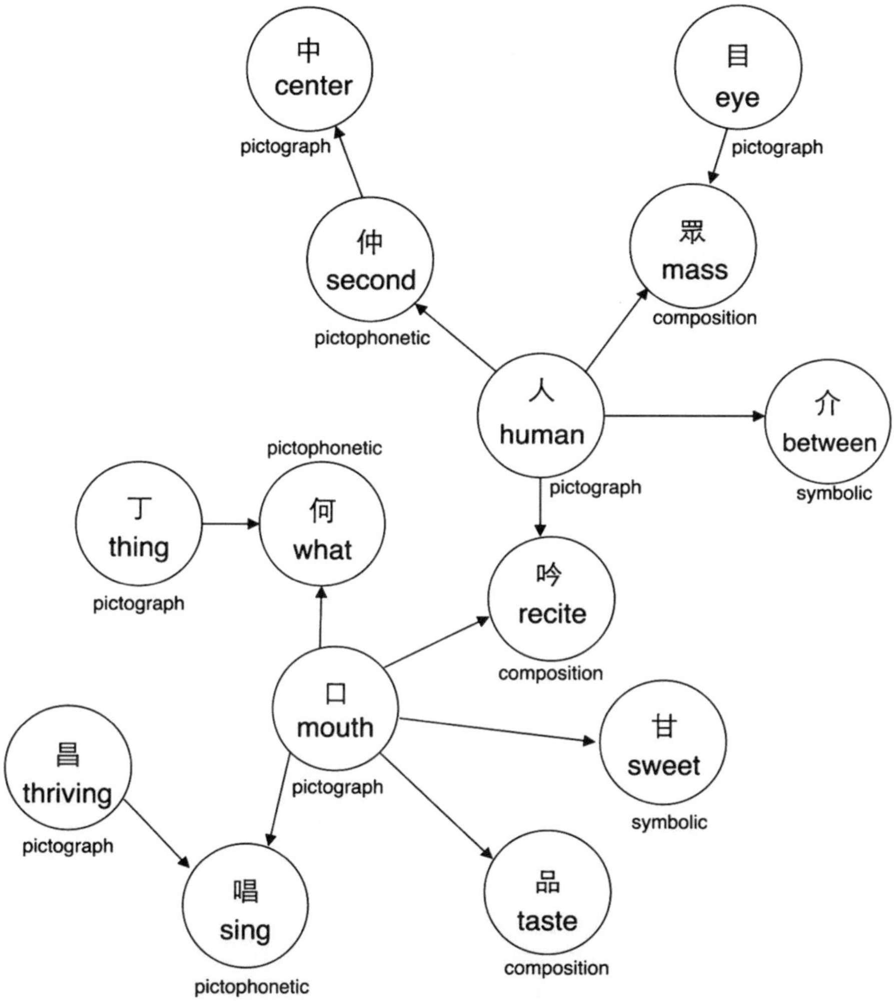


**Fig. 11.4** The network of stroke blocks "mouth" and "human" and their related words

## Encoding Pictographs

Entering alphabetical English text into a computer is a trivial task using a standard keyboard with only 26 letters. The computer keyboard is simply a descendant of traditional typewriters. However, in order to enter pictorial Chinese characters onto a computer, we must encode over 9,353 characters, or 540 building blocks, which is a non-trivial task. One early solution was to connect a traditional Chinese typewriter machine that had as many as 9,353 keys in a two-dimensional grid. This is not very

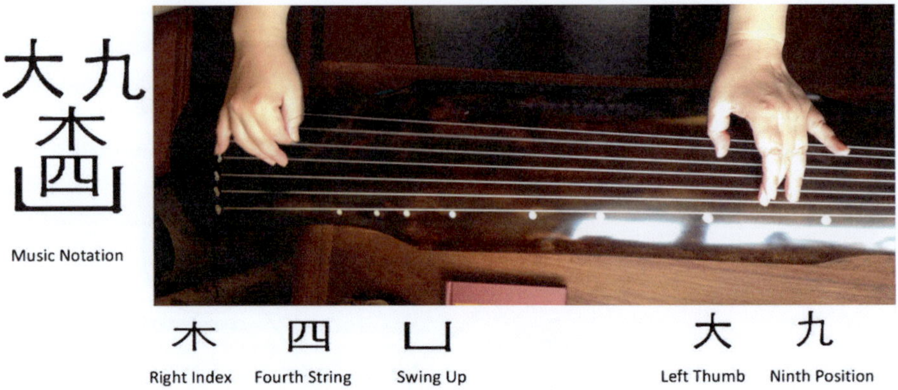


**Fig. 11.5** The ancient Chinese music notation character and the corresponding performance gesture

economical, and was quite difficult to operate because the user needed to remember and retrieve the coordinates of all the characters.

It is more desirable to enter Chinese characters using a standard computer keyboard in the smaller keyset (ASCII). There have been at least 500 encoding algorithms developed for typing Chinese characters. How do we evaluate these encoding methods and assess their efficacy? The criteria include average number of keys needed for entering a character, error rate, and mental workload. Three encoding methods have been widely used. The first method uses phonetic spelling based on "ping ying." In modern Chinese, each character has its own phonetic spelling using 3–4 alphabetic letters on average. However, not every phonetic spelling is unique. There are many words in Chinese that have the same phonetic spelling but different pictographs and meanings. As we discussed before, over 7,697 characters, or 80 % of all characters, are composed using the pictophonetic method. Therefore, there are many pictographs with the same pronunciations. Artificial intelligence pioneer Herbert A. Simon studied this phenomenon in the character recall task. He showed groups of pictographs to his subjects. Some groups had identical pronunciations; others did not. Simon found that subjects made more mistakes in the groups that had the same pronunciations than those which had different pronunciations. Simon concluded that humans encode languages in auditory forms instinctively, even though the language is pictorial! The dilemma of encoding Chinese is that the phonetic spelling and pictographic form are two separate systems in general. In order to pinpoint a particular character with a phonetic spelling, the user must add a number to select the character from a set of candidates. In many cases, the screen displays the set of candidates with index numbers for interactive text input. In reality, not everyone can spell the word correctly due to the wide variety of local dialects in China.

The second encoding method is the so-called "five strokes" method. It is based on the order of writing calligraphic strokes, such as from left to right, from top to

**Fig. 11.6** Handwriting input for Chinese characters

bottom, and so on. It is assumed that the first five strokes can be used to identify a character uniquely. The user enters the first five strokes of the character and the computer returns the exact character. This method is fast for skilled users. However, it is rather difficult for novices to remember the rules for mapping the keyboard to the strokes.

The most widely used method is the handwriting input, using gestures. Early versions of the handwriting input device involved using a stylus on a drawing pad or screen. It was not very convenient because it is much too easy to lose a stylus. The touch screen on smart phones of today revolutionized the Chinese input method. Using a finger, people can naturally "draw" a character on the screen interactively. The processor displays the candidate characters as the user inputs. Finally, the industry developed an intuitive and efficient solution for Chinese encoding for everyday use. The gesture-based input takes advantage of the sequence of strokes so that the computer may find the match even before the character is drawn completely (Fig. 11.6).

The basic anatomy of a gesture consists of multiple gesture stroke objects. Each object is made up of gesture points, including $x$ and $y$ spatial coordinates, as well as a single timestamp indicating when the point was generated. When a gesture is stored in a gesture library, it is keyed with a name. Using the order of points as the timestamp for each gesture, we can acquire an equal length vector in the form of $\mathrm{f} = (x_1, y_1, t_1, x_2, y_2, t_2, \ldots, x_\mathrm{n}, y_\mathrm{n}, t_\mathrm{n})$. For $N = 16$, it allows a 48-element vector for each gesture, which is enough resolution for feature extraction and classification. We can apply Dynamic Time Warping to match characters. Assume we have another gesture sequence with a feature vector $h_k = [x_1, y_1, z_1, \ldots, x_n, y_n, z_n]$. To compare the two gesture sequences, we want to find the minimal warping path cost in the Dynamic Time Warping function. In many practical cases, we only need to use a portion of the stroke positions. For example, we may just use the first three to five strokes and ignore the rest of the points. In some cases, we may assign weights to the points. This can be implemented in the weighted warping cost function.

Encoding Chinese characters requires twice as many resources as encoding English letters because only 1 byte of digits is needed to represent the full ASCII set versus 2 bytes needed to represent Chinese characters. There have been debates from time to time about eliminating the use of pictographic characters and commit to only using alphabetic letters. Fortunately, computer input, storage, processing, and display resources have become abundant. The advancement of technology and intuitive composition of characters enables pictographic languages like Chinese to survive in the digital age. Rather than disappear Chinese and other pictographic languages have actually been thriving in social media.

## New Age Pictographs

Rapidly growing social media in China such as Webo (micro blogger) and WeChat created platforms for new pictographs to emerge. New characters were created and spread across cyberspace, much faster than in the time before the Internet. Figure 11.7 shows two new characters that are popular in social media: an ancient horse, "Zhao Ni Mai" and an embarrassed face, "jun."

**Fig. 11.7** New Chinese characters spread over social media

The classic dictionary "Shuō Wén Jiě Zì" summarized four methods for building Chinese characters. It also includes two methods for giving new meaning to existing characters: (1) borrowing a character based on its pronunciation and (2) deriving a new meaning from a character based on analogy. In social media, users sometimes innovatively borrow the characters for their pronunciations to avoid the censorship. For example, in Chinese, a cat is pronounced identically to the sound "mao" as in the former political leader, Mao Zedong. Massive amounts of social media users have given the ancient language of Chinese some new twists.

ASCII art is another pictographic expression method based on existing alphabetic keys. On September 19, 1982, Computer Science Professor Scott Fahlman from Carnegie Mellon University posted a message to an online bulletin board that was popular among Carnegie Mellon's computer science community in the 1980s. Faculty, staff and students used these spaces as social media, discussing everything from serious issues to lost-and-found items. Given the nature of the community, many posts were humorous. But sometimes, a few readers would fail to understand a joke and they would overreact with an angry and lengthy response. This would stir up more people with even more responses, and soon the original thread of the discussion was all but forgotten. Fahlman thought it would be a good idea to give humorous posts a joke marker using whatever was available on the standard computer keyboard. He improvised a pictograph :-) representing a smiley face with a suggestion to readers to look at the symbol sideways. The pictograph is simple to type and interesting to look at. It caught on quickly around Carnegie Mellon and soon spread to other universities and research labs all over the Internet. Within a few months, dozens of smiley face variations showed up, including the face with glasses 8-) and the wink;-). Some hobbyists have gone so far to create celebrity emoticons ranging from Homer Simpson (_8(|) to John Lennon //O-O\\.[3] Other popular emoticons include:

```
:-P          tongue sticking out
:-(          sad face
:-D          laughing
X-)          mad
:-@          screaming
:-/          skeptical
*\o/*        cheerleader
o||||||o     Jeep
```

Most ASCII icons are oriented either head-turn-left or straight. Since they are text strings, it is possible to do Google searches of graphical lexicons. ASCII artwork has often been embedded in posts in social media, for example, a poster's heart pulses to illustrate the stress at work.

```
___/\___/\___    starting work
_/\_/\_/\_/\_    before deadline
_____________    after deadline
```

## *Emoji*

With rapidly-growing numbers of mobile phone users, graphical icons have become an intrinsic part of messages. An *emoji* is a small digital image or icon used to express an idea or emotion in digital communication. The word emoji is borrowed from the Japanese words *e* (picture) and *moji* (letter or character). It is equivalent to the English term emoticon, which comes from *emotion* and *icon*. There is a slight difference between emoji and emoticon. *Emoji* usually refers to a small image and *emoticon* often refers to icons composed from multiple characters. However, these

[3]William A (2007) (—: Just between you and me;-) The New York Times, July 29, 2007. http://www.nytimes.com/2007/07/29/fashion/29emoticon.html?_r=2&oref=slogin&

two forms have been getting closer and more interchangeable. For example, users can input an emoji image with a few letters from a keyboard or a trace pad. The machine automatically converts the letters into the related image. When the text box cannot display an emoji, the machine can convert the emoji image to emoticon characters. Emoji emerged in the late 1990s and the use of the word *emoji* in the English language has increased significantly since then. According to the Oxford Dictionary Corpus, although it has been found in English since 1997, its usage more than tripled from 2014 to 2015. Furthermore, the emoji called "face with tears of joy" 😂 was chosen as the word of year 2015, because it made up 20 % of all emoji used in the UK and 17 % of those used in the US.[4]

As with any language, one can start from basic emoticons and gradually move on to more sophisticated ones, even cryptic or symbolic emoticons. It is amazing that there are so many dialects of emoticons across different cultures. For example, *kaomoji* is a popular Japanese emotion style made up of Japanese characters and grammatical punctuations.[5] They are used to express emotion in texting and cyber communication. Kaomoji is a combination of the words "kao," meaning face and "moji," meaning character. Unlike Western emoticons that focus on the mouth, kaomoji focus on the eyes. A popular example is the smiling face: ^__^.

Most readers do not need to turn their head sideways to read these characters. Like Chinese characters, Japanese characters require double-byte sets for encoding and decoding. Therefore, they can potentially produce many more emoticons than Latin and Cyrillic alphabets. Kaomoji also extends individual emoticons to complex actions, objects, and even whole stories.

Social media platforms also have their own emoticons. For example, Facebook has its iconic thumbs-up emoticon to signify "likes." When you type the letter (Y), Facebook automatically converts it into the thumbs-up emoticon. Emoticons have also evolved into high-quality icons, animated cartoons, and even looping video clips, or GIFs. These advanced emoticons are available on most smart phones or in online libraries and stores.

Emoticons have in fact become a pictographic language. In 2014, the 32-year-old actress Kirsten Dunst was one of many female celebrities who discovered that a hacker had posted nude photos of her online. Naturally, she was very angry about this violation of her privacy. Dunst blamed Apple's iCloud because she had put her photos there. In a tweet, Dunst employed sarcasm when she called out Apple. She combined the pizza and poop emoji together. "Pizza" sounds like "piece of." "Poop" refers to "shit". The translation is, "a piece of shit." The phrase of emoji appears to be more elegant than plain text (Fig. 11.8).

---

[4]Oxford Dictionaries. Oxford dictionaries word of the year 2015 is . . . , retrieved on Feb. 8, 2016. http://blog.oxforddictionaries.com/2015/11/word-of-the-year-2015-emoji/

[5]Kaomoji. Japanese Emoticons. http://kaomoji.ru/en/

**Fig. 11.8** Actress Kirsten Dunst's tweet on September 1, 2014

## *Visual Passcode*

A striking characteristic of human memory is that pictures are remembered better than words. From an evolutionary point of view, this is not surprising, since the ability to remember visual environments is vital to survival. People can remember more than 2,000 pictures with at least 90 % accuracy in recognition tests over a period of several days.[6] Some autistic artists have photographic memories and can recreate vivid details of New York City after a brief helicopter tour. This phenomenon is confirmed by examining the neural correlations of memory for pictures and words in the context of episodic memory encoding to determine material-specific differences in brain activity patterns. Neuroscientists used a positron emission tomography (PET) scanner to map human brain region activity during the encoding of words and pictures of objects.[7] Researchers discovered that the encoding of pictures resulted in greater activity of bilateral visual and medial temporal cortices, compared to the encoding of words. The encoding of words was associated with increased activity in the prefrontal and temporoparietal regions related to language function. Visual memory has more effective and automatic engagement of certain areas of the brain, such as the medial temporal cortex, especially when encoding faces.

If we can remember emoji, then why don't we use them to log into a bank account or a mobile phone? Intelligent Environments, a UK technical company, found that over a third of British people have forgotten their PINs.[8] In 2015, the company launched the world's first emoji-only passcode for mobile phones. Users could compose a four emoji PIN out of 44 emoji. A traditional four digit PIN (personal identification number) has 7,290 unique permutations of non-repeating numbers. An emoji passcode can be selected from 44 emoji images, which has 3,498,308 unique permutations of non-repeating emojis. The emoji passcode has

[6]Standing L, Conezio J and Haber RN (1970) Perception and memory for pictures: single-trial learning of 2500 visual stimuli. *Psychonomic Science*. **19** (2), 73–74.

[7]Grady CL, Mcintosh AR, Rajah MN, Craik FIM (1998), Neural correlates of the episodic encoding of pictures and words. Proceedings of the National Academy of Sciences, Vol. 95, pp. 2703–2708, March 1998. http://www.pnas.org/content/95/5/2703.full.pdf

[8]Intelligent Environment (2015). Now you can log into your bank using emoji. http://www.intelligentenvironments.com/info-centre/press-releases/now-you-can-log-into-your-bank-using-emoji-1

480 times more permutations than the traditional four-digit passcode. The sequence of the emoji allows the user to compose a story in order to trigger so-called *episodic memory*. Most people prefer to remember a story rather than random numbers or words. An emoji passcode could summarize the daily routine of the user: a banana for breakfast, bike to work, pick up the kid, and have a beer in the evening. Emoji passcodes may also prevent hackers from exploiting common numerical passcodes such as birthdays. However, an Emoji passcode is not totally fool-proof. Emoji passcodes such as bed in the morning, car to work, briefcase in work, and clock at work are popular ones that are vulnerable. Chinese characters are two-dimensional combinations of building blocks, compared to one-dimensional emoji passcodes combinations. If we compose emoji icons like Chinese characters in two-dimensional space, it would increase difficulties for hackers to decode the passcode.

## *Emoji Programming Language*

Perhaps the most exciting pictographic language in cyberspace by far is the emoji programming language Emojicode. It is an object-oriented programming language that uses emoji variables yet follows conventional programming formats such as static, strong types.[9] Below is an example of Emojicode for greeting with a string "Hello, pictorial world!" The annotations are on the right hand side of each line.

```
🐇🐼🍇                              define class 🐼
🐇🐖🏁➡️🚂🍇                         define method 🏁 that returns 🚂
  😀🔤Hello, pictorial world!🔤      greeting with a string 🔤...🔤
  🍎0                               the integer 🍎 returns 0
🍉                                  end the method
🍉                                  the end
```

Contemporary programming languages such as Swift have started to support emoji variables, but Emojicode still is an amusing undertaking that is aimed for a new pictorial language. It is only in its early stage. The final goals of the project are to develop a working compiler for Android, iOS and Windows phones, along with a custom keyboard for actually composing the emoji-packed code. This is certainly a potential language for children to learn programming on a phone or desktop. For professional programmers, such as engineers or scientists, those watermelons, apples and pandas may be too distractive to their tasks, but there is a possibility to customize the emoji variables with emoticons of resistors, pumps, and reactors.

[9]Emojicode web site: http://www.emojicode.org

## Do Pictographs Fit Zipf's Law?

The linguistics professor George Zipf hypothesized that there is a universal principle in natural languages, physical, biological, and behavioral systems.[10] Zipf discovered that certain types of events are quite frequent, whereas other events are rare. For example, in English, short words, such as "the" and "a" are used frequently, but longer words, such as "mathematically" are used very rarely. If we compare a word's frequency of occurrence to its statistical rank, we see an inverse logarithmic relationship. The most frequent word occurs twice as many times as the 2nd most frequent word, three times as many as the 3rd most frequent word, and so on, creating a long tail distribution curve. Zipf's Law can be expressed as:

$$P(f) = f^{-n} \tag{11.1}$$

where P(f) denotes the probability of a word or event of rank *f*, and *n* is close to 1. Zipf's Law is the same as a Power Law. For decades, linguistics researchers have studied Zipf's Law for many languages including words in contemporary English and Chinese. For example, Zipf studied the number-frequency distributions of German, Chinese, and Gothic.[1] He used phonetic systems for comparison. However, Zipf's studies and those like it were mainly based on phonetic text and did not include visual components.

Here, let us test if Zipf's Law can be applied to figurative elements in the Neolithic rock art found in the Helan Shan Mountains. We have over 697 panels of rock art from a catalog of 884 figures. We found that the top seven figures have the probability of over 85 % of occurrences:

1. animal (235)
2. goat (163)
3. face (144)
4. symbol (66)
5. human (61)
6. pattern (52)
7. rider (35).

Figure 11.9 is the plot of the ranks of the figures and their occurrences in log-log scale. It shows the very few top ranking figures have the most occurrences, whereas the lower ranking figures are relatively rare. Using linear regression, we have, $Y = 2.91 - 1.91X$, where, we have the slope $= -1.91$, the correlation coefficient $r = -0.972$, the standard error $s = 0.17$ and the Residual Sum of Square $rss = 1.178$.

The slope of ancient rock art is steeper than the desirable Zipf slope ($-1$). The reasons behind this could be the limited figurative samples (697) and types (40) and plenty of errors in classification. For example, goat, deer, horse, and tiger all

[10] Zipf GK (1949) Human behavior and the principle of least effort, Addison-Wesley Press.

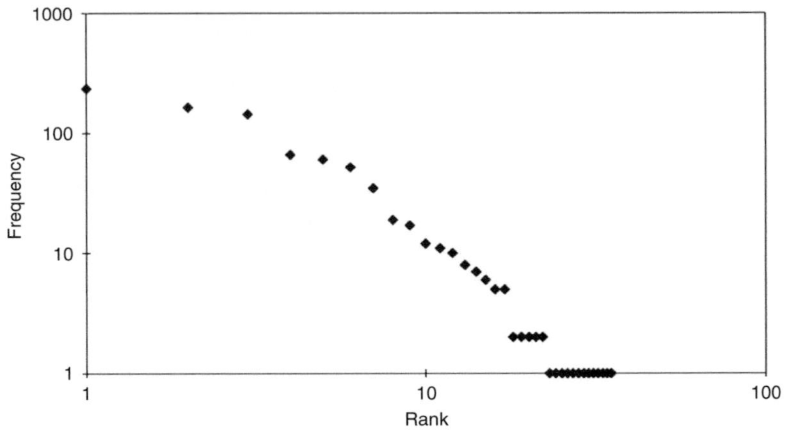


**Fig. 11.9** The ranks of figurative patterns and the frequencies of Helan Shan rock art

belong in the animal category. However, in the classification used here, "animal" is a separate type. Furthermore, some figures in the rock art interacted with each other. It is difficult to determine how to treat them as individual figures or "phrases." The low-ranked figures are rare and irregular, also resulting in errors. A steep slope also indicates the "affordance" in making rock art. Carving into rock is non-trivial work. Perhaps ancient people preferred to carve a few figures containing the most possible meaning.

Now, let us look into the building blocks of pictographic Chinese characters. We found that not all Chinese building blocks are equal. The top ten building blocks can form 4,146 words, while the ten least-used stroke groups only contribute to 59 words. This indicates that less than 2 % of the building blocks connect over 44 % of the total 9,353 words. If we banned those building blocks from use, almost half of all Chinese words would disappear!

1. water (599)
2. wood (539)
3. vegetable (534)
4. mouth (511)
5. hand (444)
6. human (362)
7. metal (336)
8. fiber (284)
9. earth (283)
10. animal (254)

Figure 11.10 shows the logarithmic distribution of the ranks of the building blocks and their frequencies in 9,353 Chinese characters.

Applying linear regression, we have the equation: $Y = 3.58 - 1.11X$, where the slope $= -1.11$, the correlation coefficient $r = -0.964$, the standard error $s = 0.126$

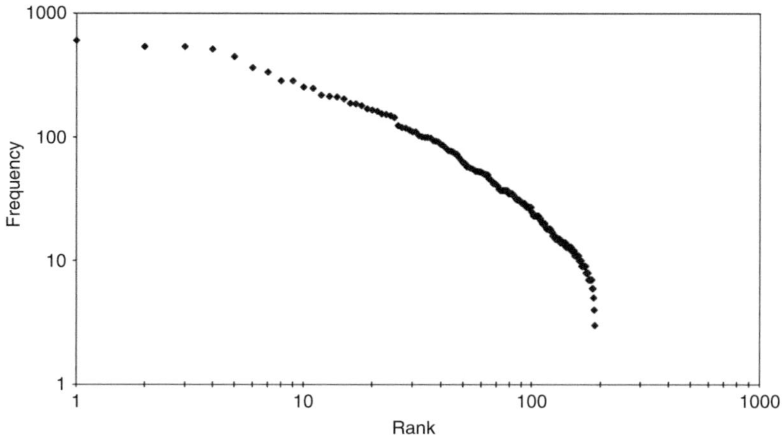


**Fig. 11.10** The rank building blocks and frequencies of their associated Chinese characters

and the Residual Sum of Square rss = 2.97. As you can see, the slope here is close to −1. This is owed to a large sample size (9,353), and systematic research and design of the character system for the past 3,000 years. The ranks and frequencies are balanced according to the least effort design principle. For example, water is vital to our lives. The building block "water" has the highest frequency, following by the building block "wood" as the second, "vegetable" the third, and so on.

Assume ancient people randomly selected building blocks to create characters. The connections for each building block would have been statistically uniform. However, people have inevitably incorporated preferences (culture) into the word creation process. In the 540 building blocks, over 236 are about humans or body parts, 106 are about artifacts, 60 are about animals, 53 are about heaven and earth, and 52 are about plants. People used anthropomorphic building blocks to describe everything in the world, including animals' eyes, noses, and toes. For example, there is no special word to describe a tiger's teeth. Purposefully selecting "nearby" building blocks reflects the "least effort" principle in word construction in Chinese characters. As a result, people have created a "scale-free" network in which a small number of "hubs" have a large number of connections to *instinctual words*, similar to the structure of today's Internet.

In Paleolithic rock art, the walls of a mountain or cave were perhaps the only communication spaces in the region. The shared "documents" were hundreds with over 40 types of figures, symbols, and patterns. Pictographic languages, such as Chinese have evolved to 9,353 characters with 540 building blocks. The volume of shared "documents" has increased significantly, ranging from turtle bone oracles and metal carvings, to rice paper media. As we discussed above, in pictographic languages, a small number of words plays a significant role. The most frequently used words are instinctual, or directly connected to our lives, such as animal, water, and wood.

## Instinctive Words in Cyber Space

Can we find equivalent instinctual words in the digital world? For over ten years, I have been trying to search for "instinctual words" in cyberspace. I started by Googling 31 commonly used words such as home, food, learning, security, sex, water, fun, etc. Then, I gradually expanded the words with 100 BBC frequently used nouns in 2015. At the beginning, I thought this was only a static study for 2003. However, the experiment continued until 2015. The changes in rank of the top words appeared as dynamic trends in cyberspace.

According to Zipf's Law, those with the highest number of hits on Google have a significantly larger gap between them than those with the fewest number of hits on Google. An interesting observation is that the top 20 % ranked words share roughly 80 % of the total hits for all words. Figure 11.11 shows a log-log plot that summarizes Helan Shan rock art, Chinese characters, and Google search results in 2003, 2009, 2013, and 2015. See Fig. 11.11 and Table 11.1.

We first found that the slopes of the linear regression lines became flatter over time. This means that the language contents became broader, and more specifically-descriptive words emerged. The low-frequency words do not fit the linear regression model very well partially due to the limited sample size.

We then look to the top five high-frequency words over different periods (see Table 11.2). We discovered interesting shifts in those "instinctual" words. For example, in the Neolithic period, people in Inner Mongolia were hunters.

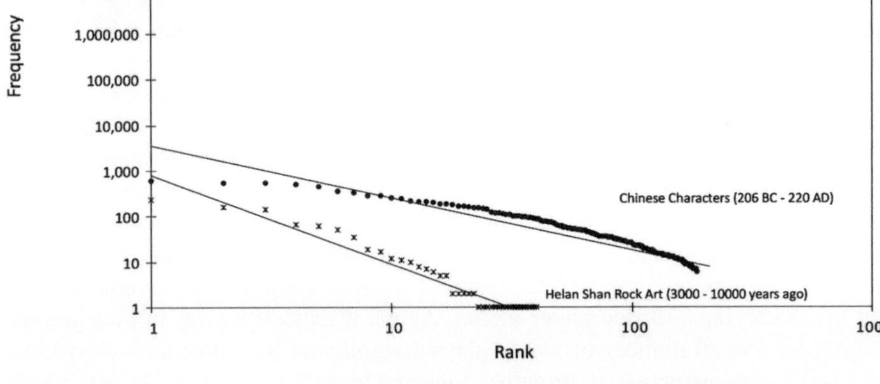


**Fig. 11.11** Helan Shan Rock Art, Chinese characters, and Google's frequently searched words all fit Zip's Law in terms of the logarithmic relationship between frequencies and ranks, but with different slopes

**Table 11.1** Linear regressions of rank-frequency samples from different periods

| Year | Linear regression | Slope | Correlation R | S | RSS |
|---|---|---|---|---|---|
| Neolithic | Y = 2.91 – 1.91X | −1.91 | −0.97 | 0.17 | 1.18 |
| 100 AD | Y = 3.57 – 1.13X | −1.13 | −0.94 | 0.13 | 2.97 |
| 2003 | Y = 9.32 – 1.34X | −1.34 | −0.94 | 0.17 | 0.61 |
| 2009 | Y = 10.0 – 1.09X | −1.09 | −0.85 | 0.25 | 2.04 |
| 2013 | Y = 10.5 – 0.95X | −0.95 | −0.88 | 0.18 | 0.88 |
| 2015 | Y = 10.7 – 0.82X | −0.82 | −0.91 | 0.15 | 1.58 |

**Table 11.2** Top 5 words in different periods

| Rank | Neolithic | 100 AD | 2003 | 2009 | 2013 | 2015 |
|---|---|---|---|---|---|---|
| 1 | Animal | Water | Home | Home | Home | **Photo** |
| 2 | Goat | Wood | Tool | News | **Music** | Street |
| 3 | Face | Vegetable | Food | People | Love | Home |
| 4 | Symbol | Mouth | Work | History | Company | Time |
| 5 | Human | Hand | Technology | **Music** | Art | News |

"Animals," or "goats" were essential food. "Face", "symbol," and "human" indicate the emergence of figurative abstract thinking. Chinese characters, on the other hand, reflect a transformation from hunting life to farming life, when "water," "wood," and "vegetable" were more essential to sustaining life, and tool-related words emerged such as "mouth" and "hand."

There are noticeable changes among the Google results of high-frequency words from 2003 to 2015. The results show that while some individual words had a significant increase or decrease in the number of Google hits, the relationship between the numbers of hits remained a logarithmic curve.[11] The differences lie mainly in the ranking of the words. From 2003 to 2009, the word "music," an instinctual form of communication, went from a ranking of 29th to a ranking of 5th – a considerable change. The increase in websites relating to music can be explained through a changing perception of music in our culture, perhaps connected to the release of the iPod. With further analysis, it is found that iPod sales peaked in 2008–2009.[12] Similar reasons for a change in position on the list can be found for each word. In some cases, the concept of a word may not have changed, but the concept may be referred to more frequently through different words, leading to a decrease in the number of sites associated with that word. Along the same lines, the word "home" encompasses many concepts, only one of which is a tangible structure. This supports the idea that language is dynamic.

[11] It's worth noting that there are different results using Google Chrome, and Mozilla Firefox, and Safari.

[12] Apple, Inc. Daily mail (2015) iPod Sales Chart 2002–2013 quarterly. Retrieved on Feb. 8, 2016. http://www.dailymail.co.uk/sciencetech/article-2547435/Death-iPod-Sales-Apples-iconic-music-player-fallen-HALF-past-year.html

In the digital world, Google search results show that "home" was the most frequently used word in online documents from 2003 to 2013. It is plausible that the word "home" is not only essential to our lives, but also has many other meanings. For example, homing in robotics means to return to the original coordinates. The home page in graphical user interfaces means the first page, or root page. iPhone and iPad products have a "home" button for returning to the top of the menu.

In 2015, Google's highest-ranked word was "photo" with 21,820,000,000 results, while the common noun that came back with the most Google results in 2015 was "time" with only 12,300,000,000 results. This illustrates a return to visual thinking in cyberspace. As multimedia storage and broadband communication capacities have increased, people are more inclined to use more "instinctual" modes of communication, such as pictures, music, and video. Overall, image content is the most dominant form of communication used on the Internet and smartphones of today. It is a renaissance of the pictographic era. While the capacity for visual media storage and communication have been improving significantly in recent decades, pictographic language input devices are still falling behind due to the limited scalability of tangible media.

A new keyboard concept aims to bring pictographic language back to the digital world. The Sonder keyboard is a set of miniature e-ink screens that can be programmed to display any icon or symbol so that keyboards are no longer bound to one set of buttons with a limited number of letters and numbers.[13] This could essentially transform any Western keyboard into a Chinese keyboard!

## Summary

As a human instinct, figurative thinking includes abstraction methods for projection, simplification, and exaggeration, all of which help us to extract salient visual features. The two-dimensional composition model of pictographic building blocks can be applied to general templates for figurative compositions such as passcodes and programming.

However, not all building blocks are created equal, and only a small number of elements contribute to a large number of occurrences in pictographs. Here, we have proved that pictographic networks follow the principle of least effort – Zipf's Law. This implies that people only use a very small set of pictorial "words" in everyday life. It is nice to have millions of emoji icons, but people only use a handful of icons on daily basis. Future instinctive computing systems may prioritize the emoticons on keyboards, search engines, and figurative programming languages.

The 10-year Google keyword survey reveals that mobile media has created a renaissance of pictographic content. Photos have taken their place as a new instinct in our social media.

---

[13] Sonder keyboard: https://www.sonderdesign.com/

# CHAPTER 12

# Machine Creativity

Creativity is the ability to make new things or think of new ideas. Creativity has been mystified as an act of the divine or genius in art, literature, music and invention. But in Nature, all species must have some kind of serendipity, or so-called "blind creativity," in order to survive.[1] As we discussed before, insects have a default mode where they demonstrate spontaneous alternation behaviors to avoid deadlocks, bias, and blind spots. Living cells maintain a delicate balance between inheritance and mutation in order to evolve; insects explore different foraging paths around a new home; even the insect-like floor cleaning robot Roomba takes different routes whenever it bumps into a wall in order to cover the entire floor. Creativity by nature is an instinctive survival behavior. The difference between a repetitive routine and creative thinking lies in the injection of some randomness. The randomness must be organized to be efficient. Creativity is controlled chaos. In this Chapter, we explore where do creative ideas come from? How do ideas crossover, mutate and develop over time? What are major challenges in machine creativity?

## The Constructionism Model

Constructionism started from the "learning-by-making" theory that people construct mental or physical models in order to understand the world around them. For example, when children begin to learn about counting numbers, they often use the help of fingers and even toes to map the abstract numbers to intuitive models. Seymour Papert was the pioneer of constructionist learning who created a computer programming language Logo for teaching children programming and mathematics.[2]

[1]Mueller ET and Dyer MG (1985) Daydreaming in human and computers. Proceedings of the ninth international joint conference on artificial intelligence, Los Angeles, CA, August 18–24, 1985.

[2]Papert S (1980) Mindstorms: children, computers, and powerful ideas. Basic Books.

© Springer-Verlag London 2016

Y. Cai, *Instinctive Computing*, DOI 10.1007/978-1-4471-7278-9_12

Like learning French or Chinese, Papert believed that it would be more efficient if the student lived in the country and tried to practice the language naturally.

Similarly, constructionism doesn't believe creative ideas come from randomness; instead creative ideas grow from making structures. Those structures are organized in forms. For example, in biology, it is about the form of living organisms and the relationships between their structures; and in linguistics, it is the study of the forms of words. We call the study of multiple forms "morphology," which comes from classic Greek "morphê" and means the study of shape or form. Morphology is about how to generate and organize ideas. Morphology is concerned with the structure and arrangement of parts of a system, and how these conform to create a whole. Such a system can be an object, a biological organism, or a social network. The conceptual forms are the scaffolding for building things that meet the expected requirements. Prototypes are simplified or scaled models and simulations that can be generated rapidly and affordably. Prototypes are observed, tested and interacted with as user experiences, which provide feedback for further prototyping and morphological processing for refinement. The constructionism model adds the prototyping process to the traditional "Generation and Test" model. Prototyping is about construction. It's a bridge between ideas and evaluation. For example, architects usually present a mockup model of a design to their clients. The clients can virtually experience the prototype, performing utilitarian tests and measurements, and determine emotional, aesthetic, and personal responses. Imagine a creative designer has ten thousand ideas in his brain or on paper. With limited recourse, he may just make a hundred physical prototypes. Then he refines the design and makes the final three or five final prototypes for his client to evaluate. With rapidly growing 3D printing and high-fidelity simulation software, virtual prototyping and virtual experience become a reality on computers. The automated morphological generation, prototyping, and experience can form a "closed loop" for generating, prototyping and evaluating millions of alternatives on a computer – a new era of machine creativity. Figure 12.1 is a diagram of the three processes in the constructionism model.

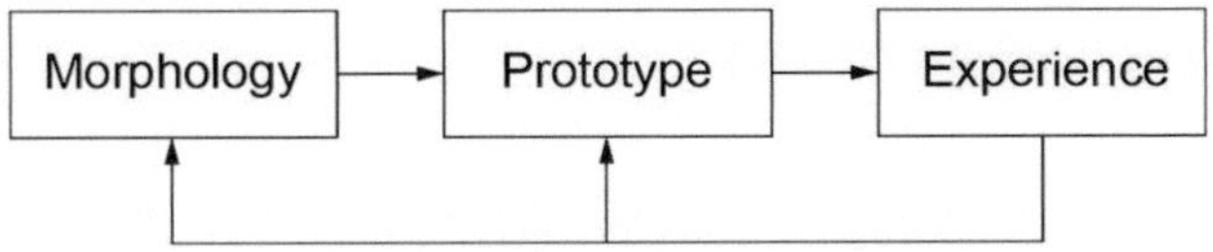


**Fig. 12.1** Constructionism model of the creativity process

## Morphological Generation

Creative ideas often come with a sort of randomness. Understanding this creative randomness factor can be traced back to early pioneers in Artificial Intelligence research in 1956. In their proposal to the Rockefeller Foundation for the famous

Dartmouth workshop,[3] John McCarthy, Marvin Minsky, Nathaniel Rochester and Claude Shannon stated that there are two approaches to making a machine exhibit originality in its solution of problems: one of these is to find how the brain manages to do creative things and copy it, for example, making a daydreaming machine. The other is to take some class of real problems that requires originality in their solution and attempt to find a way to write programs to solve them on a computer, for example, by making a robot that can think and paint autonomously like an artist. In both cases, McCarthy thought it was necessary to introduce randomness to the process, as an extension of the well-known "Monte Carlo" computational method. The method performs many random experiments on a computer to produce a rough guess of the analytical solution when it is not possible to have an exact one due to too many variables or unknown relationships. McCarthy further suggested that the mechanism which controls serial order in behavior guides the random factor so as to improve the efficiency of the imaginative processes over pure randomness.

In 1933, Swiss astronomer Fritz Zwicky invented a morphological analysis method for systematically investigating a total set of combinations. He used the method in examining the Coma galaxy cluster and was the first to infer the existence of unseen matter, what is now called "dark matter". Morphological analysis works through very simple processes, using two common principles of creativity: *decomposition* and *forced association*. Decomposition is the articulation of the components of a system into variables (or attributes) where each variable has a few possible values. Forced association is to layout the variables and their values in a matrix and generate combinatory solutions. If there are two variables, then a matrix is a common two dimensional table. The possible number of combinations is the number of rows × columns. When we increase the number of variables, the possible combinations grow exponentially. Then we have to use a computer to generate the combinatory solutions, or throw one die per variable to randomly select variables to combine. Figure 12.2 shows the morphological matrix of a furniture design as we discussed in Default Mode. As we can see, even a two by two morphological matrix may generate unexpected designs, for example, the table with an upback. It appears unusual. However, there are some uncommon applications such as privacy protection, or a drafter's desk that prevents tools from falling off the table.

The advantage of morphological analysis is its simplicity for combinatory ideas that may be "out of the box". But it often generates too many trivial or impractical solutions. As it doesn't include a prototyping stage, the Morphological Distance is introduced to evaluate the "fitness" of the solutions. The Distance can be defined as the Euclidian Distance of the sum of the morphological differences or the costs. For 3D printing materials, the distance between plastic types PVC and ABS filaments

[3]McCarthy J, Minsky ML, Rochester N, Shannon CE (1955) A proposal for the Dartmouth research project on artificial intelligence, August 31, 1955. http://www-formal.stanford.edu/jmc/history/dartmouth/dartmouth.html

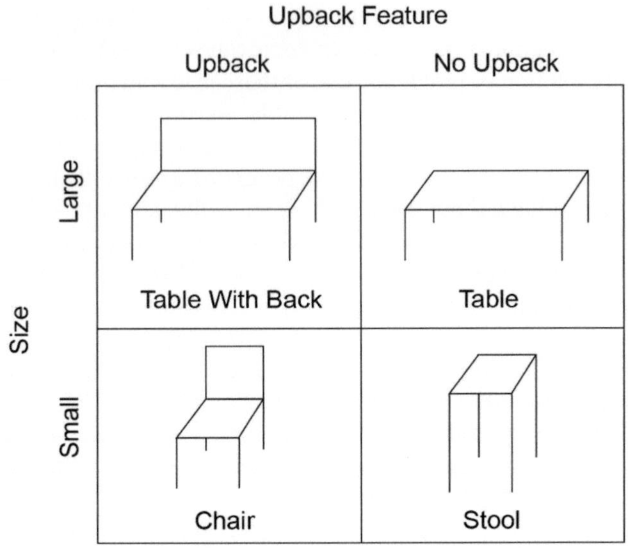


**Fig. 12.2** The morphology matrix of furniture design. Note the table with an upback is an unusual combination

could be 1 unit; and the distance between PVC and copper filaments could be 5 because their thermal properties are quite different. The Morphological Distance serves as a filter to eliminate trivial or non-realistic solutions. On the other hand, it may help to find creative solutions that are obscured by conventional thinking. Zwicky used this method to list the morphological combinations of a galaxy and found out that some unknown masses must exist to make sense of physical measurements. He called such invisible masses “dark matter.”

The Morphology Matrix certainly can be used for art production, so-called “generative art” or “algorithmic art”. The inventor of the robot Aaron is Harold Cohen, a renowned fine art professor at the University of California, San Diego (UCSD). Cohen started this project in 1973 when he visited the Artificial Intelligence Lab in Stanford University. His original inquiry was theoretical: “what are the minimal elements that make an image interesting?” But it didn’t end up with a simple answer. Instead, the project kept evolving for over two decades. Cohen started prototyping Aaron to draw scribbles like a child. The hardware was an X-Y plotting machine with an ink pen. Then he built a mobile robot that could crawl around the canvas on the floor to leave ink scribbles. Compared to prevailing generative art work, Aaron is unique because it does not paint with pixels, but with real ink or paint on an actual canvas. Aaron’s control algorithm is a rule-based expert system rather than a mathematical model such as Fractals or Cellular Automata.

Aaron evolved its painting capacity year after year as its maker taught it more difficult or complex techniques. The prototyping process itself revealed skills that professional artists have taken for granted, from free-hand drawing, to color mixing, to the rotation of 3D objects. Surprisingly, Aaron doesn't have a camera to serve as a visual sensor for feedback. Aaron is actually a "blind artist" who is trained to draw, paint and modify without looking at the objects, nor appreciating what it does, nor reflecting on the model. How far can one go with blind creativity? When Aaron generates a painting, it stores it in a file as a set of instructions for the mechanical robot to follow as it moves the brush on the paper. Aaron has a set of templates of plants, human figures, animals, pots, and other objects. Those templates are parameterized and can be deformed with control points. For example, a frontal face has about 50 control points that enables Aaron to draw a face with a variety of sizes, orientations, shapes and expressions.

Imagine one is a blind artist who has to draw, paint and modify an image from her own memory. It would be tough for both a human artist and a robot. We need a cognitive model, plus a physical model for the blind artist. The first challenge in complex drawing lies in the interactions among the objects, for example, a man in front of a tree. Aaron can remember the exact locations and outlines of the human and trees because they are all from Aaron's repertoire of templates. Aaron has to decide which lines of the tree are obscured by the man in front of it. Otherwise, the man would appear to be transparent.[4] The second challenge is color mixing. Aaron was initially designed as a drawing robot; but Cohen thought if it is intelligent enough, it should be able to paint in color. Mixing colors on the computer screen is an additive mixing of red, green and blue. But mixing colors on the paper is a subtractive mixing process. When we mix red and green paint together, each color filters out some part of the visible spectrum. Some light has been filtered out so much that we would see an almost black color. So the computer has to convert the color space from screen-based additive mixing to the paint-based subtractive mixing for painting in color. Aaron's coloring method is different from today's photorealistic ray tracing and light modeling methods. Aaron tries to find the boundaries of a closed area and "flood" it with a single uniform color like a traditional Japanese painting that contains lines and flat colors within. Aaron's latest challenge was to paint objects in a 3D space, for example, rotating the head from the frontal to a 45-degree position. Instead of building a true 3D model of the head, Cohen developed a 2.5D deformable model to present the 3D effect. In fact, this is more artistic than results that are generated from a realistic 3D model. Aaron simply selects one of 2 associated head templates and deforms it to create the 2.5 effect.

Aaron is not a copy of the artist Cohen. Rather, it is a cognitive model of a general artistic process for everyone who tries to draw, paint and sculpt. Aaron eventually has evolved to be the artistic partner of human artist Cohen. They cooperated together on art projects worldwide. My question here is: who owns the intellectual property? Aaron or Cohen? Amazingly, Cohen still signs Aaron's artwork. See

[4]Cohen H (1995) The further exploits of Aaron, Painter. SEHR, vol. 4, no. 2: Construction of the Mind. July 22, 1995: https://web.stanford.edu/group/SHR/4-2/text/cohen.html

**Fig. 12.3** The drawing made by the robotic artist Aaron, signed by his creator Harold Cohen in 1984. The drawing is exhibited on the wall of the office of the Department of Philosophy, Carnegie Mellon University

Fig. 12.3. Ray Kurzweil, the Artificial Intelligence technologist, implemented Aaron the painter on Windows as a screensaver available to public.[5] So when the computer is idle, the screen draws pictures like someone dreaming.

## Tapping into Dreams

The Spanish surrealist artist Salvador Dalí was a master of capturing subconscious dreams and turning them into surrealist paintings. Like Freud, Dalí considered dreams and imagination as central rather than marginal to human thought.[6] In his famous piece, "The Persistence of Memory" in 1931, he painted the dream-like image of melted watches hanging from a tree, table, and a horse, symbolizing a moment frozen in time. Dalí was intrigued with the vivid mental images that occurred at the boundary between sleeping and waking. He experimented with

[5]http://www.kurzweilai.net

[6]Michalko M (2012) Salvador Dali's creative thinking technique. The Creativity Post. July 17, 2012. Retrieved on March 23, 2016:
http://www.creativitypost.com/create/salvador_dalis_creative_thinking_technique

several ways of capturing those images. He would put a metal plate on the floor and then sit by a chair beside it, holding a key over the plate. When he relaxed and began to fall asleep, the key would slip from his fingers and clang on the metal plate, immediately waking him up to capture the images from his dream.[7] This phase of sleep is the so-called paradoxical sleep (PS) because of physiological similarities to waking states, including rapid eye movements (REM) and awake-like brain waves. REM is a unique phase of mammalian sleep during which the sleeper can dream vividly.

With today's brain-computer interface technologies, it is possible to replace Dalí's keys and metal plate with a phone app. In fact, there is a phone app called "Dream:ON" designed to influence a sleeper's dreams by introducing scripted soundscapes.[8] The app was launched at the Edinburgh International Science Festival in 2012 by Professor Richard J. Wiseman from the University of Hertfordshire, UK and his collaborators. It was developed and maintained by the mobile experience team YUZA based in London. The principle of the app is based on the correlation between REM and dreaming. When the user enters the REM stage of sleep where dreaming is most common, the app delivers unique audio soundscapes that the subconscious is shown to respond to. The app constantly monitors and adjusts itself to optimize the user's dream experience. The user can download soundscape files from the "Dream Store", for example, "Relaxing Rainforest", "Into the City", "Space Shuttle", "A Trip to Tokyo", "Rain through the Trees", "Ocean View" and so on. Instead of an EEG cap on the head, the app uses motion sensors to monitor the user's sleeping states, for example, awake, asleep and deep sleep. The key is to establish the correlation between movements and sleep state. During the night, the user's movement is propagated to the phone that is placed face down at the corner of the bed. The user selected soundscape is played at a carefully chosen time to coincide with the optimum point in their dream cycle. The user's subconscious will hear the soundscape, which influences his or her dream. If the user starts to wake up, the volume is automatically reduced. In the morning, the preset alarm with a specific soundscape will wake up the user. The user can view and record the dream in the Dream Diary and share the dreams with the global "Dream Catcher" experiment. Figure 12.4 shows a sample of the sleep state dynamics.

The Dream:ON project started with Wiseman's dream experiment of 20 subjects in a sleep laboratory. Subjects were shown video clips before going to sleep and exposed to audio soundscapes during their sleeping cycle. The study indicated that the future sight, or precognition of dreams is not true. That means we can't prepare what the dream will be before sleep. However, the experiment suggests

---

[7]Park West Gallery. Freud's influence on Dali's surreal "dream" painting. Blog on: http://www.parkwestgallery.com/freuds-influence-on-dalis-surreal-dream-painting/8948

[8]Dream:ON web site: http://www.dreamonapp.com

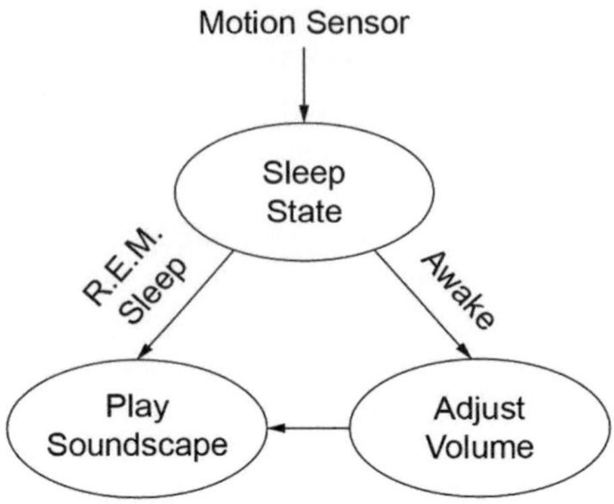


**Fig. 12.4** Dream:ON data flow

that it is possible to influence the dream with sensory stimulation subconsciously.[9] Dream:ON in fact has been a part of the world's largest dream experiment conducted by Wiseman's team. Within two years after the launch, over half a million people have downloaded the app and the team has collected more than 13 million dream reports. According to the experiment results, female users often dreamed about flying, happy, love, family, friends, and so on; male users often dreamed about flying, sex, love, happy, and so on. Flying appears to be a common dream for all. The data also show the influences of soundscapes on the emotional tone of the dream. For example, natural soundscapes triggered positive dreams, but the city soundscape produced more bizarre, disturbing dreams. The researchers concluded that Dream:ON does indeed shape dreams at a certain level. In addition, the data suggest that more bizarre dreams are being recorded on the app when there is a full moon. This correlates with neuroscientists' discovery that people experience more disturbed sleeping patterns around the time of a full moon. Wiseman's team suggests that dream reports might help psychotherapists to diagnose depression symptoms. Sleep scientists have long known that the dreams of those diagnosed with depression are especially long, frequent, and negative. It is possible that the tool can be used as a therapeutic tool to influence patient's dreams.

Just for curiosity, I conducted an experiment with Dream:ON on myself for a week. Through the night, I was not aware of any soundscapes. The app reported my sleep pattern over the nights in a diary. I did recall a couple of lucid dreams. One was about the common fear of failure: I was teaching a class and I tried to find the name of an algorithm in the textbook beginning with "PB" but I just couldn't find it, leaving the class with a long pause. The incident in a dream seems random, but the elements are realistic, such as textbook, teaching, classroom, even the cue

[9]Watt C, Wiseman RJ, Vullaume L (2015) Dream precognition and sensory incorporation: a controlled sleep laboratory study. Journal of Consciousness Studies, Vol. 22, No. 5–6, pp. 172–190, 2015.

for the word. The combination was random and unusual, but the consequence and emotional reaction were reasonable. Dreams act like a rehearsal for future plans by randomly presenting those dramatic scenarios as a warning of potential disasters.

Humans in fact spend much time engaged in thought not directly related to the current task or situation, even in daytime. This is usually called *daydreaming*. According to Sigmund Freud, daydreaming and creativity are often connected.[10] For example, daydreaming expresses our desire for freedom and to explore possibilities that have not yet happened in the real world. Science history also shows that many great inventions were inspired by another problem in a distant field, the connection being made by serendipity. The most important support for creativity perhaps lies in revisions of the dynamic memory, or episodic memory in the brain.[11] Given an episodic memory template such as "running away from a monster", a daydreaming revision could be "running away from a train", for example, if someone just saw a train on television before going to sleep. Each time, as new information is added, the movie-like episodic memory would enable richer, different, or more creative scenarios through analogy.

Erik Mueller developed a computer model called DAYDREAMER that mimics the human daydreaming process and provides an empirical tool to study the subconscious programs embedded inside our mind. He worked on this interesting problem for his dissertation and eventually turned it into a monograph.[12] DAYDREAMER takes as input simple situational descriptions, such as "the car stuck on the highway," and a procedure as output, including actions that it would perform in given situations. The program then daydreams by indexing daydreams, planning strategies, and future plans into memory. As a result, the model can learn from its experience. In addition to the episodic memory of experiences used by constraint-free planning and rehearsal, remarkably, DAYDREAMER includes an emotion component that updates its emotion states arising from goal outcomes, such as scary feelings while being chased by a monster before the daydream is over; a personality model that provides the style for guiding the constraint-free planning; and the domain knowledge of interpersonal relationships and commonsense. Mueller transcribed the daydreams from several daydreaming protocols obtained from subjects via immediate retrospective reports. He summarized four types of control goals that are common in daydreams that direct the course of daydreaming: rationalization, revenge, failure, and preparation. Once a control goal is activated,

[10] Freud S (1962) Creative writers and day-dreaming. In Freud S. The standard edition of the complete psychological works. Vol. IX, London: Hogarth.

[11] Schank RC 1982) Dynamic memory. Cambridge University Press.

[12] Mueller ET (1990) Daydreaming in Humans and Machines: A computer model of the stream of thought. Norwood, NJ.

the constraint-free planner selects and executes a script to achieve that control goal. DAYDREAMER is now open source and published on Erik Mueller's MIT web site.[13]

The daydream research accidentally continued in the Google Research group in 2015. Artificial intelligence researchers trained artificial neural networks to recognize objects in images, such as cats, dogs, cars, trees, and streets. Blessed by superb computing power, the neural networks contain a large number of "hidden layers" that represent imagery features, such as lines, orientations, shapes, colors and so on. The layers are so deep that the method is called "Deep Learning".[14] The computer system was fed with annotated objects in sample images. Because the neurons in the artificial neural network are connected to multiple layers and interact with each other as in social media, the imagery features are remixed and enhanced in strange ways. When engineers tried to visualize the intermediate results inside the deep layers, they found hallucinating images. For example, the lower layers are sensitive to edges and their orientations so that they tend to produce simple strokes and patterns. The higher layers produce more sophisticated images including contours and shadows, as the neural network weights are propagated from lower layers to the higher layers. The research team called the phenomena Deep Dream.[15]

Figure 12.5 shows an original photo of a landscape. Figure 12.6 is the result of the Deep Dream enhancement. Those simple patterns in the photos such as windows, roofs, humans, contours, are warped. The sky turns to a watery surface with spiders floating on it. This is certainly not a very pleasant image to many viewers. However, the discovery might have its scientific value because all normal perception in humans is a hallucination generated by sensory input. For example, the image on our retina is in fact upside down and distorted due to the structure of our optical elements; and when we move our eyes, the vision system actually doesn't receive anything during the movement. Most images are internally generated in the brain, not directly generated from the eyes like a camera. We actually see things that are not there all the time. To learn more about the importance of hallucination, we can read Oliver Sacks' book on hallucination.[16]

The Deep Dream reflects the hallucination experience that seems outside of our world. It is a new species of generative art that mimics neural activities inside our own neural network. Unfortunately, the perceptual experience of hallucination is not always pleasant. For example, some people are instinctively afraid of clustered holes, so-called "trypophobia," and some people are afraid of snake-like shapes.

---

[13] Mueller E website. http://xenia.media.mit.edu/~mueller/

[14] Tyka M (2015) Inceptionism: going deeper into neural networks. Google Research Blog. June 17, 2015. http://googleresearch.blogspot.com/2015/06/inceptionism-going-deeper-into-neural.html

[15] LaFrance A (2015) When robots hallucinate. The Atlantic. Sept. 3, 2015. http://www.theatlantic.com/technology/archive/2015/09/robots-hallucinate-dream/403498/

[16] Sacks O (2013) Hallucination, Vintage Books.

**Fig. 12.5** The original photo of the tower

**Fig. 12.6** The hallucination image created and rendered by Deep Dream Generator (Google Research. Deep Dream Generator: http://deepdreamgenerator.com/dream/6d3d4c41a4)

## Improvisation by Abstraction

Improvisation is to come up with an *ad-hoc* solution to a problem by adapting to a situation, despite the apparent absence of resources. During the Stone Age, ancient people used stone to make tools and rock art. They created improvised drawings on the sandstone about warriors, animals and weapons. For example, Neolithic Alpine people improvised the appearance of spears from straight line cracks on the rock, used the bowl-shaped rock to portray a lake, and used the terrain of a rock when carving a 3D map of a village. In order to improvise, we have to represent concepts at an abstract level. For example, a spear can be abstracted as a straight line with an arrow. A straight line rock crack can be viewed as a spear by just adding an arrow on one end. During the Apollo 13 crisis, the crew needed a $CO_2$ pump urgently. The ground support engineers had to improvise the pump based on what was available on the Lunar module. One problem in its abstract form is here: how to fit a circular tube to a square hole or vice versa. At the abstract level, in fact, even kids can solve the problem. For example, socks and tape were used to make a tube that could connect to any shape at either end.

Improvisation has been a creative thinking method in art, dance, and music, especially Jazz music that is often improvised during performances. American artist Andy Warhol used to say: "Life is like Jazz. It is more enjoyable if it is improvised." Warhol was one of many artists experimenting with new media in New York during the 1960s. In the summer of 1964, Andy Warhol met Billy Klüver, a scientist in Bell Labs. "I have an idea," Warhol said, "Can we make a floating light bulb?" "Sure, let's see what we can come up with," Klüver replied. Intrigued by the idea, Klüver went back to Bell Labs and discussed the problem with his colleagues. They looked into batteries, lights, a material to contain helium, etc. They did calculations on the backs of envelopes and ran simple tests. Finally, they concluded that it could not be done without the bulb being very large, because of the weight of the batteries, lights, etc. Meanwhile, Klüver's colleague found a material made by 3M that was highly impermeable to helium and could be heat-sealed easily. Klüver told Warhol that they could not make the light bulb but showed him the material they had found. When Warhol saw it, he said, "Let's make clouds." He simply folded the material over and heat-sealed it on three sides and filled it with helium. Warhol's "pillows" were born. They were called *Silver Clouds*. However, the silver pillows often stayed on the ceiling when they were installed indoors. Warhol put in a fan to blow the pillows around the room, creating a kinetic atmosphere. He further added invisible thin wires to prevent the pillows from flying outside of the room.[17] The Silver Clouds are indeed a masterpiece of kinetic art, which incorporate motion, audience participation, and the environment. It is an incredible room to bring a group of people into. Each person moves about the space in a different way. Some

[17]Klüver B. The Warhol: Silver Clouds. Andy Warhol Museum. http://www.warhol.org/uploadedFiles/Warhol_Site/Warhol_Responsive/Pages/whatson/exhibitions/EX_20100903_TE_SilverClouds.pdf

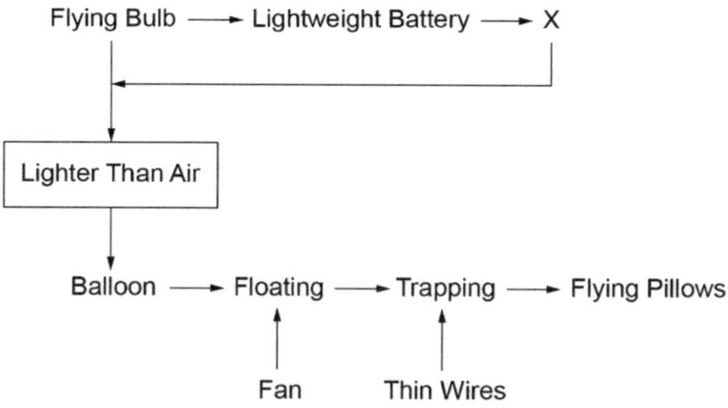


**Fig. 12.7** The mutation and development process of the ideas of Andy Warhol's kinetic art installation "Silver Clouds"

want to touch or come in contact with the clouds immediately, while others move about weaving their bodies around the floating shapes so as not to disturb their trajectories.

Let's trace the flow of ideas during the problem-solving process in Fig. 12.7, where the original goal was to create a flying light bulb. They set out to find the light battery but failed and then they went back to the drawing board. The mutation started when Klüver abstracted the concept "lighter than air" from the original concept "flying light bulb". This actually relaxed the problem constraints because there are many lighter than air things around us, such as helium balloons and soap bubbles. Another attribute of this problem is "shiny" abstracted from the concept "light bulb". There are also many ways to make an object shiny, for example, having a metallic silver coating on the surface, projecting a light onto the surface, or making the surface glow-in-the-dark. Here we can see that abstraction is the key concept in the mutation that leads to improvisational problem-solving.

## Evolving Prototypes

An idea without engineering is just a dream. Everyone has creative ideas but it takes skills and domain knowledge to select and implement the idea. Many experts normally fail many times when working in a narrow field. Prototyping is an evolutionary process that optimizes performance based on selected parameters. Like DNA, a prototype reproduces and re-engineers itself over several generations. In DNA there might be random errors. Such errors contribute to a gradual genetic drift, which seems to occur in all populations. And a few errors would have been favorable to the survival of the species. These would have been chosen by Darwinian natural selection. A Genetic Algorithms (GA) model is a simulation of DNA replication, including inheritance, mutation and crossover. Inheritance is to pass genes to the next generation. Mutation passes its errors to its generation. Crossover exchanges parts of genes between two genetic strands. For example,

Inheritance: 1 2 3 → 1 2 3
Mutation: 1 2 3 → 1 A 3
Crossover: 1 2 3 and A B C → 1 2 A and 3 B C

In order to setup a genetic algorithm, we have to map the design elements to symbolic strings that form a gene pool, and determine the percentage of the genes for mutation and crossover, as well as the fitness function for assessing the results. Then we can run the genetic algorithm to replicate those genetic strings and evaluate the fitness output. When the output meets the requirements, or the improvement of each run is less than a threshold, stop the program. We may call this "blind creativity" in nature. Without it, we probably would have not survived as a species. Genetic Algorithms can be viewed as a metaphor of the natural creative process. Furthermore, GA can be used as an optimization tool for evolving prototypes, especially those complex engineered systems we may call "artificial lives."

Combining art and science, the Dutch artist Theo Jansen created an artificial life called Strandbeests, which means "beach creatures" in English. Strandbeests are wind-powered robotic sculptures made of stiff PVC pipes, water bottles and cloth. Jansan called those plastic pipes "protein" because they are building blocks of the beach creature. Jansen wanted to create a new kind of life from scratch and totally different from existing creatures. He thought this might help humans to know the real form of life. Figure 12.8 is a sketch of one of Strandbeests.

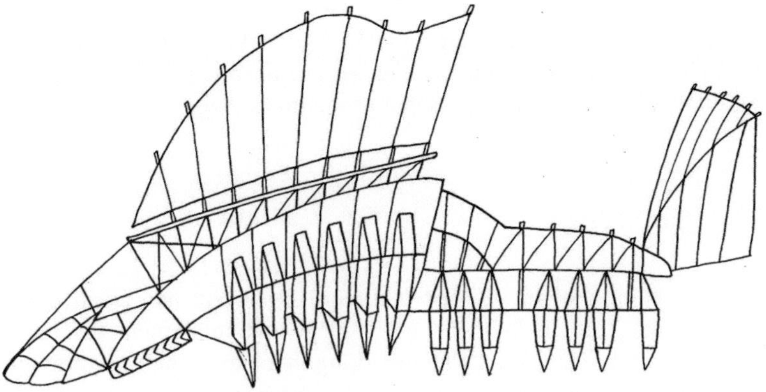

**Fig. 12.8** The artificial life Strandbeest by Dutch artist Theo Jansen

Strandbeest turned wind into energy. The winds sweep the air and store the energy in bottles, which can be used to provide energy for walking when the wind is gone. When the tide comes up, the robot will have the energy to walk away from the tide. If there is too much wind, the robot might be blown away. So the nose is directed into the wind. The hammer will put stakes in the ground.

Strandbeests have taken 20 years to evolve with the help from their inventor Jansen, and computational design tools to optimize the structure. For walking on the beach, legs are better than wheels. In order to keep the robot's hip level, the feet have been configured in a particular geometrical shape to move in the desirable trajectory: they lift up, move forward and loop backward in a horizontal line. There are total of 11 numbers to determine the leg structure. But even for a computer the number of possible ratios between 11 rods (nodes) is astronomical. Assume each rod can have 10 different lengths. Then there are $10^{11}$ possible curves. If the computer evaluated all the possibilities one by one, it would take as long as 100,000 years. Therefore, Jansen opted for the evolutionary method. He used the computer to generate 1500 legs with rods of random length. The computer then generated the walking curves to see which ones were promising. The genetic algorithm selected the best 100 out of 1500 and they were allowed into the next reproduction. Their rods were copied and combined into 1500 new legs that were evaluated by the walking curves to see how similar they were to the ideal curve (the goal function). This went through multiple generations and finally resulted in 11 numbers that produced optimal length rods. Figure 12.9 shows the leg structure and the 11 "holy" numbers for rods.[18] Here, Jansen and the computer masterfully co-designed the sophisticated walking leg with 11 parameters. This proves that computer-assisted creativity can speed up

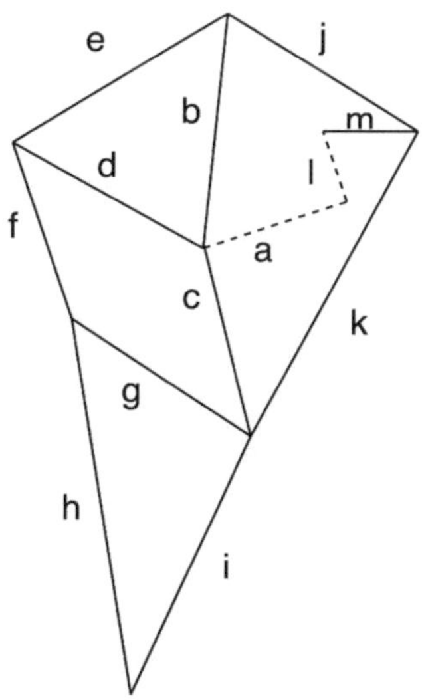


**Fig. 12.9** The optimal solution for the leg configuration. There are 11 "holy" numbers as given by the inventor Theo Jansen: $a = 38$, $b = 41.5$, $c = 39.3$, $d = 40.1$, $e = 55.8$, $f = 39.4$, $g = 36.7$, $h = 65.7$, $i = 49$, $j = 50$, $k = 61.9$, in addition, rotational rods $l = 7.8$ and $m = 15$

[18]Theo Jansen's Strandbeest. http://www.strandbeest.com/beests_leg.php

the evolution of complex prototypes, from 100,000 years or more to a few days or weeks. After an artificial life dies, new lives can be created with similar genes and variations. The evolutionary algorithm can be deployed again and again with new parameters and input values.

Evolutionary algorithms have been widely used in technical innovation, including prototyping and evaluation. Compared to the evolution of artistic prototypes created by Theo Jansen, technical innovations often have more constraints and multi-dimensional fitness functions. The automated computer based evolution method has been successfully applied to the design of an X-band antenna for NASA's Space Technology 5 (ST5) mission. The X-band is a segment of the microwave radio region of the electromagnetic spectrum. Portions of the X-band are assigned exclusively for NASA Deep Space Network (DSN) communication between the satellite and ground DSN facilities. The ST5 mission consists of three miniaturized satellites. Each satellite has two antennas, one centered on the top and the other centered on the bottom of each spacecraft. The ST5 mission antenna is a single arm with wire-like antenna aperture. It is a challenging design problem because the satellite is spinning at roughly 40 rpm and it is important that the antennas have uniform gain patterns in the azimuth, but a single-arm antenna is inherently asymmetric.

According to the study,[19] two evolutionary algorithms to automatically design an X-band antenna were used. The first used a vector of real-valued parameters without branches and the second used a tree-structured generative representation for constructing the antenna with branches. The evolutionary algorithms had to meet the key ST5 antenna requirements, including transmit frequencies, receiving frequency, the ratio between the highest voltage and the lowest voltage in the signal envelope along a transmission line (VSWR), gain pattern, input impedance, diameter, height and weight. The two algorithms use two different fitness functions. After setting up the evolving structures, system requirements and fitness functions, the algorithms start to evolve generations with different configurations. With the parameterized evolutionary algorithm, a population of 50 individuals was maintained, of which 50 % were kept from generation to generation. The mutation rate was 1 %, with a Gaussian mutation standard deviation of 10 % of the value range. Normally, the average fitness score was stagnant for 10 generations, but the best score was stagnant for about 40 generations. After that, no significant improvement would be found.

Finally, one of the evolved antennas, ST5-33.142.7 met the mission requirements and is in use on each of the three ST5 spacecraft, launched on March 22, 2006. They became the first computer-evolved antenna hardware in space. Remarkably, the evolutionary results were not just an incremental variation of existing antenna shapes, and have unusual structures that expert antenna designers would not be likely to produce. Figure 12.10 shows the evolving sequence of antenna structures

[19]Hornby GS, Lohn JD, Linden DS (2011) Computer-automated evolution of an X-band antenna for NASA's space technology 5 mission. Evolutionary Computation. Vol. 19, No. 1, pp. 1–23. 2011, MIT Press.

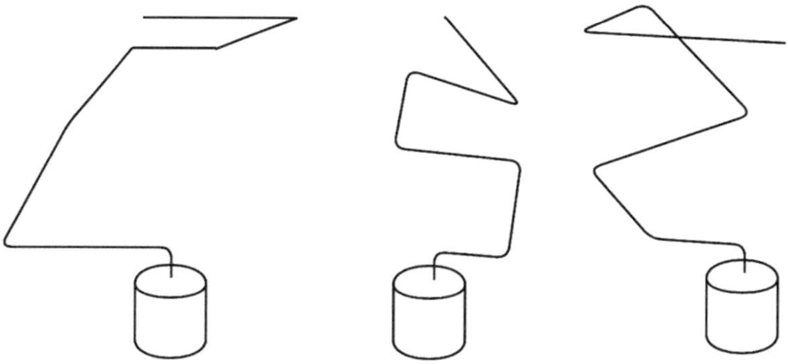


**Fig. 12.10** The evolving sequence of antenna structures that led to the antenna ST5-33.142.7 for active NASA ST5 mission (Redrawn from the original image courtesy of Jason Lohn)

leading to the ST5-33.142.7 antenna. Although the evolutionary algorithms were not consciously producing totally new shapes, the results are in fact creative and elegant. They have different appearance and optimal performance. This indicates that evolutionary algorithms have potential for creative design, but they cannot work alone by themselves. They do need expertise to specify the evolving structures, requirements and fitness functions. In addition, it is important to evaluate the behavior of antenna prototypes. In this case, engineers use electromagnetic models to estimate the output of the antenna such as pattern quality.

## Virtual Experience

American chemist Linus Pauling said that simulation is the third scientific method in addition to observation and experiment. For machine creativity, it is necessary to have an automated method for evaluation that simulates physical reality and human experience. While simulation, modeling and virtual reality have not been viewed as creativity tools alone, they are certainly filters for non-optimal solutions. Normally, it takes a lot of time for humans to evaluate a large quantity of prototypes such as the 1500 leg structures for Strandbeest and millions of configurations for the ST5 antennas. Computerized virtual experience simulators are able to close the loop on automated assessment without human intervention, which speeds up the prototype evolution process. There are two potential benefits: human experts may spend more time on seeding more original prototypes for algorithms to evolve; and the algorithms themselves evolve novel, working prototypes that have *never* been seen before.

Bose is a popular brand for its compact sound systems. The desktop sound system is designed to replicate the acoustic sound effect of a concert hall that is normally generated from large sound systems, all in a compact design. The secret

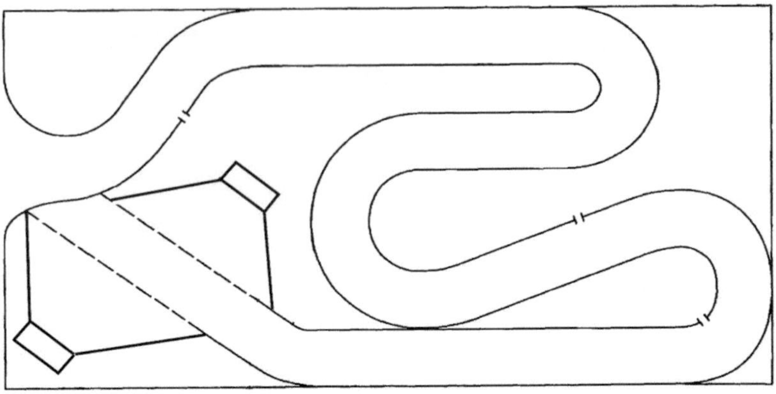

**Fig. 12.11** One of the folded wave guide designs in Bose's sound system, redrawn from the US Patent: US8615097 B2, 2008

inside the Bose systems is a unique 26 inch-long folded acoustic wave guide, which is a series of passages from the speaker driver to the speaker grill. The acoustic wave guide was invented by then MIT professor Amar Bose who devoted his life to studying acoustics and psychoacoustics, the study of the human perception of sound. He eventually debunked most of the prevailing wisdom on high-fidelity sound reproduction by inventing the folded wave guide system. His team did numerous simulations and testing in order to evolve the wave guide to the shape it has today, both aesthetically pleasing and with acoustically high-fidelity.[20] See Fig. 12.11.

In order to simulate the experience of a concert hall, Bose's company developed a software tool, called Auditioner that allows acoustic engineers to hear precisely what the proposed audio system would sound like from any seat in a large venue even before building construction begins.[21] The program has been used to design public address systems in many acoustically challenging places, such as the grand mosque at Mecca which, full of reverberating marble, often failed conventional audio solutions.

## Summary

Creativity is like finding a needle in a haystack. Computers can not only generate a haystack quickly, but also can find the needle with evolutionary algorithms. It is debatable whether a computer can invent things *consciously* in the process, or if it

[20]Parker RP et al (2008) Waveguide electroacoustical transducing. US Patent: US8615097 B2, 2008.

[21]Clynes T (2013) The curious genius of Amar Bose. Popular Science. No. 7, 2013.

really enjoys the process. But studies show that a computer is capable of co-creating new things with human experts, from paintings, to sculptures, to inventing, even though the computer does not do this for pleasure. With a virtual experience function in place, a computer may be motivated to create working prototypes autonomously.

In the proposed constructionist model of creativity, prototyping is the key to bridging ideas and user experience. Throughout the case studies in this chapter, we have seen the footprints of prototypes in all cases: the painting robot Aaron, Warhol's Silver Clouds, ST5 antennas, Strandbeests, and Bose wave guides. Without a prototype, ideas are just dreams. Ironically, even dreams have prototypes. There are a lot of books, even dictionaries about types of dreams. Many dreams follow the same patterns. By modifying the variables or sequences of a dream template, a computer can produce daydreams for rehearsal and future planning.

Morphological abstraction is surprisingly important in the creative process. This is almost counter-intuitive because most people believe that creative people don't think in abstract terms. Morphological abstraction is a set of representational features of a structure or concept. It still has a basic form, not numbers. For example, Aaron uses control points for locating or moving the objects; Jansen used a network of rods to abstract Strandbeests' leg structure; Klüver and Warhol used "lighter than air" to mutate their original "flying bulb" concept. Abstraction facilitates conceptual mutation and improvised problem-solving.

Finally, emerging virtual experience systems allow creative prototypes to evolve and evaluate themselves autonomously. The ultimate challenge here is to enable the computer to learn our human subconscious value systems such as aesthetics, morality, and empathy. That way a machine can evolve from "blind creativity" to "conscious creativity".

# Part III
# Evolving Cybernetics

# CHAPTER 13

# Self-Awareness

According to ethnologists, humans are simply survival machines programmed to preserve and replicate our genes. In order to secure this process, we have developed self-awareness to identify ourselves and to avoid life-threatening situations at multiple levels, from our cells to our existence as organisms. In this Chapter, we explore self-defense systems in nature and cyberspace, the behavioral models of malicious programs, biomorphic defense models, and collective consciousness in the Internet of Things.

## Self-Awareness Model

Self-awareness is a set of basic instincts that appear in genes, organisms, and at social behavioral levels. Across all levels of self-awareness, there are at least four instinctive functions of self-defense: *self-replication*, *self-recognition*, *control and communication*, and *stealth*. Figure 13.1 illustrates the components of the self-awareness behavioral model.

*Self-replication* is the most instinctive behavior found in nature. In his book "The Selfish Gene," British ethnologist Richard Dawkins states that genes spread throughout a population in order to further their own chance of replication for survival. Organisms are built by genes. Therefore, organisms are machines created by genes to increase the chances of the gene being replicated.[1] Genes also send protein messengers that instruct other proteins to build structures in body and the brain. In different organisms, the same gene does not always produce the exact same structure, creating diverse physical differences. There are also errors in the replication process. If genes had been making exact copies, then they would have never changed and complex life forms would never have evolved from simple organisms.

[1] Dawkins R (1976) The selfish gene. Oxford University Press.

© Springer-Verlag London 2016

Y. Cai, *Instinctive Computing*, DOI 10.1007/978-1-4471-7278-9_13

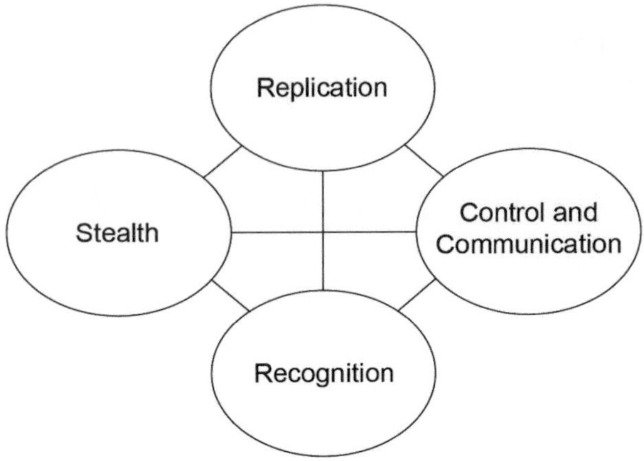


**Fig. 13.1** Self-awareness model: recognition, control and communication, replication, and stealth behaviors

There must be a great number of errors in replication to enable genes to evolve and adapt to the conditions where they live. In order to propagate a population (or a message), cyberspace makes reproduction easier and low in cost. When we repost a photo on Facebook or retweet a message, we are engaging in a reproductive process.

*Self-recognition* is about the capability of identifying oneself, members of a family, or an acceptable group to share resources with. Self or non-self discrimination is so critical in nature that there are many specialized authentication programs for distinguishing family from invaders. These range from facial geometry, DNA, smell, sound, shape, to gesture. However, Buddhism and neuroscience can both agree that there is no consistent self. The brain and the body are constantly changing. There is no absolute borderline between self and non-self.[2] Therefore, a self-recognition algorithm must constantly adapt to the changing world in the wake of emerging techniques for improving quality of life, as well as techniques for deception and concealment.

*Control and Communication* ($C^2$) includes monitoring and changing dynamic states, sending signals, and coordinating actions. At the biological level, this includes control and communication in an immune system for a coordinated defense, cellular health monitoring, and self-purging for energy conservation. At the cyber-physical level, this includes command and control in network traffic, remote control, vehicle-to-vehicle communication, and autonomous driving. In order to secure a functioning system, frequent, periodic checkups are often implemented to self-test the system. For example, each morning when we wake up, the first thing we do is try to move our body, open our eyes, and listen to sounds as a part of self-awareness about our health condition. Soldiers performing landmine detection

[2] Ouellette J (2014) Ten things I learned about me. Slate, posted on January 30, 2014. http://www.slate.com/articles/health_and_science/science/2014/01/the_science_of_the_self_what_can_neuroscience_and_psychology_tell_us.html

often use a metal chisel to test the sensitivity of the metal detector because it may have changed due to hot weather. Frequent chisel tests enable the soldier to fine-tune the sensor to function within an optimal range.

*Stealth* reflects self-awareness about the observation of self by predators. It includes strategies for hiding, deception, camouflage, and erasing to cover up trails. Nearly all animals utilize deception for protection. The most common example of deception in nature is when the color or texture of the animal blends into the background. This is exemplary of camouflage. Behavior may play a key role, and smell, taste, and sound are also often involved.[3] Here, we concentrate on stealth behaviors in cyber attacks by malicious software. In the following chapters titled "Stealth" and "Visual Privacy," we will further discuss particular algorithms in a broader scope.

## Self-Recognition in Immune Systems

A living organism typically has two elements: a set of instructions that tell the system how to sustain and reproduce itself, and a mechanism to carry out the instructions. In biology, these two parts are genes and metabolism. However, these functions can also exist in non-biological systems such as computers. A computer virus is a program that makes copies of itself in the memory of a computer and can infect other computers. According to Steven Hawkins,[4] this fits the definition of a living system. Like a biological virus, a computer virus is a simple, degenerate entity. While it contains only instructions or genes, it does not have any metabolism of its own. It acts like a parasite that relies on the metabolism of the host. When we turn off the power to a computer, its metabolism is cut off. If the virus resides in short-term memory, then it will die. If the virus resides in long-term memory (e.g. a hard drive), then it will freeze its activity until the power is back.

In 1882, Russian biologist Élie Metchnikoff pierced a starfish larva with a rose thorn under a microscope. The following day, he observed a few cells trying to engulf the thorn. The starfish has remained largely unchanged since it evolved some 500 million years ago, suggesting that organisms have innate immunity. The innate immune system consists of cells and proteins that are always ready to fight microorganisms at the site of infection. Metchnikoff is credited with the discovery of white blood cells (macrophages) that engulf any invaders. Macrophages are a major component of the innate immune response.

The immune system has further evolved to provide *adaptive immunity*. When invader pathogens bypass innate defenders, the adaptive system is called into action. The adaptive immune system functions as a kind of "second brain" because it can store memories of past experiences in the war on invaders, and it generates responses

---

[3]Gregory RL and Gombrich EH, eds. (1973) Illusion in nature and art. Charles Scribner's Sons, New York.

[4]Hawking S. Life in the universe. hawking.org.uk. http://www.hawking.org.uk/life-in-the-universe.html

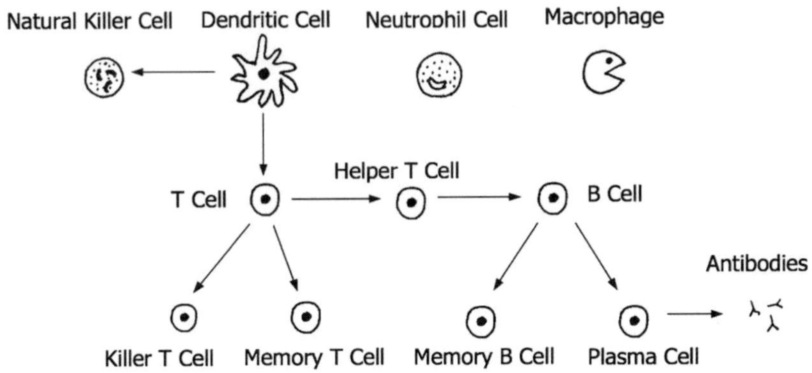


**Fig. 13.2** The human immune system. In contrast to today's security systems that are mainly a "castle-like" defense, the immune system's defense is from the inside

to new and novel antigen patterns. Assume a man steps on a rusty nail. The first barrier of the immune system (skin) is breached. Nearby bacteria enter the wound. They start to double their numbers about every 20 min, which is why dentists advise us to brush our teeth within 20 min after a meal. When the replication of the bacteria reaches a certain level, nearby macrophages "smell" the bacteria and move toward the smell through a process called "chemotaxis." They can swallow up to 100 intruders at a time. When macrophages fight for too long, they call for backup by releasing messenger proteins that communicate location and urgency. Neutrophils then travel through the blood stream to the battlefield. When neutrophils are not enough to stop the invasion, the more intelligent immune system jumps in. Dendritic cells act like a computer that reacts to wireless signals from the battlefield and starts collecting samples of enemies for machine learning. They decide whether to call in anti-viral or anti-bacterial forces. Dendritic cells are literally *toolmakers* which activate T cells stored in the nearest lymph node. When T cells become activated, they quickly duplicate thousands of times. Remarkably, some become memory T cells that stay in the lymph node and will make the body immune to this enemy throughout life. Some travel to the field of battle to help out. A third group goes on to travel to the center of the lymph node to activate their partner B cells that produce antibodies. See Fig. 13.2.

## *T Cell Receptors*

T cells are defense cells that are generated from bone marrow and migrate to the thymus above the heart, which is like a "school," in order to mature and to be "educated." Hence, we call those who have graduated from the thymus "T cells." The learning behavior of T cells form a part of the adaptive immune system, which means that they can recognize and memorize different infection patterns they have experienced in the past. The pattern recognition process is accomplished by a peptide bond, a chemical bond between proteins. In this process, T cells bind to a protein attached to a disease-causing organism (e.g. a bacterium, virus, or cancer cell), called a pathogen and attempts to remember its shape for later use. Occasionally, T cells make mistakes and memorize the shape of a protein attached to our own cells.[5] When this happens, T cells attack healthy cells as enemies. This is known as auto-immunity. Examples of autoimmune diseases include rheumatoid arthritis, celiac disease, and psoriasis.

To prevent T cells from attacking healthy cells called "self," the adaptive immune system has two screening processes: they are called positive selection and negative selection. When new T cells are replicated in the thymus, they are exposed to a large collection of their own healthy proteins as training samples. If the new T cells do not interact with them at all, then they are not motivated. These ones fail the exam and are left to die. If they bind moderately or strongly, then they are allowed to move onto the next level – negative selection, which separates the T cells that interact moderately from those that interact too strongly. If T cells bond tightly with the protein of healthy self cells, then they would memorize the shape of that protein and attack the body's own healthy cells down the road. Therefore, these T cells are also ordered to sacrifice themselves and are eaten by other immune cells. The thymus "school" is so competitive that only one fifth of T cells survive on average.

The few T cells that successfully graduate from the thymus are the immune system's critical security agents, attacking pathogens while leaving the body's own cells intact. The most amazing talent of T cells is their self-recognition capacity. How can T cells distinguish friends from invaders?[6] Studies show that molecules located on the surface of the T cell, known as T cell receptors (TCR), can recognize the antigens on target cells. In immunology, an antigen is any substance that induces an immune response. The receptors can be trained in the negative selection process to distinguish those threats from the body's own cells to avoid triggering an unwanted immune response.[7] It is estimated that every person has the ability to generate $10^{15}$–$10^{18}$ different T-cell receptor (TCR) molecules. Since the genomes (inherited DNA) of many species have been sequenced and found to contain only 30,000–40,000 genes, how do so few genes produce so many different antigen receptor molecules? TCR genes use a unique strategy involving combinations of genes to achieve the degree of diversity required for the immune response. For example they can rearrange themselves within the genome of a differentiating cell.

[5]Dzhagalov et al. (2013) Negative selective T. Accessible Bioscience. http://accessiblebiosci.blogspot.com/2013/08/negative-selectivi-t-dzhagalov-et-al.html

[6]Toon J (2011) How T cell recognize the bad guys. Futurity, Jan. 24, 2011. http://www.futurity.org/how-t-cells-recognize-the-bad-guys/

[7]Jian N, Huang J, Edward LJ, Liu B, Zhang Y, Beal CD, Evavold BD, Zhu C (2011) Two-stage cooperative T cell receptor-peptide major histocompatibility complex-CD8 trimolecular interactions amplify antigen discrimination, Journal of Immunity, vol. 34, No. 1, p13–23, January 2011.

T cells are perfect examples of self-awareness at the microscopic level. Self and non-self discrimination is a critical recognition function for immune systems in order to avoid auto-immunity. T cells need to recognize and memorize the shape of proteins attached to invaders that are stealthy and deceptive enough to sneak around the body undetected. None of these functions can be achieved without control and communication, such as urgently releasing protein messengers, water, or antibodies into the blood stream.

## *Artificial Immune Systems*

Artificial immune systems (AIS) aim to simulate the self-defense and self-healing behaviors found in natural immune systems on a computer. Let us begin with the recognition function of T cells as discussed above. When a foreign invader attacks the body, it attaches a protein to healthy cells. Can T cells detect any change in a protein sequence? Assume a protein shape can be simplified as a binary digital string. Our own normal health patterns are defined as the "self" in a very long chain of code containing a lot of zeroes and ones. Any modification of the code that exceeds an allowable variation would be defined as an invasive pattern, or "non-self." To any computer, memorizing a long string of the binary code of "self" is not a problem. However, to compare an input string with the string "self" is non-trivial. It is quite computationally expensive when the length of the code is enormous. We could cut the long string into small segments for T cells to adapt and remember, for example, 4 digits for each segment. Assume the string to be protected is "000001011111." The string then can be segmented into "0000," "0101," and "1111." In order to distinguish from self and others, we need a matching rule. Here, we may define that the two strings match when digits are identical in four contiguous positions (Fig. 13.3).[8]

The self or non-self discrimination algorithm works in the following way: First, start with replicating immature T cells with a random string segment assigned to each T cell. Then, test if the random string segments match the self string segments. If the random string segment does not match any of the "self" string segments, then it is saved in the repertoire of detectors. If the random string segment does match a "self" string segment, then it is eliminated.

This algorithm is a simple simulation of the self-recognition function in T cells. It can be applied to detect changes in computer systems, such as the boot sector of a computer where computer viruses often invade and then launch infectious attacks. Conventional methods include digital signatures and file authentication programs. However, these are vulnerable to being counterfeited or bypassed. For example, the checksum method for change detection can be detected and added into the malicious

[8]Forrest S and Perelson AS (1994) Self-nonself discrimination in a computer. Proceedings of IEEE Computer Society on Research on Security and Privacy, 1994.

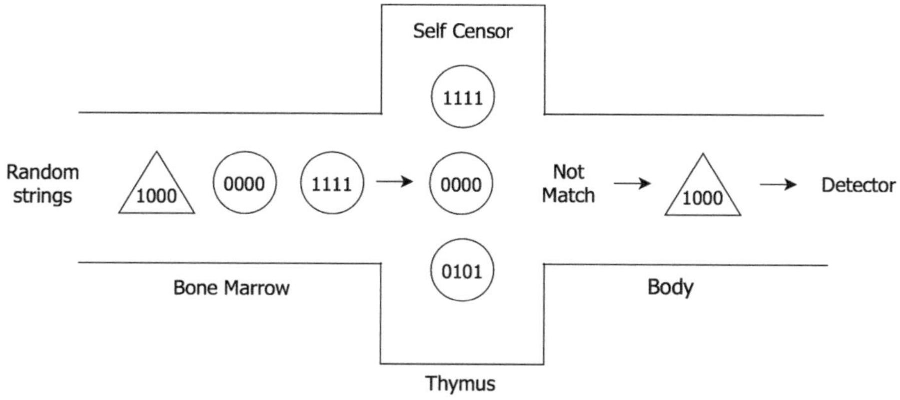


**Fig. 13.3** The negative selection process for T cells

code. The artificial immune system approach ensures that each copy of the detection algorithm is unique and can be applied to multiple sites. Therefore, if one site were compromised, other sites would likely survive. However, this algorithm is not foolproof if a virus is crafted from a subset of the "self" string in different order. The algorithm would not be able to detect this deceptive virus. Besides, a major limitation of the random generation approach appears to be the computational difficulty of generating valid detectors, which grows exponentially with the size of the "self." The algorithm can be improved by determining the total number of unmatched strings that exist for the defined "self" string, then selecting some of them to generate a diverse set of detectors for monitoring data patterns. In this way, the algorithm can run in linear time with respect to the size of the "self" string.[9] This method can be further improved with genetic algorithms rather than testing randomly generated candidates for detectors. This method enables detectors to evolve based on the "survival of the fittest" principle. The most qualified parents are able to produce qualified offspring with allowable mutations.[10]

Aside from T cell behavioral simulation, researchers have also developed B cell population simulation algorithms as a part of AIS for pattern recognition. The applications for models based on immune system principles are increasing in fields of science and engineering such as computer security, virus detection, anomaly detection in time series data, fault diagnosis, structural optimization, and the recognition of promotion in DNA sequences.[11]

[9]Dasgupta D and Forrest S (1998) An anomaly detection algorithm inspired by the immune system. In Dasgupta D. (ed.). Artificial Immune Systems and Their Applications. Springer.

[10]Forrest S, Javornik B, Smith R and Perelson AS (1993) Using genetic algorithms to explore pattern recognition in the immune system. Evolutionary Computation, vol.1, no. 3, pp.191–211, 1993

[11]Dasgupta D. An overview of artificial immune systems and their applications. In Dasgupta D, eds. (1998) Artificial Immune Systems and Their Applications. Springer 1998

Current computer security systems present certain immunity behaviors. However, compared to natural immune systems, these functions are rudimentary and limited. For example, the prevailing anti-virus algorithms use signatures or heuristics to scan for viruses that were discovered before. This kind of method would not be able to detect viruses that have never been seen before.

Natural immune systems can recognize malicious behaviors in addition to anomalous signatures. For example, our immune system can train its self-defense mechanism using "beneficial disturbances." Beneficial bacteria such as the commensal bacteria in the gut train the immune system to better distinguish between disease-causing pathogens and harmless antigens, which may help prevent the development of allergies. Similarly, these "good" bacteria may influence the immune system's sensitivity to antigens, potentially helping to prevent autoimmune diseases – conditions in which the immune system attacks the body's own tissues. The bacteria also produce useful antibodies and trigger the expression of intestinal proteins, which cause the immune system to repair internal injuries.

A recent study showed that receiving a single tattoo can temporarily lower the body's immunity because the body responds to the invasive ink. Researchers took saliva samples from volunteers at tattoo shops and measured levels of immunoglobulin A, a stress hormone known to suppress immune response.[12] Results show that immunoglobulin A dropped significantly in those receiving their first tattoos. After the initial stress response, the body returns to an equilibrium. However, if one continues to have more tattoos and stresses the body over and over again, the decrease in immunoglobulin A is markedly less. Instead of returning to the same set point, the immune system adapts its internal set points and moves towards a higher resistance to foreign agents. Perhaps having twenty tattoos might help the body to fight off common infections.[13] This appears to be a novel machine-learning algorithm that could allow computers to learn from the behaviors of daily non-lethal incidents, to adapt detection thresholds, and to predict future attack patterns.

Furthermore, unknown code or data can be contained or quarantined in a protected area so as not to interfere with the operation of the computer. This is similar to the immune response of macrophages that attempt to engulf invaders. A virtual machine can be created to run the infected program and analyze its behavior. The virtual machine would act as a poison food taster did for royal families in ancient times. Before royal members ate their meals, the professional poison taster would sample the food and drink to ensure safety. In a computer, if any suspicious behaviors were found (poison detected), the tester would send the encrypted samples to the admin machine or a "central virus analysis" system for further analysis or to update the epidemic warning. The central virus analysis system

---

[12]Christopher D. Lynn et al. (2016) Tattooing to "Toughen up": Tattoo experience and secretory immunoglobulin A, American Journal of Human Biology. March 4, 2016.

[13]University of Alabama in Tuscaloosa (2016) Want to avoid a cold? Try a tattoo or twenty, says researcher. Medical Press. Online post on March 9, 2016. http://medicalxpress.com/news/2016-03-cold-tattoo-twenty.html

is similar to the CDC (Centers for Disease Control and Prevention) in the US where many specialized laboratories culture infectious viruses, observe their behaviors, and produce vaccines. In nature, it could take anywhere from minutes to days to elevate the infection of a virus or bacteria. In a digital space, it only takes a few microseconds to infect millions of computers across the Internet. Artificial immune systems must evolve from cell level self-awareness to collective situation awareness with more advanced real-time behavioral analysis techniques.

## Identifying by Behaviors

Self-recognition is a by-product of self-awareness. It is critical for an immune system to avoid auto-immunity. It is also important to the authentication of users in computer security and privacy. At the cell level, self-recognition relies on the peptide bonds of proteins, which is a slow, chemical chain reaction process. At the organism level, this change recognition method is not enough. For example, two genetically identical twins may have different behaviors in daily life. Therefore, we need to further investigate how to identify self or others by behavior.

### *The Mirror Test*

In 1970, psychologist Gordon Gallup Jr. developed a behavioral experiment called the "mirror test," or "mirror self-recognition test (MSR)" to determine whether an animal can recognize itself in the mirror. The MSR test is the classic way to measure self-awareness. Gallup marked a visible colored dot on the animal's body while the animal was sleeping. The test then determines if the animal can use its reflection to touch the mark on its body. Despite hundreds of animals having been tested, only a very few species pass the mark test of mirror self-recognition (MSR). Aside from humans, only the four great apes, bottlenose dolphins, and Asian elephants have passed the test. Even human children under 18 months old typically fail this test. The majority of children begin to use the mirror's reflection to investigate the mark on their noses by age 2. At the other end of the human life span, people with Alzheimer's disease lose the ability to recognize themselves in mirrors.[14] Gallup argues that mirror self-recognition is an indicator of self-awareness in its rudimentary form, which is the ability to become the object of ones' own attention. However, most organisms respond to themselves in the mirror as if confronted by another organism. The neuropsychological evidence also suggests that patients with damage to the frontal cortex have an impaired ability to recognize their own

[14]Gallup GG, Anderson JR, and Shillito DJ (2002) The mirror test. In Bekoff B et al. (eds). The Cognitive Animal. The MIT Press, 2002.

faces. The frontal cortex appears to be involved in self-recognition, self-evaluation, episodic memory, introspection, humor, and mental state attribution.

The next question is can a computer pass the mirror self-recognition test? In some very simple cases, it is possible. Assume the computer tests whether an input string matches the "self" string stored in the memory in a reverse order like a reflection in a mirror. The algorithm can simply push the letters of the input string into a First-In-Last-Out (FILO) stack, and then perform a standard string match. This is just a one-dimensional string-matching problem for a computer. If we ask a mobile robot to recognize itself, then it would be a three-dimensional matching problem that becomes more complicated. Now, a robot has no curiosity to study itself and instinctually touch any marks on its body. Robots may come from the same manufacturer with the same appearance, but with different control software installed. Besides, even if each robot had a unique identification passcode, it is possible for the code to be stolen and compromised. In this case, we must look into behavioral patterns, including gesture, sound, geographical profile, communication patterns, and the consequences of actions.

## *Voices, Keystrokes, Locations, and Gestures*

Behavior-based recognition methods analyze a user's dynamic behaviors and can capture behavioral patterns continuously – not just at the startup time – and may be accurate enough to trigger an alarm to another system or person to double check the situation. Behavior-based recognition methods take multi-dimensional measurements and time series into consideration, providing more verification factors than a simple passcode or static appearance. Taking sound recognition as an example, many different motorcycles look similar in photos. However, virtually no two engines sound exactly alike, especially the sound of a Harley Davidson V-Twin motor, which can usually be identified by the unique rhythm of its famous "potato-potato-potato" sound.[15] The reason behind this sound is because of the unique V-shaped angular configuration of the engine pistons and a single pin connecting the pistons, giving a distinct, non-symmetrical rhythm, not only in sound, but also in vibration.

Similarly, human voices are unique because virtually no two vocal tract structures are identical. There are also significant differences in the anatomical structures of vocal tracts between men and women. Voice recognition can be viewed as soft biometrics, or a secondary identification factor, as opposed to hard biometrics such as fingerprints, retina scans, and DNA.[16] The problem with speaker verification is

[15]CycleFish.com. What gives a Harley-Davidson its unique sound? Posted on August 4, 2011. http://www.cyclefish.com/RoadCaptain/blog/1311/

[16]McGehee F (1937) The reliability of the identification of human voice. J. of Psychology, vol. 17, pp. 249–271, 1937.

in determining whether the input voice pattern matches the training voice patterns associated with the identified user. This is useful when other biometric data such as facial imagery or fingerprints are not available. For example, one of Colombia's biggest drug cartel kingpins Juan Carlos Ramirez Abadia radically altered his facial appearance with multiple plastic surgeries. In the end, his own voice gave him away, thanks to voice fingerprint analysis. The U.S. Drug Enforcement Agency made a match based on his telephone conversations taped by Brazilian police and his known voice recorded by Colombian officials.[17]

Automatic speaker verification appears to be easier than speech recognition because it does not need to extract content from the sound waveform. In reality, it is still a challenging problem because of the diversity of a speaker's vocal characteristics, environmental noise, languages, emotion, and multi-tasking interactions, including voice message, telephone conversation, and remote customer service.

In order to characterize a speaker's voice, we must extract voice features. Ideally, these features should be: efficient in representing speaker-dependent information, easy to measure, stable over time, occurring naturally and frequently in speech, changing little from one speaking environment to another, and not easy to mimic. Pitch, the fundamental frequency of vocal cord vibrations, is known to be an important feature to characterize speech. It is largely resistant to noise and channel distortions. The average pitch for a speaker varies considerably from one individual to another, but by itself is not sufficient to distinguish between many speakers. More descriptive pitch parameters include the entire pitch contour that covers a sentence-length utterance.[18] In addition to pitch, the most commonly used feature has been the Mel-Frequency Cepstral Coefficient (MFCC) as we discussed in the previous chapter "Sensuality." Once the voice features are extracted machine learning models can be used, for sample clustering and classification, including Hidden Markov Models (HMM), Gaussian Mixture Models (GMM), Support Vector Models (SVM), and Vector Quantization (VQ). Unfortunately, most existing speaker verification algorithms require long training and testing time and are designed for specific tasks that are either text-independent or text-dependent.

An Android platform-based speaker verification algorithm was developed in the author's lab for multitasking interactions such as text-dependent and text-independent cases.[19] The algorithm contains two parts: training and testing. In the training process, the phone records one 1-min sound clip of the speaker as a training sample. This training sample is automatically cut into 5 parts and each clip is processed separately. The algorithm first calculates the pitches and pitch contours

[17]AP Press (2007). Alleged drug kingpin with altered face arrested, August 10, 2007. http://www.nbcnews.com/id/20217024/ns/world_news-americas/t/alleged-drug-kingpin-altered-face-arrested/#.VvycZ7RMLMU

[18]Atal BS (1976) Automatic recognition of speakers from their voices. Proceedings of IEEE, vol. 64, no. 4 April, 1976

[19]Cai Y, Li X, Gong ZJ, Codina TR (2014) Speaker verification for multi-task interactions. Journal of Interacting with Computer, vol. 26, no. 2, pp. 135–144, 2014

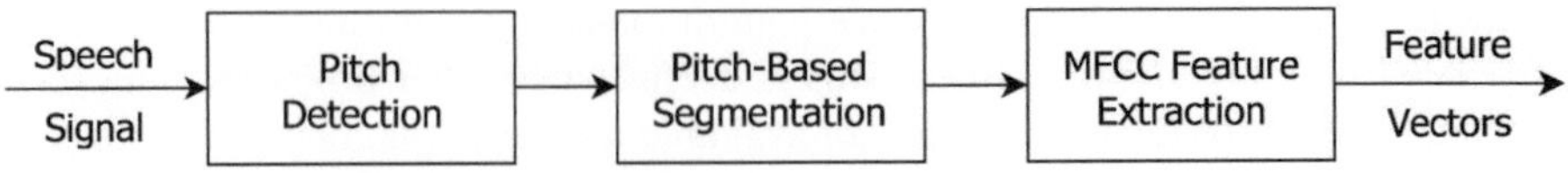


**Fig. 13.4** The speaker verification system

in each clip and removes the silent parts. Then, it calculates the MFCC features from the rest of the clip. The features are clustered using the Vector Quantization method to determine a self-adaptive threshold based on the similarity of the training samples. During the testing process, one 10-second sound clip is recorded by phone as the test sample. The system then performs the same process as the training process: analyzing the patches and pitch contours, removing the silent parts, and calculating the MFCC. Then it classifies the features in the test data. The pitch features are examined first. If the pitches are very different, the system will output "not the same person." If the pitch difference is not very large, the classification goes on to check the similarity measured using Vector Quantization or Mahalanobis distance. If the distance is lower than the self-adaptive threshold, then it is the same person. Otherwise, it is not the same person. Figure 13.4 is a diagram of the basic flow of the related processes.

In the text-independent speaker verification test in a normal office environment, 10 male and 10 female subjects were asked to speak random sentences for up to 10 s. The results show that Mahalanobis distance has a better verification rate than the VQ method. In the text-dependent speaker verification test, 20 male and 20 female subjects were asked to speak two short sentences: "This is [name] signing for FedEx. Today is [date]." The results show that the VQ has a better verification rate than Mahalanobis distance. Overall, the VQ method is also much faster compared to Mahalanobis distance. The speed increases by a factor of 10, which is almost real-time processing. However, it is not as accurate as Mahalanobis distance, which takes about 20 seconds.

Furthermore, the lab experiment data show that the test sample length has an impact on the positive detection rate. Based on 10 subjects for the text-independent task, with test sample lengths in 30 seconds, 1 minute, 1 minute and 30 seconds, and 3 minutes. The results show that 1-minute is a turning point. Below 1 minute, the positive detection rate drops significantly. Above the 1-minute threshold, the positive detection rate is above 80 % and continues to rise gradually.

The empirical study shows that the speaker verification algorithms work better when the person is speaking a native language, and that the positive detection rate can reach 100 %. Cross-language verification normally produces poor results. For example, the positive detection rate drops to 50–75 %, based on the testing of subjects who have trilingual skills in English-German-Chinese or Spanish-Catalan-English. In addition, an accent in a spoken language may also contribute negatively to the verification. There are potential applications for such an algorithm in non-critical and frequent tasks. For example, automated customer services tailoring their dialogue based on recognition of returning customers' voices, especially gender identification. The pitch-based gender classification is accurate enough for this kind of application.

Keystroke dynamics is another behavioral personal identification method that measures a user's typing style, including dwell time (a key press) and flight time (between the key release and the next key pressed down). Keystroke dynamics can be traced back to the Morse code telegraphy era. Frequent users normally could identify an individual by listening to the rhythm of "dots" and "dashes." Military intelligence used the rhythmic patterns to distinguish ally from enemy. In modern times, keystroke dynamics can be a second factor for authentication in addition to a password. If a hacker stole a password, he would still have to type the password in a similar rhythm as the authorized user.[20] Keystroke dynamics can be analyzed from keystroke logging. Each keystroke is time stamped twice: once when the key is pressed down and once when the key is released. The algorithm calculates keystroke features such as dwell time and flight time. For a given typing period, these features can be stored in a feature vector. Assume a user enters a password many times, and a few feature vectors are used to form a keystroke profile for that user. After enough training, the computer may switch to the monitoring mode by comparing the keystroke feature vector and the stored keystroke profile. The similarity can be measured by Euclidean or Mahalanobis distances, or even more sophisticated Dynamic Time Warping in order to align the peaks and valleys of signal curves. If a match is found, then the entry is verified; otherwise, an alert will be issued.

However, keystroke dynamics are not always reliable as a detection method. For example, the resolution of keystroke timing measurement is sensitive to the type of operating system. Keystroke dynamics is also associated with various types of keyboards. I remember my experience of typing English text on a German keyboard. I had long pauses when looking for the right keys. Overall, keystroke dynamics reflects a person's physiology, behavior, emotion, culture, and habits.

New technologies bring more challenges to keystroke research. For example, Microsoft released a new smartphone keyboard that enables a finger to slide from key to key without needing a pressing gesture for every key. Google provides the Instant Search function that is able to monitor the letters typed in the search text box in real-time. This potentially enables Google to capture keystroke dynamics (e.g. flight time between keys). Furthermore, 3D touch screens are able to capture a user's keystroke forces, adding one more dimension to the feature vector.

Geographic profiling is another behavior-based methodology that analyzes the spatiotemporal occupancies of a person of interest. Compared to voice recognition and keystroke dynamics, geographic profiling is absolutely behavioral and dynamic. Geographic profiling was originally used in criminal investigations, such as the investigation of serial murders, rape, arson, robbery, and bombing through the sequence of connected crime locations.[21] However, geographic profiling has also

---

[20]Maxion R (2012) Making experiments dependable. The Next Wave. Vol. 19, no. 2, 2012

[21]Rossmo DK (1999) Geographical profiling. CRC Press.

been applied to other areas because of rapidly growing geographic data streams. For example, Internet Protocol (IP) addresses are normally associated with the service provider's device addresses. Smartphones send out the GPS coordinates of the user whenever a user makes a phone call or a navigation system is activated. Gasoline stations and electronic tollbooths transmit drivers' location data to banks that issue credit cards. The automated license plate recognition systems on highways or parking lots send the toll bill to the driver's home address. A geographic profile of a user can be collected over a period of time and can be updated continuously with changing patterns. The simplest form of a geographic profile is the geographic centroid of the collected coordinates and the average travel radius of the person. Just like animals, our daily travel patterns are not random and are fairly consistent in terms of the geographic centroid and radius. Sometimes, we travel outside of our normal radius by a long distance and the travel is flagged. For example, a credit card company denied my charge at a gasoline station near Malpensa International Airport, Milan, Italy, because it detected an unusual traveling pattern compared to my normal geographic profile stored in the bank's database. The bank sent me a text message to verify the charge and listed a toll-free telephone number for me to call to unlock the credit card in Italy. The text message and phone call created two-factor authentication for security purposes. However, geographic profiling is invasive to personal privacy. For example, credit card companies often recommend that users inform them about their planned locations before any international trips so they can avoid false alarms during their travels.

Perhaps the most exciting behavioral authentication method is the gesture-based method that incorporates body motion sensing. The new mobile phone app called "AirSig" takes advantage of motion sensors on mobile phones such as accelerometers, gyroscopic sensors, and magnetic sensors to capture the user's sequence of gestures.[22] By waving the phone in the air, for example, spelling the name "Cai," the user can log in to his or her account with or without a password. Human gesture is not easy to mimic exactly, even if it is videotaped, because of it has at least eight degrees-of-freedom (DOF) plus time stamps. According to AirSig, the possibility for counterfeiting an air signature is about $2^{-2000}$. However, it is also a challenge to match the user's own gestures in such a high-dimensional feature space. In contrast to the voice, keystroke dynamics, and locations mainly used for passive recognition and verification, gesture-based authentication must be performed actively by the user. Gesture-based methods have created a new direction for behavioral authentication, especially for mobile payment, digital signatures or approval, and the Internet of Things. If you see someone dancing with a phone in her hand, please be calm. She may be just signing her credit card at that moment.

[22]AirSig: http://www.airsig.com

## *Are You Human or Robot?*

For decades, computer users have written their email addresses in an image in order to deter spam emails or unwanted commercial posts, while assuming it was difficult for spam robots to harvest email addresses from images. To further defeat spam robots that may be equipped with advanced optical character recognition algorithms, the letters in the image are distorted by waves or noise. A popular algorithm used today is called CAPTCHA, which stands for "Completely Automated Public Turing Test to Tell Computers and Humans Apart," and assumes hackers do not know much about computer vision. Ironically, Google researchers have cracked their own CAPTCHA with 99.8 % accuracy.[23] So far, CAPTCHA has been an effective tool to distinguish robots from humans. However, there are some problems. Besides its exposed vulnerability, CAPTCHA's usability is not perfect. The wavy text often confuses and frustrates human users during the login process.

Emerging behavior-based methods are designed to simplify the test by adding dynamic interactions. For example, facial recognition algorithms have been used for user identification in financial transactions. However, algorithms can be fooled using a stolen image. A new behavioral approach called "swiping face" (an analogy of "swiping card") implemented by Alibaba, asks the user to change his or her expression from normal to smiling, or to blink the left or right eye, in order to capture changing facial landmarks. The server can then determine whether the face is a still image or a live face. Other approaches include using the mouse to click the box next to the question "Are you a robot?" or typing a number in the box for "$4 \times 5 = ?$" The bottom line of such behavior-based approaches is to make an action easy for humans but difficult for robots.

## Malware Behaviors

User authentication methods are often used in human-computer interactions. They may not always inhibit malicious software, or "malware," that is present inside the system. Malware is any software used to disrupt computer operations and produce unwanted results. According to a survey on March 16, 2011, the top three malware categories were Trojan horses (69.99 %), viruses (16.82 %), and worms (7.77 %). These three types contributed about 94.58 % of the entire malware population. A Trojan horse is a malicious program embedded inside a seemingly legitimate and benign application that lures the user to run it. The typical behavior of a Trojan horse is to open connections to a command and control server. Once the connection is made, the machine becomes a remote controlled "zombie" machine. An example

[23]Kastrenakes J (2014) Google develops computer vision accurate enough to solve its own CAPTCHAs. The Verge on April 16, 2014: http://www.theverge.com/2014/4/16/5621538/google-algorithm-can-solve-recaptcha-almost-every-time

could be a free online game app that attracts users to download it, and then steals the user's credit card information immediately after installation. Trojans do not self-replicate. They normally spread through user interaction, such as an email attachment or downloading and running a file from the Internet. A computer virus, on the other hand, is a program that replicates and inserts itself into other programs or files for a malicious purpose. A worm can self-replicate, but it is an "autonomous vehicle" that crawls over a network to infect other computers. Relatively speaking, a virus requires the user to run an infected program in order for the virus to spread, whereas a worm spreads itself.[24] However, there are no absolute borders between different malware types. Many malware products are actually hybrids of more than one type. For example, a Trojan horse may hide inside a USB drive that could release a virus and worm to infect the files on the host machine and then infect other machines in the local network. Behavior-based malware analysis systems are designed to detect anomalous, even unseen behaviors.

## *Self-Awareness in Malware*

Malware has evolved from innocent hacking programs to highly sophisticated software engineering products. Despite the many different categories of malware,[25] contemporary malware fits four characteristics of our self-awareness behavioral model: control and communication, stealth, self-replication, and self-recognition. *It is truly an arms race when both attacker and defender have self-awareness behaviors.*

*Control and Communication ($C^2$)* Control and communication behaviors are common in malware such as Trojans, viruses, worms, bots, and spyware. After the host is infected, it becomes a "zombie" machine controlled by a remote attacker. Multiple "zombies" can form a botnet that delivers a coordinated attack such as the distributed denial of service (DDOS) attack. Hundreds or thousands of bots can create a data traffic surge to crash a server. Remote controlled bots also act as ransomware that can delete important data files such as medical records in an emergency room. A majority of infections occur in files, usually when performing basic operations like open, write, create, move, and delete. Suspicious behaviors include performing those operations in greater numbers than expected or targeting specific files.

*Stealth* Once a malicious program is installed on a system, it is designed to stay concealed in order to avoid detection. Rootkits are a collection of software that enables this concealment by modifying the host's operating system to prevent it

[24]Britannica.com. Computer virus – Encyclopedia Britannica. Retrieved 28 April 2013.

[25]OmniSecu. Different types of computer viruses. http://www.omnisecu.com/security/types-of-computer-viruses.php

from being visible in the system's list of processes. Viruses are designed to hide from the operating system or anti-virus software by making changes to file sizes or the directory structure. *Polymorphic* viruses can change their forms in order to avoid detection. They encrypt themselves, which is a behavior known as *mutation*. The behavioral hiding patterns also include a child process in the malware infection tree that deletes the static file image of an ancestor process, and is considered a possible attempt to evade detection.[26]

*Self-Replication* Popular malware such as viruses and worms copy themselves into a new file or into an existing file on the host machine or other machines in the network. Viral self-replication behavior can be detected by monitoring the operating system function calls using tools such as *trace* to track down the process identifications and their relationships in real-time. For example, if the process references itself, then it is likely it is trying to self-replicate.[27] However, the particular operating systems and the local host limit this method. Fast-spreading worm behaviors can be detected through peer-to-peer cooperation. The host-based detection method exploits a worm's temporal consistency and similarity in the worm's invocations of system calls as it propagates. Similarly, patterns in how an influenza epidemic spreads can be detected by observing the dynamic patterns of symptoms and pharmacy visits of a household and neighbors.[28] Visualization models can be used to further track the patterns of malware spreading. For example, one could build malware infection trees in a directional graph by identifying processes and files related to a specific malware sample. Precise time stamps certainly help to reconstruct the dynamics of the viral spreading patterns.[29]

*Self-Recognition* Advanced malware is often designed to recognize itself by remembering the infected machine so that it will not infect it twice. Malware can also detect "non-self," or enemies such as anti-virus programs and try to delete them before being killed. This behavior in the digital world mimics primitive self-recognition behaviors in nature. Detecting anti-virus processes is an indication of a malicious program. This behavior can be used as a "honey pot" to lure the malware and trap it in a confined area for further investigation. See Fig. 13.5.

[26]Morales J (2013) Prioritizing malware analysis. SEI Blog. November 4, 2013. https://insights.sei.cmu.edu/sei_blog/2013/11/prioritizing-malware-analysis.html

[27]Morales JA, Clarke PJ and Deng Y (2010) Identification of file infecting viruses through detection of self-reference replication. Journal of Computer Virology, vol. 6, pp. 161–180, 2010

[28]Malan DJ and Smith MD (2005) Host-based detection of worms through peer-to-peer cooperation. WORM'05 Proceedings of the 2005 ACM workshop on Rapid Malcode, pp. 72–80, 2005.

[29]Morales JA, Main M, Luo W, Xu S, and Sandhu R (2011) Building malware infection trees. The 6th International Conference on Malicious and Unwanted Software (MALWARE), IEEE Xplore:27 Dec 2011.

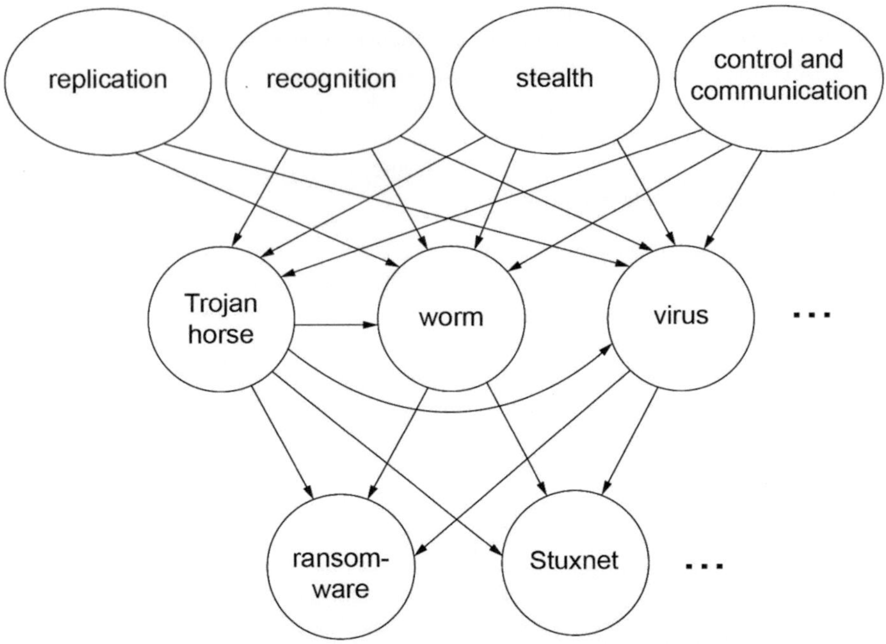


**Fig. 13.5** Self-awareness behaviors in malware at genetic, organism, and species levels

## *Malware Behavior Profiling*

Current prevailing methods for describing malicious code behaviors are state machine, time line, event diagram, or text. We need a more unified model to represent the timing and sequence of behavioral "vocabularies." Malicious software is not a simple-minded attack, but a sequence of well-choreographed known behaviors with certain mutations. Malware behavior profiling creates an abstract model for describing the sequence of behavioral patterns of an event or an attack. In many cases, we can use familiar behavioral vocabularies and "grammar" to describe a novel or complex attack, and share this knowledge with the broader community. In addition, abstract models allow researchers to simulate the event and test assumptions on a computer without implementing an actual system, which could save significant time and cost.

A Petri net is a potential model for representing dynamic, discrete behavioral sequences. It was first introduced by Carl Adam Petri in 1962,[30] and the time factor was added 10 years later. Petri nets have been further extended to Colored Petri Nets (CP-nets) with data and modules. A basic Petri net is a graph that

[30]Petri CA (1962) Dissertation: Kommunikation mit Automaten, Institut für Instrumentelle Mathematik, Bonn, 1962.

contains two types of nodes: *places* (circles) and *transactions* (bars). *Tokens* (dots) are moved from place to place by the firing of transactions of the net, which reflects dynamic properties of system behaviors. Petri nets are particularly useful in modeling systems with concurrent and asynchronous events, which are quite common in the real world. For example, when multiple users try to print on the same printer concurrently, or race conditions occur when multiple users try to push elevator buttons for different floors at the same time, or a deadlock occurs when a user tries to set up a new password but he needs to first recall the old, forgotten password. Many conventional modeling methods such as Finite Element State Machine (FSM) often fail to represent such concurrency. Petri nets can help us to simulate concurrent malicious attacks at an abstract level without an actual target machine. They have the potential to model complex cyber-physical attack sequences, where digital events are linked to physical events that are carefully coordinated.

Let us investigate malicious behavior in a ransomware case. In the summer of 2015, an attacker hacked into a database and then sent a ransom message through a series of emails between July 25 and 26. The first email was sent on July 25, 2015 at 01:43 UTC from 190.88.135.5 located in Curacao with the domain name memorabilia.xyz; then at 06:17 UTC, from 85.217.235.46, located in Bulgaria with the domain name conclusion.xyz; at 11:42 UTC from 163.22.18.56 located in Taiwan with the domain name sanyo.xyz; and again at 16:32 UTC, from 201.139.80.92 located in Argentina. Finally, on July 26, two ransom emails were sent: at 07:06 UTC, from 178.159.229.230 located in Ukraine with the domain name Levitra.xyz; and at 19:21 UTC, from 120.151.28.96, from Australia with the domain name instrument.xyz. The following is a part of one of the ransom notes (Fig. 13.6):

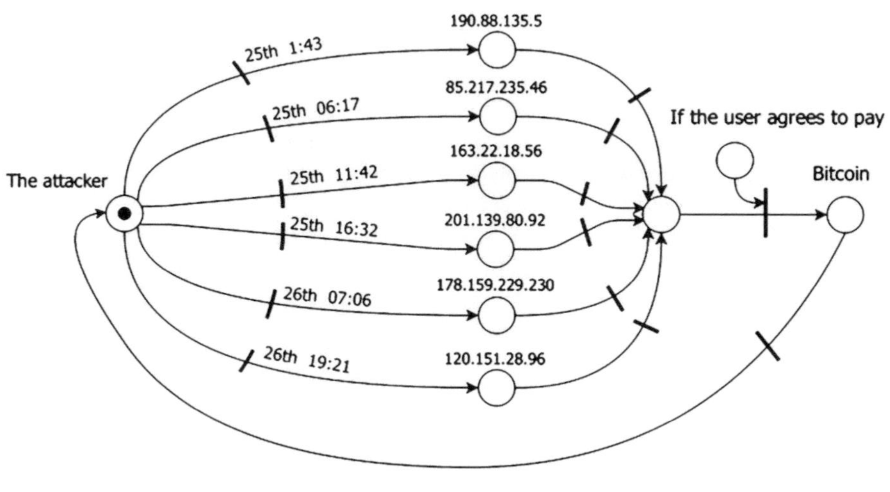


**Fig. 13.6** The ransomware behavior profile indicates the behavior patterns of control and communication and stealth

> We are providing a chance to solve this case. You make a payment to the BTC (Bitcoin) address. The time ends in the next 24 hours. You can get bitcoins at an exchange like Expresscoin.com, Localbitcoins.com, Clevercoin.com, Coincorner.com, Coincafe.com, Coinbase.com, Circle.com or a Bitcoin ATM machine Coinatmradar.com. You need to send exactly 1 bitcoin (BTC) to the following BTC address:
> 1FNQ8zdGbWThucxzwNiKkm76M8vCnYYUu4.

Petri nets are valuable for modeling cyber-physical malware behaviors such as Stuxnet, which is a revolutionary malicious computer worm that was designed to target a particular physical system through an industrial local computer network and Programmable Logic Controllers (PLC). A typical industrial local network is not necessarily connected to the Internet. A cluster of computers in a plant are usually employed as a user interface for monitoring sensors and actuators in real time (so-called SCADA, or Supervisory Control And Data Acquisition), and they communicate to the PLC local bus that is directly connected to sensors, switches, motors, and other devices. The SCADA computer systems also store the latest copy of the PLC code and libraries that can be downloaded to the PLC. Stuxnet has three modules: a Trojan horse that automatically executes the payload and distributes the worm to other computers on the network; a worm that executes all routines for the attack and handles communication back to the remote server; and a rootkit for hiding all malicious files and processes to prevent detection. Stuxnet is a combination of all four primitive self-awareness behaviors: self-replication, self-recognition, stealth, and control and communication.

Stuxnet was reportedly released to the target environment through an infected USB flash drive as a Trojan. The worm then replicated itself across the network. How easy was it for people to pick up a USB flash drive and plug it into a computer? Researchers from the University of Illinois, Urbana Champaign dropped 297 flash drives on a large university campus, including parking lots, hallways, libraries, and cafeterias. They found that almost half of the flash drives (48 %) ended up being used on a computer, and almost all of them (98 %) were picked up from the dropped places, potentially causing widespread computer infection later on. Interestingly, those who plugged in the flash drive were usually trying to find information about the owner so that they could return the lost flash drive. In most cases, plugging the flash drive into a computer is the only way to find out who the owner is.[31]

Stuxnet is also capable of recognizing itself and its target (Siemens S7-300 system) with the PLC programming toolkit Step 7, which is a Windows-based program for writing PLC executable code, sending settings to the PLC, and intercepting data from the PLC. Stuxnet erases itself when it realizes it is not in the targeted network. When it infects a network via a USB stick, it is supposed to only spread to three additional computers and erase itself after a certain number of days.[32]

---

[31] Tischer M, Durumeric Z, Foster S and Duan S (2016) Users really do plug in USB drives they found. Proceedings of 37th IEEE Symposium of Security and Privacy, San Jose, May 2016.

[32] Falliere N, Murchu LO and Chien E (2011) W32.Stuxnet Dossier. Version 1.4. Feb. 2011. Online document from Symantec.

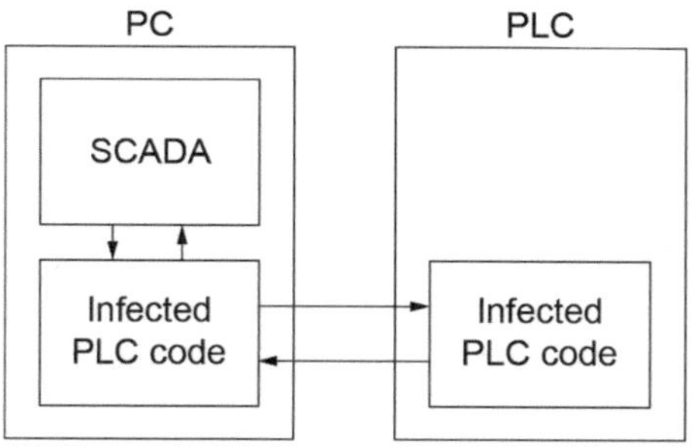


**Fig. 13.7** The Man-in-the-Middle (MITM) behavior in Stuxnet

Stuxnet uses a sophisticated behavioral stealth method, called "man-in-the-middle (MITM)" to hide its attack in the PLC. The worm hijacks the critical communication library between the SCADA and PLC called s7otbxbx.dll, which is a Dynamic Link Library (DLL) that is executed when the SCADA and PLC are running. The malware takes snapshots of the normal operating parameters and values for a period of time and then plays them back during the attack. Therefore, the operator will only see earlier normal parameters values instead of the actual ones. See Fig. 13.7. This is similar to a scene in the 1994 movie, "Speed", where a bomber installs an explosive device on a city bus, and monitors it through a wireless hidden camera. In the movie, the bomb would be activated if the bus went over 50 mph (80 km/h) and would detonate when the speed dropped below 50 mph. The police lieutenant "Mac" McMahon jumps on the running bus travelling over 50 mph, gets a local news crew to record footage of the bus, and then has it rebroadcast in a continuous loop to fool the bomber while the passengers are safely unloaded onto an airport bus before all the fuel ran out.

Stuxnet and "Speed" share the same "record-and-playback" behavioral pattern. This often happens when a user overly trusts a system or solely depends on a system such as SCADA. The "man-in-the-middle" risks can be reduced by frequent check ups and multiple inspections of diagnostic channels. For example, a PLC can be accessed from a laptop through engineering software such as the Ladder Logic Programming language or toolkits that are provided by the PLC manufacturer. By directly reading the measurements from the sensors and motors, the user may discover inconsistencies between what the SCADA displays and what is the raw data. Unfortunately, it is common practice in industry to consolidate tools into one of a kind in order to reduce associated costs. However, this sacrifices the diversity of inspection channels.

Stuxnet also uses a control and communication server for lengthy operations on a grand scale. Stuxnet has hijacked two legitimate certificates to connect the victim machines and the attacker's command center. Whenever Stuxnet malware infects a target, it sends version and status information back to the attacker. The worm can be executed autonomously and semi-automatically by remote control. The command center can order the worm to hibernate, download and execute the malicious code, or self-destroy. A control and communication server might be identified by abnormal traffic patterns, for example, a SCADA terminal sending a mysterious message to a server outside of the firewall.

The most challenging forensic analysis on Stuxnet is to understand the physical sabotage behaviors behind the control logic of the PLC code. Security experts around the world have been trying to piece together the puzzle of the attack sequences, mostly using State Machines. Here, we explore the Petri net version of the attack sequences in the PLC code. Using an abstract temporal model, Petri net allows us to include timing factors, to standardize malicious behavior profiling, and to incorporate physical simulation languages such as ladder logic that is widely used in PLC programming, testing, and monitoring. Figure 13.8 is the Petri net of the PLC behavior sequence. The initial state is Start. The transitioning time from start to State Record takes about 13–90 days. During this time, Stuxnet monitors and examines the target system to determine whether it is in a particular state of operation to meet the criteria for the attack. After all criteria are met, Stuxnet moves to record snapshots of the operational data. After 2 h in states of Attacks, Stuxnet sabotages the target system by slowing down or speeding up motors to different rates at different times. Stuxnet replays the recorded data to fool the SCADA displays in front of the operators. The two attacks have a 15-min or 50-min break in between. Finally, Stuxnet resets various fields before transitioning back to the initial state and starts a new cycle.

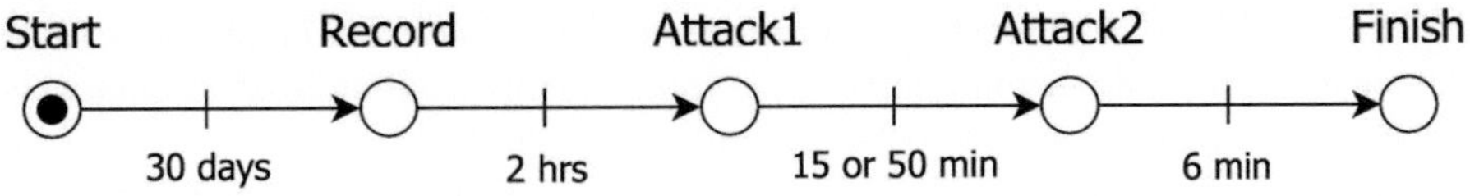


**Fig. 13.8** A simplified Stuxnet PLC sabotage behavioral sequence in Petri net

Stuxnet teaches us about the comprehensive malware behaviors: control and communication, stealth, self-recognition and self-replication, which covers almost all major self-awareness functions. The Petri net helps us to model the behavioral sequences such as the periods for waiting, recording, attacking and resetting. This allows analysts to share their knowledge about the attack behaviors and simulate the events on a computer without running the actual hardware or software.

## Visualizing Malicious Behaviors

Malware behaviors often hide inside an ocean of ASCII and binary data that are difficult to see. Visualization transforms data into visual artifacts, including graphs, animations, and video games. There are two purposes of visualization. It can be a tool for analysts who want to see the whole picture before diving into details. It is also a media for reaching out to a broader set of viewers with an intuitive interface. Visualization does not attempt to replace conventional analytical tools; rather, it helps analysts in building a visual vocabulary of malicious behaviors, articulating the processes, and eventually automating the detection process. In the following sections, we explore how to apply visualization models to analyze network traffic data and potential insider threats.

## *The SQL Attack*

Visualizing malicious behaviors in a network traffic dataset is a challenge because of fast packet traffic between many connected IP addresses. A prototype of a 3D visualization model has been developed at the Visual Intelligence Studio, CyLab, at Carnegie Mellon University. The model uses a force-directed graph to represent dynamic interactions between network nodes. The more frequent the traffic volume is between nodes, the closer they are. The model also uses directional belts to display color-coded network protocols, application types, and packet sizes. The visualization tool is designed to give an overview of the network situation during a given observation period. Pre-processing and post-processing are critical to visual analytics. Pre-processing sets up filters that narrow down the displayed entities so that they do not overwhelm the user. Post-processing sets up filters for the visual data so the user can investigate particular events or patterns. Post-processing is similar to tracking. Usually, it is most effective when using multiple tools, such as WireShark, to investigate in further detail.

Consider the "Capture The Flag" dataset. These are results from a student hacking competition that lasted for hours, creating over 150 GB of packet capture (PCAP) data, including source and destination IP addresses, protocols, packet sizes, network states, and packet contents. The 3D graph shows a strange behavioral pattern between the local server and a foreign IP address. Their dynamic locations are very close, indicating a frequent interaction. Both sides sent and received HTML and TCP ACKS with a uniform rhythm. See Fig. 13.9. It could have been normal traffic, but why would they mirror each other with small packets in such a fixed rhythm? This suspicious interaction behavior pattern led the researchers to look into the packet contents of the two addresses using WireShark. After filtering out all other addresses, the researchers were able to piece the puzzle together: it was an SQL attack, in which the attacker, located in Spain (based on the IP address), sent a sequence of SQL commands in order to obtain the script stored on the server. The attacker sent one letter at a time to reverse engineer the script content, assuming the attacker happened to know the script file name. The following are examples of WireShark's interpretation of the attacker's input and the server's responses.

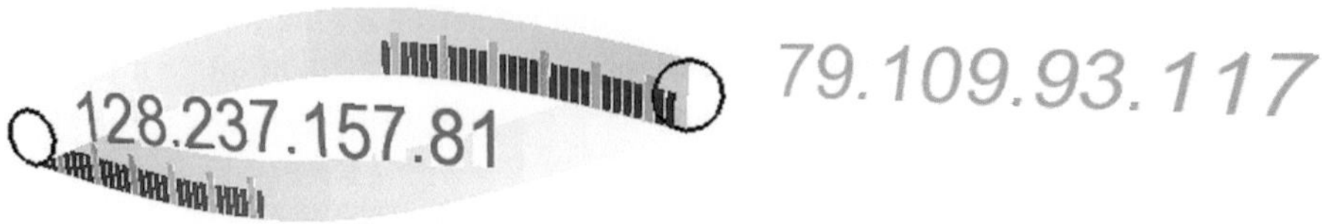


**Fig. 13.9** Visualization of an SQL attack. The HTML packets (*dark vertical strips*) and TCP ACK packets (*light vertical strips*) alternate in a uniform pattern between the local server and the foreign IP address

The attacker submitted a semi column ';' in ASCII value 59 to see if it matched the character at position 107/var/www/problem1.php in the script on the server.

```
Form item: "username" = """" OR (SELECT ascii(subtrs(load_file(0x2F7661722F
                      7777772F70726F626C656D312E706870),107,1))=59) OR "a"="b"
Form item: "password" = "a"
Form item: "submit" = "Enviar consulta"
```

The server replied with a general login failure message. Below is the content of the packet in HTML.

```
<html><body>Wrong username/password.<br />\n
<form method="post" name="form1">\n
\tusername: <input type="textbox" name="username" maxlength="10" /><br />\n
\tpassword: <input type="textbox" name="password" maxlength="10" /><br />\n
\t<input type="submit" name="submit" />\n
</form>\n
<p>\n
Notices:<br />\n
This system is now using the advanced SHA1 encryption function. Call the helpdesk
if you need to change your password.\n
<p>\n
\n
<body></html>
```

The attacker kept trying more characters until a match was found at ASCII value 120, which is the letter 'x.' The server replied with a welcome message.

```
<html><body>Welcome  "OR (SELECT ascii(subtrs(load_file(0x2F7661722F7777772F
                       70726F626C656312E706870),107,1))=120) OR "a"="b.<br />\
<form method="post" name="form1">\n
\tusername: <input type="textbox" name="username" maxlength="10" /><br />\n
\tpassword: <input type="textbox" name="password" maxlength="10" /><br />\n
\t<input type="submit" name="submit" />\n
</form>\n
<p>\n
Notices:<br />\n
This system is now using the advanced SHA1 encryption function. Call the helpdesk
if you need to change your password.\n
<p>\n
\n
<body></html>
```

## *Insider Threats*

Our body has a kind of self-awareness of anomalous behaviors, for example, the feeling of a shaking hand or a runny nose. We then go to see the doctor to check our symptoms. For an organization, anomalous behaviors may be connected to insider threats. The challenge here is how to define normal and abnormal because everything is dynamic. Norms change over time. Visualization plays an interesting role in revealing spatiotemporal behavioral patterns over time. Analysts can animate the sequences and create alerts suspicious behaviors.

For example, given the dataset of an enterprise system that has over 11 access points that require a user ID to gain access, the 2-month log data would show group and individual behavioral patterns. The force-directed graph is used to represent the average number of visits and frequency between access points. The more frequent visits between the two nodes, the closer they are. The intensity of the path indicates the number of visits, and the path can be coded for individuals or for a group. In Fig. 13.10, the left graph shows the individual's paths in May superimposed on the accumulated paths of the group of users. The right graph shows the individual's paths in June superimposed on the accumulated paths of the group. As we can see, the individual's paths were a part of the group's overall path pattern. However, the individual's path pattern in June was significantly different from the paths in May because of the frequent visits to a particular access point. In order to visually compare the motion histogram, the graphs should be aligned at the same position whenever the data or time stamps are the same. This ensures the repeatability of the visual rendering.

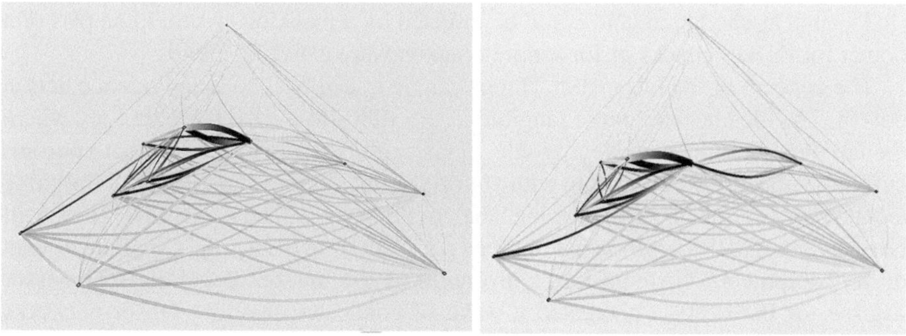

**Fig. 13.10** The *left graph* shows the individual's paths in May superimposed on the accumulated paths of the group of users; the *right graph* shows that the individual's paths in June superimposed on the accumulated paths of the group

## Collective Consciousness

On February 28, 1983 at 11:03 pm, the flow rate of the two largest water tunnels in New York City suddenly jumped by 6.7 million gallons each, 3 min after the broadcast of the last episode of the popular CBS television comedy, "M-A-S-H." The water flow continued at an abnormally high rate for about half an hour. This translates into about one million New Yorkers who went to the bathroom after the 2-hour show.[33] The connected water pipes and tunnels in New York City collectively produced useful information about the size of the audience that likely watched the last episode. Three decades later, ideas about collective consciousness can even go further.

The Internet of Toilets (IoT) has become a reality. Smart toilets originated in Japan. They are equipped with basic functions such as automatic heating, washing, and drying. Manufacturers are continuing to add more features. For instance, in order to automatically upgrade the control and measurement capabilities of the embedded system, it is convenient to add WiFi capability as a backdoor for the vendors. Furthermore, the smart toilet can be programmed for health informatics with additional sensors for personal health screening and monitoring, including weight, body temperature, glucose levels, bleeding, STDs, and other biomedical data. Health data can be stored in a cloud server provided by the toilet manufacturer for a monthly fee. Eventually, health agencies such as Center for Disease Control (CDC) might be able to access this data for epidemic disease detection and prevention. While the scenarios are technically feasible, they also present potential threats to security and privacy. For example, a 90-year old grandfather may just use the factory default password for the product instead of making a new one. His toilet's IP address would be shown in the search engines for the Internet of Things (IoT), such as shodan.io, and could be infected by a malicious worm to be part of a botnet for DDOS attacks or for sending ransomware extortion emails.

The concept of the Internet of Things can be traced back to many science fiction novels. The idea became more tangible in the 1980s at the MIT Media Lab, where the lab director, Nicholas Negroponte, envisioned that "you will find that your left cuff link will be communicating with your right cuff link via satellite. With flat panel technology, every license plate, wine label, or price tag will be a 'display.' There will be many more MIPS (million instructions per second) in the nation's appliances than on its computers."[34] Media Lab's invention of the low-power wireless processor ZigBee, or IEEE 804.11, created a wave of connected wireless sensor networks called Motes or Mesh Networks.

When hundreds, thousands, or even millions of sensors work together, it creates a *collective intelligence* about events. For example, Google can use search engine query data to track influenza-like illnesses in a population due to the relative

[33]UPI (1983). The last episode of 'MASH' was a royal flush. March 14, 1983. http://www.upi.com/Archives/1983/03/14/The-last-episode-of-M-A-S-H-was-a-royal-flush/3459416466000/

[34]Brand S (1988) The media lab. Penguin Books, Inc.

frequency of certain queries, with a high correlation with the percentage of visits to the doctor in which a patient presents influenza-like symptoms. Google's tracking system can accurately estimate the influenza activity level within a lag of 1 day.[35] Sharing data across the Internet enables the collective intelligence of events in a region, country, or even the world. There is no doubt that this collective intelligence will only grow deeper and broader.

However, individuals must have self-awareness to work in a community. Google uses web crawlers that have a built-in sense of self-awareness to feed its search engine with web pages. These crawlers have been designed to take pauses once in a while to yield the data traffic to the server. Otherwise, these crawlers would be too greedy and will be banned by the Internet community.

*Collective consciousness is an equilibrium between collective intelligence and individual self-awareness*, which is important to user privacy and the well-being of the Internet as a whole. For example, people have the right to keep their toilet usage and biomedical data private. The toilet should not stream private data to a stranger's cloud data server or onto another computer without permission.

Social insects and animals in nature tend to balance their self-awareness and collective intelligence. For example, many organisms use different frequencies to avoid predators, hunt for food and communicate with family and friends (Chap. 14). Diversity is key to keeping communication channels separate from one another and localized to a small region, or to keep messages brief or even maintain silence when predators are nearby. Figure 13.11 shows six diverse types: (a) IP-addressed devices, (b) industrial control systems such as PLC networks, (c) Bluetooth devices, (d) IP-addressed devices that are connected to a cloud server, such as Google Glass, (e) peer-to-peer connected mesh networks, and (f) vehicle-to-vehicle networks. As we can see, there are many architectures of the Internet of Things for a variety of purposes. By making systems diverse and self-aware, security vulnerabilities can be significantly reduced.

For example, the IP-addressed devices such as closed-circuit television (CCTV), Google Glass, as well as emerging smart toilets and autonomous driving vehicles are vulnerable to being hijacked and used as botnets. A self-aware program or security settings are necessary to prevent data leaking beyond the border for security and privacy purposes. A troubling example would be an infant monitoring camera being hacked and a strange man contacting the monitored infant.

Industrial control systems such as PLCs have their own local networks, but they are normally connected to SCADA systems for operation, maintenance, and development. To prevent "man-in-the-middle" attacks, a regular check up with lower-level inspection tools such as sensor readers and ladder logic programmers is necessary. The SCADA PCs may cut down their online time to reduce the risk, similar to the "silence" strategy seen in nature.

---

[35]Ginsberg J. et al. (2009) Detecting influenza epidemics using search engine query data. Nature, vol. 457, pp. 1012–1014.

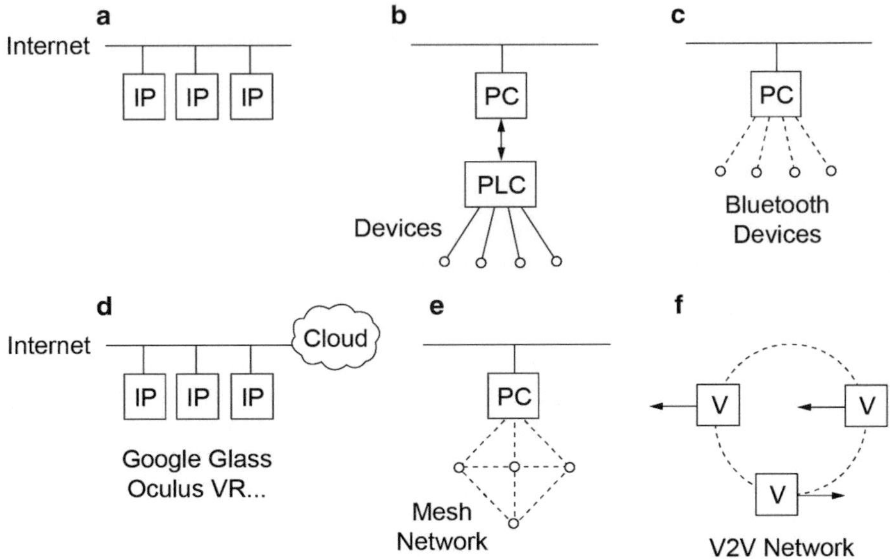


**Fig. 13.11** The variety of species of the Internet of Things (IoT): (**a**) IP-addressed devices, (**b**) industrial networks, (**c**) Bluetooth devices, (**d**) IP-addressed devices attached to clouds, (**e**) mesh networks, and (**f**) vehicle-to-vehicle communication systems

Personal electronic devices normally can be connected to a computer via WiFi, Bluetooth, or medical wireless channels. The more channels to be connected, the more troubles lie ahead. A physical attack on a wireless device can be malicious. One such example is a microchip that was invented for female birth control. The device is designed for implantation under the skin. It dispenses 30 micrograms per day of hormonal contraceptives. To conceive, women turn off the implant with a remote control; another click of the remote restarts it. The chip has to exhibit self-recognition and stealth to keep the implant wearer's data secure and to prevent it from being hacked.

Mesh or Mote networks must be connected to a server computer for programming and monitoring, which can be infected by worms or viruses. Amazingly, peer-to-peer sensor networks such as Crossbow's Motes behave much like worms themselves. Using the Linux-based TinyOS, the server and sensors can communicate by hopping through the nodes along the shortest path.[36] The server computer can upgrade the OS while sensors are still working. The OS software can spread across the whole network by hopping from one node to another like the spread of a virus. Motes' battery life is supposed to last up to an entire year. However, simple malicious attacks such as bursts of status inquiries can shorten the battery life and

[36]Crossbow 2005). Wireless sensor networks seminar, San Jose, April 19–20, 2005.

cause the sensor network to die out. A self-awareness program might be able to recognize the inquiry, monitor battery life, and respond accordingly.

Finally, vehicle to vehicle communication networks are based on a special frequency channel that is designated for communication purposes. It is not easy to modify the chip directly, but it is possible to fake the transmission signal (e.g. to simulate vehicles surrounding a target from all directions to force it to stop). Such hacking can be prevented by implementing multiple self-awareness channels beyond standard V2V signals.

## Summary

Self-awareness is a basic instinct present in genes, organisms, and groups. There are at least four self-awareness behaviors: self-replication, self-recognition, control and communication, and stealth.

The human immune system is a self-aware system that recognizes self and non-self, replicates defensive cells through self-replication and self-recognition, and coordinated attacks against invaders. In an immune system, all cells are security guards that can adapt to unseen invaders. The identity is verified through protein shapes from inside sources rather than a secret code given at the gate.

Behavior-based identification methods such as voice, keystroke dynamics, location, and gesture are promising alternatives to conventional identification methods. These behavior-based detection methods have the potential to detect unseen malware from learned behavioral patterns such as self-replication, self-recognition, stealth, and control and communication. Malicious behavioral patterns can be represented using temporal models such as Petri nets and visualization methods.

Finally, unique and diverse security functions are key requirements for designing the Internet of Things to prevent malicious remote attacks.

# CHAPTER 14

# Stealth

When we talk about hiding information, we often refer to data encryption. It is a mathematical way to convert signals to seemingly random noise. Only the authorized receivers have the key to unlock the data and read the signals. However, encryption itself also reveals a signal – the file contains some secrets that the sender wants to hide. Once the suspicious package is located, analysts may use a powerful computer to run decryption algorithms to unlock the secret data.

Is there any way one can hide information without raising attention? Let's look into how animals hide signals. In this Chapter, we explore stealthy algorithms that are inspired by secret communication in nature and the digital variations of stealth algorithms.

## White Noise

The simplest hiding algorithm perhaps is silence. Keep quiet when a suspected intrusion occurs. Many creatures instinctually use the silence strategy to avoid predators with low energy and ease. However, silence does not come for free. It interrupts normal communication and work. Another way is to generate noise so that predators can't detect individuals. For example, frogs sing together to communicate and they also generate synchronized noises to avoid being identified individually. When predators approach, they all keep quiet. Therefore, no one would be singled out. Modern sound engineering can produce customized soundscape environments on demand. A *white noise* machine is a device that generates sound across all frequencies. White noise is used mainly to ensure privacy or promote concentration because it covers the largest spectral range. Imagine turning on thousands of TVs simultaneously so that people in another room will not overhear a

© Springer-Verlag London 2016

Y. Cai, *Instinctive Computing*, DOI 10.1007/978-1-4471-7278-9_14

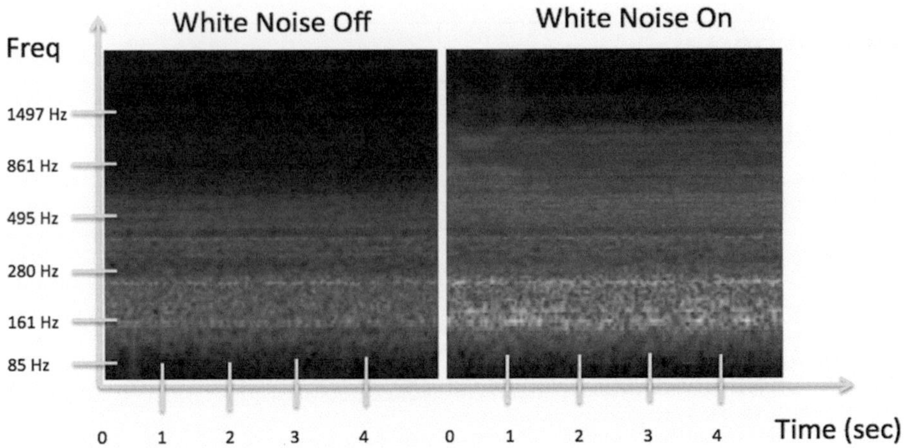


**Fig. 14.1** Spectrograms of the office environment with the white noise machine off (*left*) and on (*right*)

private conversation. Devices often produce colors of noise, rather than actual white noise, which has a harsh sound. For example, *pink noise* is a blend of high and low frequencies. *Brown noise* utilizes lower sound frequencies to generate a deep ambient rumble.

Let us consider a real-world case study: a contemporary office building installed a white noise machine to help with privacy in the office. Figure 14.1 shows the comparative spectrograms that the employee measured when the white noise machine was turned off and on. The spectrograms show that the frequency of the constantly bursting white noise ranges from 161 to 280 Hz. The noise is in a range similar to that of refrigerators and microwaves running simultaneously.

In the digital world, software for anonymous communication uses white noise to conceal the identity of users. But when the white noise is off, it is easy to single out individual users. For example, a student from Harvard was afraid of the final exam. So he emailed a bomb hoax in the morning, half an hour before the final exam. He sent the email from his laptop computer using TOR and Guerrilla Mail, a free Internet application that creates temporary and anonymous email addresses. Theoretically, no one could trace the email back to him. However, the FBI and police still found him. How? TOR is not a foolproof tool. It assumes there are many users in the same area. Unfortunately, only the student used TOR on that day. The rest of the local users were silent. It was easy to single him out.[1]

[1]AP (2013) Harvard student made bomb threats to get out of exam, cops say. CBS News, December 17, 2013. http://www.cbsnews.com/news/harvard-student-made-bomb-threats-to-get-out-of-exam-cops-say/

## Dedicated Channels

Our ability to detect vibrations in the air is referred to as our sense of hearing. Many animals can detect the presence of vibrations in the environment around them using a variety of vibration-detection structures.[2] This includes vibrations spread through a contact such as walking on a plant, or vibrations in the air or water around them.

Animals use separate channels or dialect signals for secrecy, so other animals can't jam their survival channel. A human's audible range is typically 20–20,000 Hz. Sounds outside of this audible range will not be detected by our ears, but can be detected by many animals. Elephants produce infrasound, or so-called long-wavelength sounds below 20 Hz to communicate with other members of their herd.[3,4] These low-frequency sounds travel great distances through the air, and through the ground, enabling elephant groups to maintain contact over a range of tens of kilometers.[5,6] The low-frequency sounds are below the audible range of human ears. On the other hand, bats, dolphins and some other species use ultrasound above the frequency of audible sound (20,000–100,000 Hz) for a variety of purposes, including communication, predator avoidance, and echolocation.[7] Bats use ultrasonic sound for navigation. They are able to catch flying insects while flying full speed in a totally dark environment.

In addition to sending out pulses of sound at different pulse rates, some animals also modulate the frequencies within their emitted sound pulses. For example, bats use frequency-sweeping clicks in the range between 10 and 60 kHz for both communication and navigation. By sweeping through frequencies bats can differentiate between echoes returned from objects at different distances. To get feedback on the location of prey or obstacles at a faster rate, the frequency of their pulses increases from between 8 to 15 Hz and 150 to 200 Hz when there is a tricky maneuver to be made.[8]

Since dolphins don't have vocal cords, they use clicks which emanate from small knobs near their blowhole. They have their own distinctive dialect of calls to secure communication. Dolphins vary their sound frequency from 1 Hz to 170 kHz

---

[2]Burnett S (2011) Perceptual worlds and sensory ecology. Nature.com. http://www.nature.com/scitable/knowledge/library/perceptual-worlds-and-sensory-ecology-22141730

[3]Langbauer WR, Jr et al (1991) African elephants respond to distant playbacks of low-frequency conspecific calls. Journal of Experimental Biology **157,** 35–46, 1991.

[4]Poole JH et al (1988) The social contexts of some very low frequency calls of African elephants. Behavioral Ecology and Sociobiology **22,** 385–392, 1988.

[5]McComb K et al (2003) Long-distance communication of acoustic cues to social identity in African elephants. Animal Behaviour **65,** 317–329, 2003.

[6]Pappas TC, Motamedi M, Christensen BN (2004) Unique temperature-activated neurons from pit viper thermosensors. American Journal of Physiology – Cell Physiology **287,** 1219–1228, 2004.

[7]Ultrasonic Sound. http://hyperphysics.phy-astr.gsu.edu/hbase/sound/usound.html

[8]Sales G and Pye D (1974) Ultrasonic Communication by Animals, Springer.

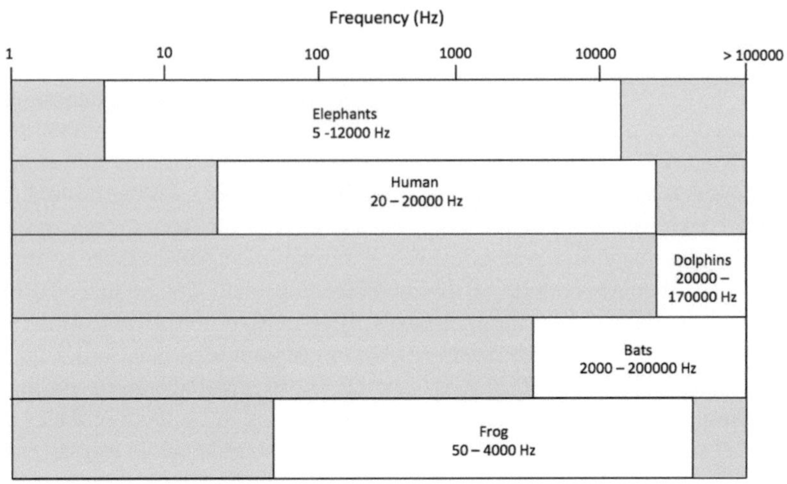


**Fig. 14.2** Range of sound frequency that can be detected by a variety of animals (After Stephan Burnett's Nature Education, 2011)

through a wide variety of clicks. The low-frequency sounds are used for territorial and mating calls, while the high-frequency sounds are used for better echolocation because the imaging resolution is proportional to the frequency of the imaging wave. Amazingly, typical frogs' sounds are audible, but some of them are able to produce ultrasonic sounds to communicate over lower frequency background noise. Figure 14.2 shows a diagram of the multiple frequency channels among humans, elephants, bats and dolphins.

LiFi (Light Fidelity) is a 5G visible light communication system that uses light from light-emitting diodes (LEDs) as a medium to deliver mobile, networked communication in a similar way as WiFi. Visible light communication works by switching bulbs on and off within nanoseconds, which is too quick to be noticed by human eyes. The bulb itself can be dimmed, so it's almost invisible to human eyes, but still functional. Line-of-sight is necessary for Li-Fi communication. On the other hand, this may be a weakness, but it is also an advantage for stealthy communication. The light waves cannot penetrate walls, so the range of the system is shorter, but it is more secure from hacking.[9]

Wi-Fi can work over one of two spectrum bands: 2.4 GHz or 5 GHz. The tradeoffs between them have to do with interference, range and speed. Cellphones use the band between 2 and 4 GHz, sharing it with WiFi and some satellite devices. Figure 14.3 shows the frequency distributions of modern digital communication systems.

The vehicle-to-vehicle communication system (V2V) uses the dedicated channel for short-range communications (DSRC), a standard set by FCC. It works in the

[9]Wikipedia: Li-Fi. http://en.wikipedia.org/wiki/Li-Fi

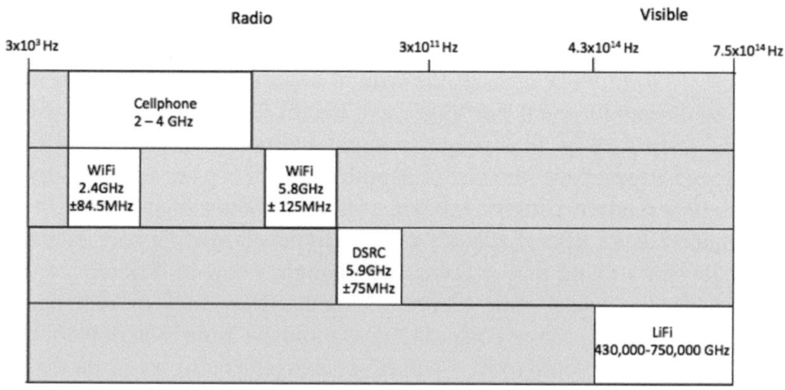


**Fig. 14.3** Frequency distributions of modern digital communication systems: cellphones, WiFi, Vehicle-to-Vehicle communication (DSRC) and LED-based light communication (LiFi)

5.9 GHz band with a bandwidth of 75 MHz and an approximate range of 1000 m. When implemented, the first generation of V2V systems will warn the driver but not take control of the car. Later implementations will include improvements to braking or steering around obstacles and eventually merge with self-driving cars. V2V will be a mesh network, meaning every node (car, smart traffic signal, etc.) will be able to send, capture and retransmit signals. Five to ten hops on the network would gather traffic conditions a mile ahead.

## Frequency Hopping

For a long time, torpedoes were remotely controlled by a long electrical wire to ensure that the communication channel between the commander and torpedo would not break. It was hard to make a wire long enough, even ten miles was too short, leaving the torpedo to chart its own course. Wireless radio waves were better than wire for controlling weapons. But radio waves had a serious flaw: enemies could access the same radio signal and jam it. During WWII, there was a high demand to find a simple and reliable way to communicate, but made it difficult for the enemy to decode the content.

Hedy Lamarr was a glamorous Hollywood star, born Hedwig Eva Maria Kiesler in Vienna, Austria in 1914. She was famous in the movie Ecstasy in 1933. The beautiful actress was also an inventor.[10] During WWII, she met pianist George Antheil at a party in Hollywood and they became friends. Antheil was a pioneer of experimental music and very familiar with mechanical equipment. Lamarr talked

[10]Couey A (1997) About CDMA spread spectrum. http://people.seas.harvard.edu/~jones/cscie129/nu_lectures/lecture7/hedy/lemarr.htm

to Antheil about her idea for guided torpedoes and its control system to support the war against the Nazis. Although Lamarr had no formal training in military weapons, she was exposed to military ideas as the wife of a weapons manufacturer.

Lamarr told Antheil about her idea for a Secret Communications System that could guide torpedoes to their target without being intercepted by the enemy, by sending messages between the transmitter and receiver over multiple radio frequencies in a random pattern. Lamarr used an analogy of piano keys to radio waves. A piano has 88 keys. Each key on a piano has a specific harmonic acoustic frequency. Just as a piano player holds and changes notes at different intervals to make a melody, the secret radio communication system held and changed radio frequencies to make a secret code. Signals could be transmitted without being detected. The message would move so quickly across the radio spectrum that anyone tuning in to a particular frequency would only hear a blip, and would be unable to intercept the message.

The invention was based on radio frequencies that changed at irregular periods that were synchronized between the transmitter and receiver. Lamarr thought Antheil could help to solve the synchronization problem, as Antheil had experience with sound synchronization for the soundtrack of twelve player pianos,[11] an airplane propeller and a symphony in his famous silent movie *Ballet Mecanique* with the filmmaker Fernand Léger in 1920.[12] Antheil's solution was to use two paper rolls with the same patterns of punched holes installed in the transmitter and receiver. If the two rolls were started at the same time, they would maintain the synchronization right on down to when the torpedo hit the ship. In fact, the two inventors designed their system to use eighty-eight frequencies – exactly the number of keys on a piano. See Fig. 14.4.

The two inventors submitted their patent proposal in 1941. On August 11, 1942, Lamarr and Antheil were awarded US Patent Number 2,292,387 for the Secret Communications System. However, the technology was not implemented until decades later. In the 1950s, engineers from Sylvania Electronic Systems Division began to experiment with the ideas in the Secret Communication System patent, using digital components in place of the paper rolls. They developed an electronic spread-spectrum system that handled secure communications for the US during the Cuban Missile Crisis in 1962.

Since then, the term "*spread spectrum*" began to be widely used. Initially, spread spectrum remained a military communications technology. In the mid-1980s, the US military declassified spread-spectrum technology, and the commercial sector began to develop it for consumer electronics. Today, it's an increasingly important component of mobile telephony. CDMA (Code Division Multiple Access) technology uses spread spectrum. In the US, since the Federal Communications Commission

---

[11] Erwin D (2008) Boogie Woogie on the player piano. YouTube. http://www.youtube.com/watch?v=ydcRAMZl0l0

[12] Supercinema77. The movie *Ballet Mecanique on Youtube.* http://www.youtube.com/watch?v=H_bboH9p1Ys

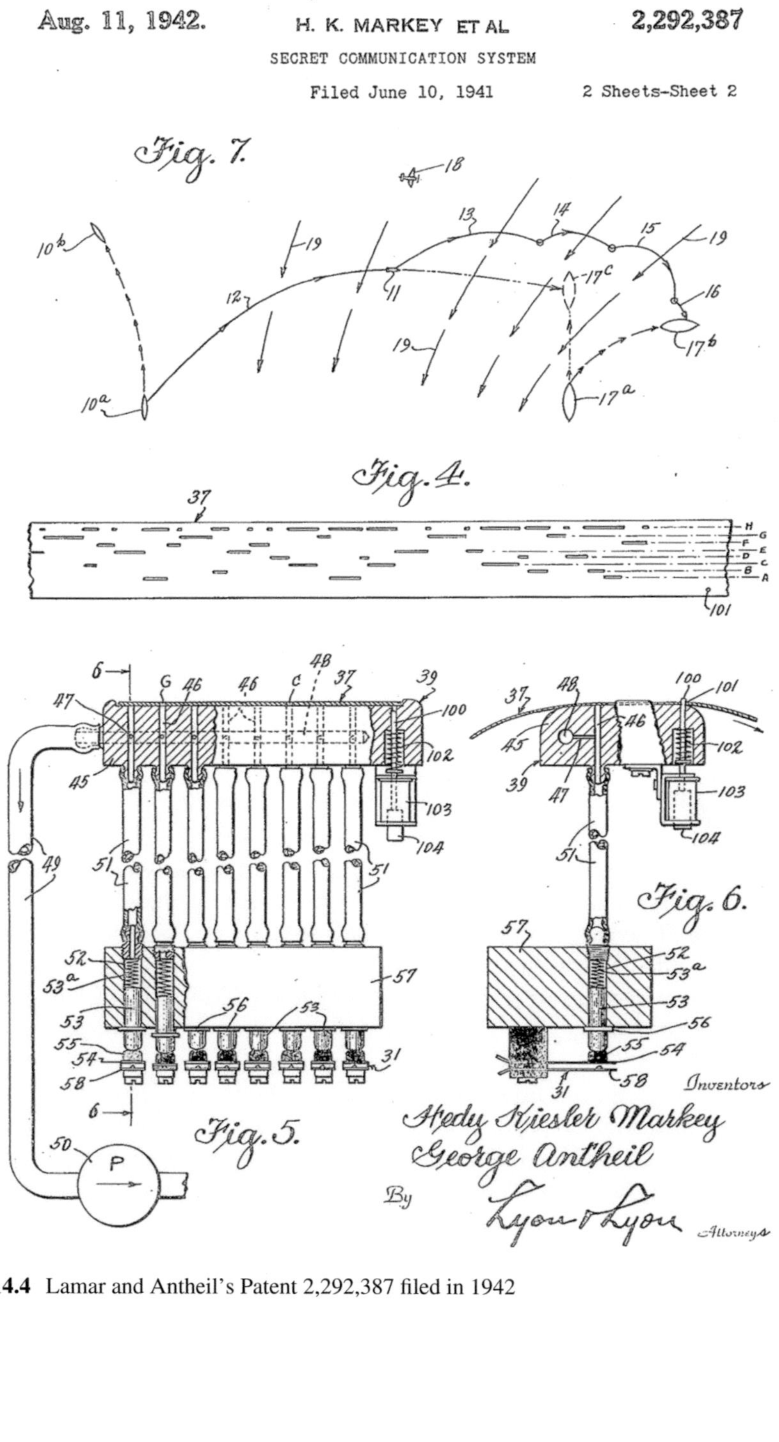


**Fig. 14.4** Lamar and Antheil's Patent 2,292,387 filed in 1942

(FCC) amended rules to allow frequency hopping spread spectrum systems in the unregulated 2.4 GHz band, many consumer devices in that band have employed various spread-spectrum modes. Spread spectrum has proven highly useful in cellular telephones, because its inherent encryption guarantees better privacy for cellular phone users. The technology has also proven to be an extremely efficient method for using radio waves. Rather than requiring each transmission to use its own frequency, spread spectrum enables people to simultaneously communicate over the same bands of spectrum without appreciable interference. Thus, as more people buy cellular phones, the increasing demand for spectrum can be accommodated by sharing the same frequencies.

Frequency-hopping spread spectrum (FHSS) is also applied in short distance communication systems such as Bluetooth or Adaptive Frequency-hopping spread spectrum (AFH). The variation of FHSS improves resistance to radio frequency interference by avoiding using crowded frequencies in the hopping sequence. The key idea behind AFH is to use only the "good" frequencies, by avoiding the "bad" frequency channels that are experiencing frequency selective fading, or being actively jammed. The AFH system can detect good or bad channels. To give further intelligence to the frequency management, the *cognitive radio* emerged which can be programmed and configured dynamically. Such a radio automatically detects available channels in the wireless spectrum, then accordingly changes its transmission or reception parameters to allow more concurrent wireless communication in a given spectrum band at one location.

## Steganography

Steganography is a Greek word meaning concealed writing. It is an ancient way to hide a secret message in an invisible ink or a micro dot in a letter or a cover image. The advantage of steganography is that the intended secret message does not attract attention to itself. In contrast, cryptography often arouses suspicion with plainly visible encrypted messages.[13] For example, a man may shave off his hair, tattoo a secret but readable message on his skull and wait his hair to grow back to conceal the hidden message. This is a kind of steganography. The man may also tattoo an encrypted message on his face. No matter how unbreakable the code is, people would notice there is a message on his face. This is the limitation of cryptography. But the two can be combined in many cases. For example, the man may tattoo an encrypted message on his skull, or tattoo an encrypted message on his face with an invisible ink that only displays under an ultraviolet light.

In digital steganography, communication may include hidden messages inside of images, sounds, videos, games, and network packets. Social media files are ideal for steganographic transmission because of their large size. For example, a sender may

[13]Cipher Machines. Retrieved on May 24, 2016. http://ciphermachines.com/types.html

embed a copyright signature in an image file by adjusting the color of every 100th pixel to correspond to a letter in the alphabet, a change so subtle that it is not visibly noticeable. In the following section, we present an overview of steganographic coding methods for images, videos and network packets.

## *The Least Significant Bits (LSB)*

A pixel in an image normally contains either 8-bits or 24-bits of information. The 8-bit resolution is good enough for representing gray-scale images, which has 255 shades of gray. The 24-bits resolution is appropriate for representing color images, which consist of Red, Green, and Blue channels with 8-bit resolution for each, which can represent up to 16,581,375 colors.

Now let's take one pixel out of a gray-scale image. It is an 8-digit number in the binary form: 00110011. We can break the number into two parts: the left most four digits are the most significant bits (MSB); and the right most four digits are the least significant bits (LSB). Any changes in the most significant bits will make a noticeable difference. On the other hand, any changes in the least significant bits won't make a perceptively noticeable difference. As the changes in the least significant bits won't have much visual impact, let's embed the secret message or image into the cover image. Assume we have the pixel with its value 00110011 as a cover and the secret pixel with its value 11001010. Let's embed the four digits of the most significant bits from the secret pixel into the four digits of the least significant bits from the cover pixel. Then we have the steganographic pixel with the hidden data: 00111100.

$$0011\underline{0011}\,(\text{cover}) + \underline{1100}1010\,(\text{secret}) = 00111100\,(\text{with hidden data})$$

Figure 14.5 shows the steganography image and the hidden secret image. This embedding bits method is called the Least Significant Bit method, or Padding method. It is quite simple to implement. However, it's easy to detect, by just checking the least significant bits.

How do we measure the image distortion due to hidden messages? We can manually compare the before and after images. But that is subjective. For computerized image quality evaluation, we can view the steganography as a noisy image and use the Peak Signal to Noise Ratio (PSNR) for an objective measurement. Assume the maximum possible pixel value of the cover image is MAX. When the pixels are represented in 8 bits per sample, it is 255. Also assume we have the mean squared errors (MSE) of pixels of the cover image and the resultant steganographic image. We have,

$$PSNR = 10\log_{10}\frac{MAX^2}{MSE} \tag{14.1}$$

**Fig. 14.5** The steganography encoded image (*left*) and the hidden image (*right*)

PSNR simulates human's logarithmic response to the visual differences between the two images. It has been widely used for measuring the performance of image compression algorithms. The higher the PSNR value the better the image quality. Typical values for the PSNR in video compression are between 30 and 50 dB, provided the bit depth is 8 bits. For 16-bit data typical values are between 60 and 80 dB. Acceptable values for wireless transmission quality loss should be between 20 and 25 dB. Therefore, the PSNR of the steganographic image cannot be lower than the typical range to avoid the noticeable imagery distortion. In Fig. 14.5, the PSNR of the image with the hidden image on the left of is 36.57, which is in the range of a low-end video compression.

## *Image Transformation*

An image transform is a mathematical transformation that converts pixel values into particular transform coefficients. The idea of the image transform methods is to hide secret information in the least significant image transform coefficients. Common image transforms include the Discrete Cosine Transform (DCT) and discrete wavelet transform (DWT). Both are widely used for image compression. So far, over 70 % of online images are compressed in JPEG format, which is based on DCT algorithms. DWT, on the other hand, is the compression method for the JPEG2000 image format. DWT has a better compression ratio and better image quality over DCT. However, DWT is computationally expensive, compared

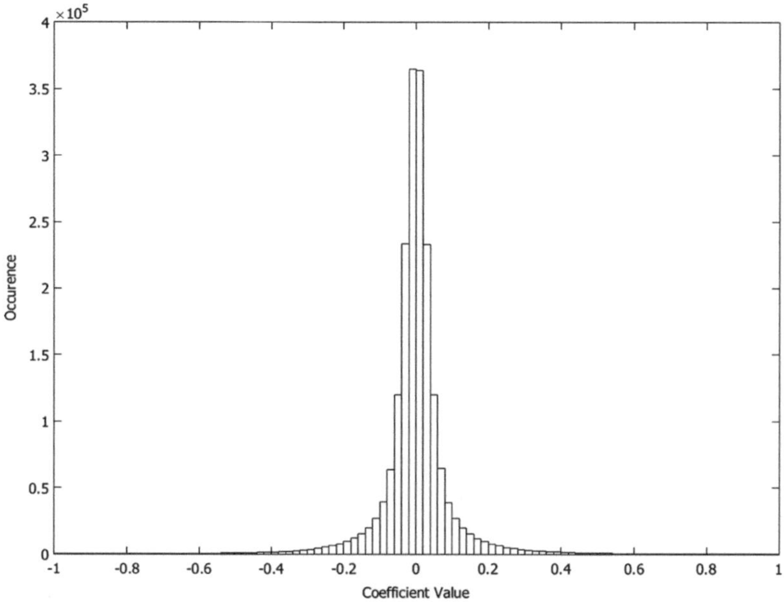


**Fig. 14.6** DCT coefficients in an uncompressed image contain many least significant coefficients, which can be set to zero, leaving plenty room for hiding secret messages or images

to DCT. Both DCT and DWT can be used for steganographic coding.[14] But from a popularity point of view, the DCT-based image steganography makes more sense at this moment.[15]

The DCT decomposes an image into many cosine waves. Each wave has a coefficient. Those coefficients form a two-dimensional matrix to represent the pixel intensity variation frequencies in the vertical and horizontal directions. Normally, a very small number of low-frequency coefficients have significantly large values, but the rest of the coefficients, typically the high-frequency coefficients, have rather small values, close to zero. Figure 14.6 is a histogram of the DCT coefficients of the image on the left of Fig. 14.5.

If we set the small coefficients to zero, it will compress a lot of the coefficients, which leads to a deep compression of the image after the inverse discrete cosine transform (IDCT) is performed. Eliminating a lot of near zero coefficients normally makes little noticeable changes to our eyes. In fact, our vision is less sensitive

[14]Desai J, Hemalatha S, Shishira SR (2014) Comparison between DCT and DWT Steganography Algorithms" International Journal of Advanced Information Science and Technology (IJAIST). Vol. 24, No. 24, April 2014.

[15]Patel H and Preeti D (2012) Steganography technique based on DCT coefficients. International Journal of Engineering Research and Applications. Vol. 2. No. 1, 2012: 713–717

to the subtle high-frequency intensity details. This is the so-called "psycho-visual redundancy" in image compression. The so-called "*zero-hiding*" method replaces the zero value coefficients with a secret message or image. The pseudo code for the DCT-based zero-hidden is presented below:

```
Compute DCT of the cover image
Set the DCT coefficients to zero if less than a threshold
Replace those zeroes with the hidden message
Compute inverse DCT to get the steganographic image
```

## *Video Steganography*

If we made an analogy based on image steganography being similar to hiding a person in a car, then video steganography would be like hiding a person in a train with many more rooms to hide. Obviously, video media provides more rooms for hidden messages and they are more difficult to discover. A video contains visual, audio and metadata channels. It is possible to hide messages in every channel. Here we only focus on the visual channel because it has the largest capacity for embedding messages, compared to metadata and audio channels.[16]

In order to implement video steganography, we need to understand video compression algorithms. A video consists of a number of frames of images. Because video processing is time-sensitive, and often happens in real-time, most algorithms use macro-block as a processing unit, which contains $4 \times 4$, $8 \times 8$, or $16 \times 16$ pixels per unit. *Video compression removes color spectral, spatial, and temporal redundancies, which can be replaced with hidden messages.*

Removing color spectral redundancy is based on the fact that human eyes are sensitive to brightness changes and less sensitive to color changes. Studies show that over 90 % of objects can be recognized just from grey-scale photos. So a typical video compression algorithm would decompose the full-color RGB channels into one intensity (grey-scale) channel and two color channels. For example, European systems commonly use the YCrCb format, in which Y is the intensity, and Cr and Cb are the red and blue channels. North American systems use the YIQ format, in which Y is the intensity channel, and I and Q are color channels. Multimedia MPEG systems use YUV format, where Y is the intensity channel and U and V are the color channels. Our eyes are sensitive to the intensity channel. Let's keep it uncompressed. The color channels can be compressed to 50 % (4:2:2) of their original resolutions, or even 25 % (4:1:1) of it. In light of this, it is logical to hide secret messages in those color channels because it is rather difficult to perceive the visual differences after hiding messages.

[16] Al-Frajat, AK et al (2010) Hiding data in video file: an overview. J. of Applied Science, Vol. 10, No. 5, 2010: 1644–1649.

Removing spatial redundancy is an image compression within a frame. This is similar to DCT and DWT image transform compression algorithms. However, even after individual frames are compressed, transmitting all the frames is still not efficient because the differences between adjacent frames are subtle, even unnoticeable to naked eyes. In extreme cases such as surveillance camera videos from an office building at night, many frames are just empty background without any change. So it doesn't make sense to send every frame without a temporal compression.

To remove temporal redundancy, the inter-frame differences are modeled to create highly compressed virtual frames. Let's call the original frame the I-frame. Assume we can track the macro-blocks between the I-frame and the next frame and the movement of macro-blocks can be represented by motion vectors, which include angles and lengths. The content in the next frame can then be predicted from the motion vectors. We call this reconstructed frame a P-frame, or a predicted frame. So the encoded video started from an I-frame, and then a few P-Frames, and then an I-frame again, followed by P-frames. However, after a few P-frames, the predicted locations of macro-blocks might have drifted off from the actual locations. So we need to insert B-frames, or bi-directional frames to interpolate between the P-frame and the next I-frame for smoother transmission. Now we have a typical frame sequence:

$$I \cdot B \cdot B \cdot P \cdot B \cdot B \cdot P \cdot B \cdot B \ldots$$

Figure 14.7 shows how the P-frame is reconstructed from the macro-blocks in the I-frame and motion vectors.

Video formats such as MPEG-4 use file structures called containers to store the bit streams produced by the encoder. Any video data file on the disk can be viewed as a sequence of bytes. This sequence of data is arranged as a collection of boxes, including file type (ftyp), media data (mdat), and metadata. The file type box (ftyp)

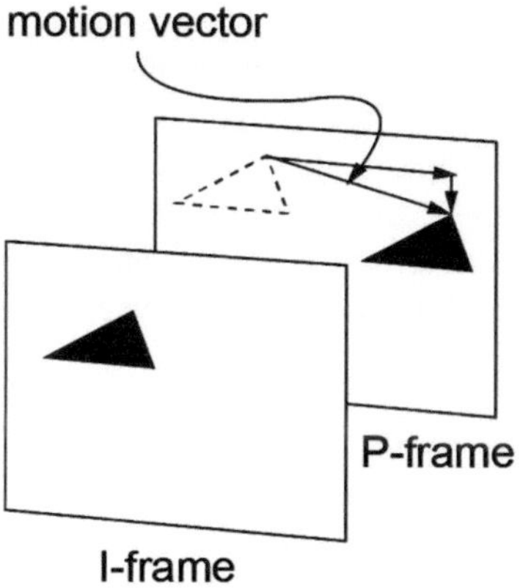


**Fig. 14.7** The P-frame can be reconstructed from the macro-blocks in the I-frame and Motion Vectors

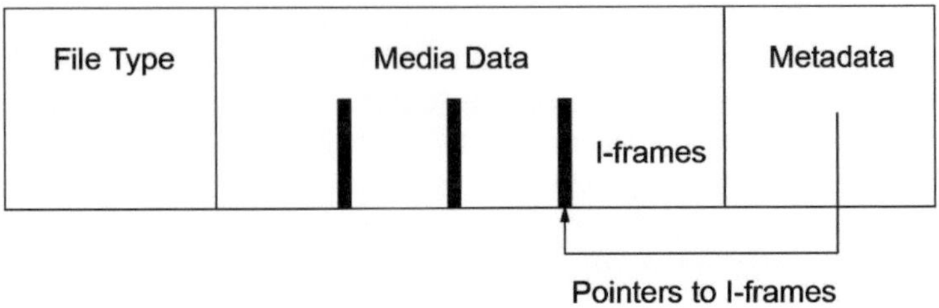


**Fig. 14.8** The video file MPEG-4 data structure

tells the movie file type such as m4v. The media data box (mdat) stores actual image transform coefficients such as DCT coefficients and motion vectors that are associated with I, P and B-frames. The metadata box contains a movie header box, or moov that stores video frame track metadata such as closed captions (CC) and sound track data. The metadata box also contains the sample table sync samples (stss) with pointers to I-frames stored in the media data box, and sorts them in a list so that the video player can run video frame by frame. Figure 14.8 is a simplified diagram of the data structure of the popular video format MP4 (MPEG-4), which has been used in most mobile media players.

We can hide messages in the I-frames in video files. Given an MP4 video file, we can locate and process the key frames (I-frames) with the metadata and actual media data. In the I-frame, we calculate the DCT coefficients of the block. For all DCT coefficients below a threshold, replace the least significant bits of the pixels by hidden data bits. Then, we make a key that stores the locations of the pixels whose least significant bits were changed. Here we store one bit of hidden data in one pixel, if the DCT coefficient of that pixel location is less than a threshold. The key is required during the extraction phase and acts like a passcode for the recipient of the video. Thus, we convey this passcode along with the video with hidden data to the recipient, who is then able to completely decipher the secret message. The extraction process works in just the reverse way: locate the I-frames in the video data file and use the key passcode to locate the pixels within an I-frame where data has been hidden. Then extract the least significant bit from the pixel location. Finally, reconstruct the hidden message in the form of text or image.[17]

The advantage of this method is that we do not need to decode the video into I, P and B-frames separately. Everything is done in one place and stored back into the file directly. Thus we can avoid the image quality losses in decoding and re-encoding I, P and B-frames. Furthermore, it prevents potential errors in extracting the reassembled I-frames using libraries such as FFMPEG. If the extracted I-frame is not the original I-frame, then the hidden data would be lost.

We can also hide data in the motion vectors, which are used to encode and reconstruct both the predictive P-frames and bidirectional B-frames in the

[17]Pandit H and Wable C (2015) Video steganography, CMU ECE 18-798 Project Report. Fall 2015.

compressed video.[18] Given a video data file, we can separate the video into frames. Each frame can be decomposed into one intensity and two color channels. We can select one channel for hiding a message. We then divide the single frame image into macro-blocks and calculate the motion vectors that include magnitude and angle, starting with the second frame. If the magnitude is larger than a threshold, then the motion vector is considered to be a candidate for hiding data. A subset of the motion vectors are selected if their associated macro-block prediction error measured in PSNR is below an initial threshold value – so viewers won't pay attention to the area that is already blurry. Also, if the angle is acute, the data is embedded in the horizontal component of the motion vector; and if the angle is obtuse, the data is embedded in the vertical component of the motion vector. Once we have finalized the candidate motion vectors (CMV), replace the least significant bit (LSB) of both magnitude and angle by bits of the secret message. In the decryption phase, we calculate the motion vectors of each frame starting from the second frame P-frames and B-frames. Calculate the magnitude and angle of each motion vector to find out if any data is present in them. If the magnitude is larger than the threshold, continue the decryption process for the current frame and extract the data using the least significant bit part accordingly.[19]

Similar to the steganography method discussed above, this algorithm can directly hide messages in a compressed video so that it makes detection hard. The algorithm optimizes the suitable value of the threshold for selecting the embeddable macro-blocks so that it maximizes the data hiding capacity. It is also rather fast to encode and decode the messages within motion vectors. However, changing motion vectors may drop the video quality and not all the motion vectors that have large magnitudes are associated with macro-blocks of high prediction error. There are motion vectors whose magnitude is small but their associated macro-block prediction error is high.

## *Network Steganography*

In contrast to conventional steganography that hides information inside individual digital files including images, sounds and videos, *network steganography* instead uses the network protocol itself, which makes it more difficult to detect hidden data. Network steganography hacks the Internet Protocol (IP) in data communication, voice or text based on the Internet. All the data traffic on the Internet travels via a method called packet switching, which chops digital data into tiny packets, and sends them over a network shared by many users. The IP follows the standards for

[18]Pan F, Xiang L, Yang XY and Guo Y (2010) Video steganography using motion vector and linear block codes. Proceedings of IEEE International Conference on Software Engineering and Service Sciences (ICSESS), 2010.

[19]Penmetsa S and Yendluri D (2015) Video steganography, CMU-ECE-18-798 Project Report, Fall 2015.

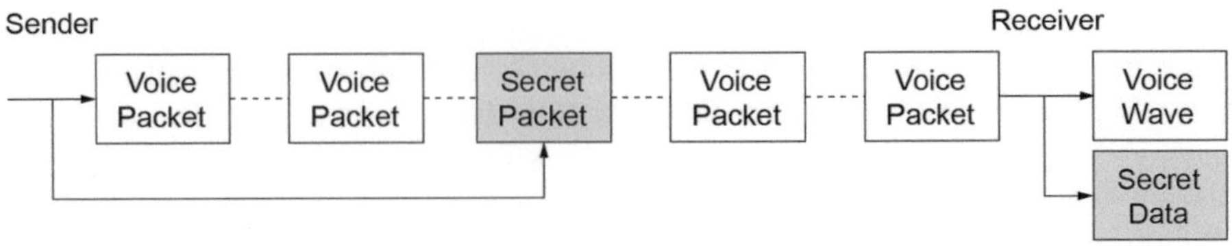


**Fig. 14.9** Network steganography over IP

packaging those packets, for example, packets don't need to reach their destination in any particular order. This is fine for order-insensitive data like emails or static web pages but problematic for voice or video data.

Voice over Internet Protocol (VoIP) has been developed to meet the challenge of streaming live audio data online. The algorithm breaks the audio data into packets and routes them over the Internet like a regular IP. Since telephone calls must occur in real-time, VoIP uses the Real-Time Transport Protocol (RTP) and the User Datagram Protocol (UDP). RTP provides timestamps for packet synchronization and sequence numbers for packet loss and reordering detection. So it gets time-sensitive video and audio data to its destination in order. UDP takes care of fast packet streaming without jiggling. In contrast to the conventional protocol TCP that requires sending packets again if there are any errors or missing packets occur, causing annoying jiggling effects, UDP doesn't bother to check whether the data are reliable, intact, or even in order. It just keeps sending or receiving packets, despite packets that are sometimes out of sequence. Relatively speaking, UDP is not so reliable, but it provides much smoother audio streaming over wireless networks. Thanks to the psycho-audio phenomenon – our perceptions have a pretty good tolerance for missing a few packets in an audio stream.

The unreliability of the UDP may result in several types of error, including data corruption and lost data packets. Steganography exploits these errors and tries to hide secret data spread among many IP packets. When the receiver's computer gets a packet, it checks for errors using that packet's checksum, which is a numerical value assigned to a packet for authenticating the integrity of the data. Normally, if the checksum is wrong, the computer discards that packet. But if a receiver has the steganography program installed, it won't discard these intentionally wrong checksums – instead, it will know that these are precisely the data packets to scan for the secret message. Because packets don't stay anywhere except in the recipient's computer, there is no easy way to detect them unless an analyst recorded all the traffic data and could sit down to download a suspicious image or audio file for investigation.[20] See Fig. 14.9.

[20]Lubacz J et al (2010) Vice Over IP: the voice over IP steganography threat. IEEE Spectrum, posted Jan. 29, 2010. http://spectrum.ieee.org/telecom/internet/vice-over-ip-the-voip-steganography-threat

In addition to VoIP, there are more than a hundred techniques that can transfer secret data using meta information, such as header elements. There are also many options for the hidden data carrier. For example, to identify the data packets associated with the silence between words. These can then be packed with covert data.

## Steganalysis

Steganalysis is the study of detecting the signs of steganography in data. The first place to check is file size. For example, a normal compressed image in JPEG format is relatively small. If someone sent a 700-megabyte picture of a dog, then it might arouse suspicion. In order to find possible hidden messages in image transform-based steganography, such as DCT coefficient zero-hiding, we can analyze the histogram of the DCT coefficients. Normally in a compressed image, the DCT coefficient values below the threshold are dropped to zero. If the zeroes are replaced with hidden data, then the histogram of the DCT coefficients would show a shape similar to that of an uncompressed image.

Also if the hidden messages are in alphabetic text, then we can check out the repeated patterns. For example, in a natural image, the distribution of the pixel intensity is rather random. When the hidden words occur, certain set of letters would have recurrent patterns. Statistic modeling can help detect these repeated patterns and detect potential steganography.

## Summary

Stealth is a survival instinct used in nature to avoid predators or competitors. It is seen in cells, bacteria, insects and animals. From an instinctive computing point of view, security and privacy work together seamlessly for survival.

The white noise method approach for survival is to join in with multiple signal sources to avoid being singled out. Dedicated communication channels help animals and electronic devices communicate efficiently without interference or being discovered. Frequency Hopping switches frequencies on the fly to avoid being discovered or jammed.

Steganography conceals words and images inside images or sound signals without visible changes to the cover image or sound. Secret messages can be hidden in the least significant bits (LSB) of pixels, or in the image transform coefficient values (e.g. DCT and DWT) that are below a detection threshold. Video media provide much more space for steganography, including room for hiding data in the intra-frame (I-frame) based and motion vectors in the forward predictive frames (P-frames) and bidirectional frames (B-frames). Furthermore, network steganography embeds secret data inside the network protocols, where it is even more difficult to discover. However, steganography is not totally fool-proof. The tell-tale signs can be revealed by analyzing the file size, the histogram of the transform coefficient values and statistical modeling of the repeated patterns.

# CHAPTER 15

# Visual Privacy

Our sense of privacy comes from cavemen because we are still the hunter-gatherers of 100,000 years ago. We always feel uncomfortable if someone is staring at us because we are hunters and we don't want to be hunted. Privacy is also about territorial rights. We take private space as a measure of security. A private space is normally the most vulnerable place. Anything that enters such a space is seen as a threat or a discomfort, triggering emotional, even aggressive reactions. In this Chapter, we discuss visual privacy for cavemen and cavewomen in the digital age, including personal territories in physical and cyber spaces, privacy technologies, and privacy research methodology.

## Evolving Personal Space

Private space in human society is more sophisticated than in animal communities. People normally regard the region surrounding a person as so-called *personal space*. Like animals' territorial rights, most people feel discomfort, anger, or anxiety when their personal space is invaded. Allowing another person to enter one's personal space is a privilege of a relationship between those people. In his 1966 book, *The Hidden Dimension*,[1] anthropologist Edward T. Hall introduced the concept of *proxemics* in personal space, which describes the subjective dimensions surrounding each person, called *personal reaction bubbles*, or *interpersonal space* and the physical distances they try to keep from other people. Researchers have revealed that there are four different zones of interpersonal space. The intimate zone is reserved for lovers, children, and close family members. The personal zone is used for conversations with friends, to chat with associates, and in group discussions. The social zone is reserved for strangers, newly formed groups, and new acquaintances.

[1]Hall ET (1966) The hidden dimension. Anchor Books. ISBN 0-385-08476-5.

© Springer-Verlag London 2016

Y. Cai, *Instinctive Computing*, DOI 10.1007/978-1-4471-7278-9_15

The public space is used for open spaces with random interactions such as plazas, speeches, and theaters, which are subjected to local laws, traditions, and regulations.

Personal space is highly variable according to subtle cultural rules. In Roman times, politicians would bathe together during meetings. Their boundaries were divided by social status. In the book *History of Private Life*, Philippe Ariès and Georges Duby explored the changes of personal space over time. With regard to a person's social standing, personal space has not changed much over the years, perhaps due to human instinct. A person of more affluent status expects a larger personal space.[2]

In modern society, it is difficult to maintain personal space in crowded areas, such as public transit systems or elevators. When normal space requirements simply cannot be met, individuals will *adapt* to the situation but still try to avoid *physical contact* and *eye contact*, which are intolerable entries into their personal space indicating harassment and challenge. Figure 15.1 shows Hall's personal space diagram. It also introduces new levels of personal space within cyberspace. Comparing the two diagrams, we can see the level of intrusion in cyberspace is much stronger than before, in terms of broadness and depth.

Cyberspace also extends the concept of personal space beyond physical dimensions. Different digital communication modes, such as email, social media, SMS, and Skype, cover different personal spaces. Due to their ubiquitous nature, smartphones have been used widely in public spaces. Ironically, mobile phones also reveal personal data, including location, contacts, and contents. One of the by-products of this is the *Bluetooth Sensor*. The wireless signal emitted from a driver's Bluetooth headpiece can be collected by highway administrative agencies for measuring the vehicle traveling speed based on the *Doppler effect* of the wireless signal waves.

The invasion of robots is happening as we speak. What if a mobile robot enters our personal space? Shall we then redefine it as a human-robot space?

The rapid propagation of surveillance cameras on highways, streets, and in shopping centers creates privacy concerns. In China, highway CCTV camera systems sometimes flash lights onto vehicles to record drivers' faces. In Europe, license plate recognition (LPR) systems have been used to detect speeding and for collecting rush hour tolls. However, many countries have restrictions on camera use in public places. For example, in Germany, the use of CCTV cameras in line-of-sight of residential areas is restricted.

Some camera products come with a warning light to indicate that the camera is in operation. For example, the webcams on personal computers have a green light to indicate their "on" state. The light cannot be turned off from the software because it is wired directly from the chip. However, hackers have been able to bypass the chip to turn the light off.

[2]Ariès P and Duby G, eds. (2001) A History of Private Life. Belknap Press. ISBN 978–0674399747.

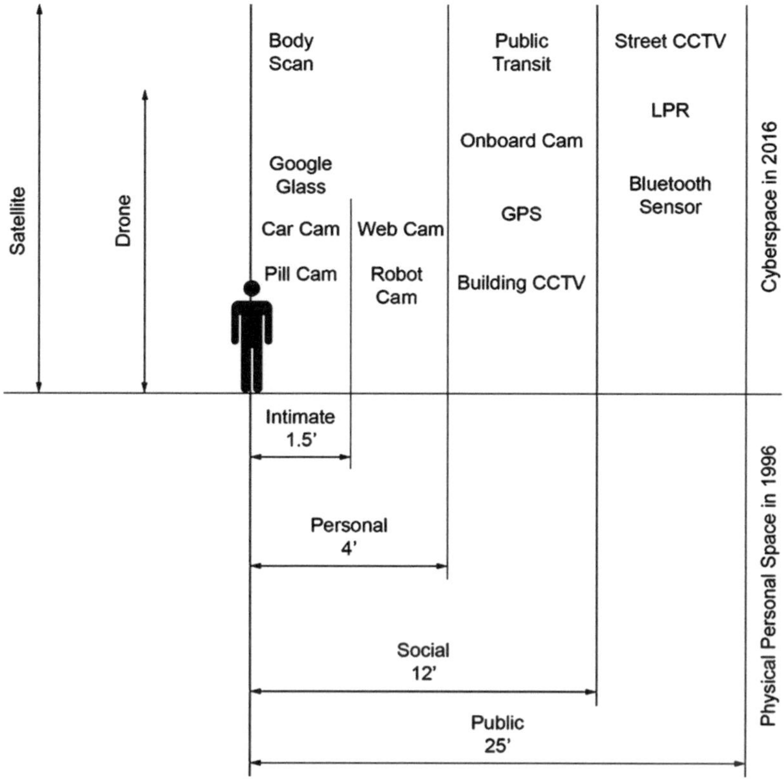


**Fig. 15.1** Hall's physical personal space (*bottom*) in 1966 and personal space within the realm of cyberspace (*top*) in 2016. The measurements are in feet

Many new gadgets are actually quite invasive, such as Google Glass, which is a wearable camera and display built into eyeglasses. Many bars, classrooms, theaters, and restaurants have banned the use of these glasses, simply because it invades peoples' privacy. Because we are simply the cavemen and cavewomen of old time, we will always feel uncomfortable if someone watches us and tries to record our every movements and every words.

The caveman principle shows us why many things are never well-received by the public. Instinctive computing, on the other hand, attempts to find alternatives that balance our multiple instinctual needs, including security and privacy. Can we hit two birds with just one stone?

## One-Way Mirrors and Smart Glass

The invention of glass was a giant leap forward for humanity, enabling people to view their surroundings from the inside of buildings. Glass also reveals the inside of structures to outside viewers. In 1945, American architect Philip Johnson designed his Glass House in New Canaan, CT, which is an iconic work of architecture because of its innovative use of glass walls and its seamless integration into the surrounding landscape.[3] In fact, Johnson himself lived in the Glass House from 1949 until his death in 2005. On the opposite side of the spectrum, American architect Frank Lloyd Wright designed the famous house at *Falling Water* like a cave with low ceilings and modestly sized windows. He once said, "Philip Johnson is a highbrow. A highbrow is a man educated beyond his capacity. His house is a box of glass – not shelter. The meaning of the word shelter includes privacy."

*One-way mirrors* enable insiders to see outsiders, but not the other way around. Physically, a one-way mirror is a piece of glass coated with a thin and almost transparent layer of metal. The result is a mirrored surface that reflects only some light and is transparent to the rest. A one-way mirror is typically used as a normal mirror in a brightly lit environment (e.g. on the street in daylight), with a much darker environment on the other side (e.g. a room without light). People on the brightly lit side see their own reflection, like a normal mirror. People on the dark side see through it, like a transparent window. One-way mirrors can be used for hiding security cameras where the camera is placed inside a mirrored enclosure without being discovered. British artist Monica Bonvicini created the public installation, "Don't Miss a Sec." The piece was a functional toilet that featured a cubed structure with one-way mirrors. Visitors can use the public bathroom while observing commuters pass by who are not able to see the people occupying the bathroom.[4]

*Smart glass* is also called switchable glass or switchable window. It refers to glass or glazing that changes light transmission properties under the application of voltage, light, or heat. *Electrochromic windows* are commonly used for changing opacity levels. This technology is based on the phenomenon of electrochromism, where a metal oxide material (such as NiO) will reversibly change opacity or color when electric charge is applied.[5] Most electrochromes are all-solid systems with multiple layers, including glass substrates, transparent electrodes, and electrochromic mixtures (see Fig. 15.2). They have a number of potential uses including displays, smart mirrors and windows, active optical filters, and computer data storage.[6] InterCity Express (ICE) high speed trains in Europe use electrochromatic glass panels between the passenger compartments and the driver's cabin. The standard mode is clear, and can be switched by the driver to frosted or translucent modes, mainly to protect the privacy of the operators and to conceal unwanted sights from passenger view. Electrochromic windows are also used in the Boeing 787 Dreamliner aircraft.

---

[3]The Glass House. http://theglasshouse.org/

[4]Carlile J (2004) A new way to view London from a toilet. Weird news on NBC News. http://www.nbcnews.com/id/4326340/ns/world_news-weird_news/t/new-way-view-london-toilet/#.VBUGgWhIXaU

[5]University of Gent. Electrochromic displays. http://lcp.elis.ugent.be/tutorials/tut_echrom

[6]Salaneck WR, Clark DT, Samuelson EJ (1991) Science and Applications of Conducting Polymers, IOP Publishing Limited.

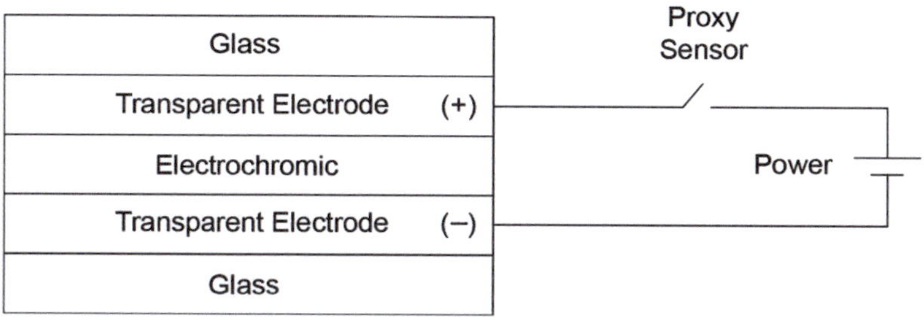


**Fig. 15.2** Structure of the electrochromic display

Perhaps, the most remarkable application of smart glass is for visual privacy on Singapore's Bukit Panjang Light Rail Transit (LRT), which is an eight-mile long people mover in downtown Singapore. It is a fully-automated state-of-the-art transit system. In the northwest region of the city, population and housing density is relatively sparse. LRT trains are used to bring residents of these areas closer to their homes from the main train stations. As a result, the trains drive close to residential buildings. Visual privacy was a concern here, since commuters on the trains would have been able to easily see into other people's living rooms. To protect residents' privacy, electrochromic windows were installed on all vehicles. Whenever the train reaches the edge of a residential area, an electromagnetic signal is sent from the rail track to the vehicle control system. A voltage is applied to the smart windows, and the windows appear frosted. The windows are returned to their transparent state after the train has left the residential area. The process is so ubiquitous that many passengers do not even notice the change. Figure 15.3 shows the smart mirror in action while the train passes a residential area.

It is worth noting that the engineering design was implemented in 2000, many years before privacy issues became a highly-debated subject. One-way mirrors and smart glass are not only physical devices but also design metaphors for solving visual privacy problems. One-way mirrors enable one-directional opacity and transparency. Smart glass switches between opacity and transparency on both sides.

**Fig. 15.3** Smart Windows on the Bukit Panjang Light Rail Transit trains in Singapore

## Webcam Privacy

Growing webcam-based technologies enable telepresence such as live video banking. This is especially useful in remote places where bank branch offices are not easily accessible. Videoconferencing facilitates face-to-face communication, but it can also invade one's personal space. People do not want to show their home or office backgrounds to the bank manager on the other screen. How can a user's privacy be protected in a webcam video stream? The key involves segmentation of the foreground and the background.

The first thought is the "green screen" solution. A bright green colored cloth is placed in the background of the user. Software can remove the green colored pixels and replace them with another background image of the user's choosing, such as a beach scene. The reason behind using a green color cloth is that people do not normally have a lot of green color on their face, unless the user is very sick. However, the green screen method is not foolproof. If the user were to wear a green shirt, then parts of the foreground would be mistaken as the background.

Ideally, it would be convenient if the algorithm could segment the speaker from the background in a natural environment without having to use a green cloth or white wall. Assume the background does not move. The first step is to catch the image of the background without humans. Imagine the intensity distribution of the background is like a three-dimensional terrain, where the height is represented by the intensity of the pixel. We can approximate the intensity distribution with multiple Gaussian functions to approximate the terrain of the background intensity. We call this a Gaussian Mixture Model (GMM). Once the background model is created, we can subtract the foreground from the background whenever there is a human presented in the background. Figure 15.4 shows a subject with the original background and with the modified version.

**Fig. 15.4** The original webcam video from a bedroom (*left*) and with the replacement of background of a beach (*right*)

Background modeling is not perfect. To improve segmentation quality, we can use an RGB-D camera, such as the Kinect, to obtain depth information. With simple thresholding, we can segment the foreground and background based on the depth measurement. We can also use two webcams to create a photometric stereo model for the depth information. In addition, we can use a face detection algorithm to ensure that the face is always in the foreground.

## The Body Scan Problem[7,8]

The human body is one of the most vulnerable systems in nature and privacy is one of its most vulnerable aspects. Our bodies are highly visible with the rapidly growing technology of visual media, such as video surveillance, thermal imaging, microwave body imaging, and medical imaging. *Visual privacy* is a sensitive topic because it literally deals with humans' private parts.

The growing demand of three-dimensional holographic imaging systems has created significant interest in many disciplines. Current devices operate using a coherent millimeter wave transceiver to reflect signals off the human body and any objects carried on it. They then use holographic reconstruction algorithms to reconstruct an image of the person. These devices penetrate items that are less dense, such as clothing and hair. Unlike the latest metal detectors, this system can also detect non-metal threats or contraband, including plastics, liquids, drugs, and ceramic weapons hidden under clothing. However, these high-resolution scanned images reveal intimate bodily details and have raised serious privacy concerns.

Most of the research and development of human scanning systems has occurred within unpopular projects in a few private companies or in government laboratories. As a result, most of the technologies are either device-specific or proprietary, which has slowed down the overall advancement of privacy technology for 3D body scanning systems. It presents a bold challenge to the field of Computer Science. As Artificial Intelligence (AI) has been applied to many security systems, we ask the question, "Can we use AI to protect our privacy in *visual media*?" (Fig. 15.5).

### *Privacy-Aware Rendering*

Visual privacy is a complex issue involving computer vision, graphics, aesthetics, and even social values. So far, there is no quantitative measurement of how

[7]Law J and Cai Y (2006) Feature Hiding in 3D Human Body Scans, Journal of Information Visualization, Vol. 5, Number. 4, 2006.

[8]Laws J and Cai Y (2006) A Privacy Algorithm for 3D Human Body Scans, in LNCS 3994, Springer.

**Fig. 15.5** The human scanning system was tested at an airport

people sense the visual presentation of 3D holographic human bodies.[9,10] Here, we experiment with several rendering methods that blur or obscure private parts of the human body. For example, we can increase the mesh size to blur certain areas, or we can change the transparency of the skin texture to blur details in an area. We can set the scale of blurriness so that users can evaluate their preferences (see Figs. 15.6 and 15.7). In this study, we want to design an empirical model of visual privacy assessment, including sensitivity analysis under different conditions. We aim to prove that people's opinions towards visual privacy are dynamic, peer-influenced, and even creative.

[9]McMakin DL, Sheen DM, Collins HD, Hall TE, and Severtsen RH (1995) Wideband, millimeter-wave, holographic surveillance systems. EUROPTO International Symposium on Law Enforcement Technologies: Identification Technologies and Traffic Safety, Munich, FRG, SPIE, vol 2092, pp. 131–141, 1995.

[10]Sheen DM, McMakin DL, Collins HD, and Hall TE (1993) Weapon detection using a wideband millimeter-wave linear array imaging technique. EUROPTO International Symposium on Substance Identification Technologies, Innsbruck, Austria, SPIE, vol. 2092, pp. 536–547, 1993.

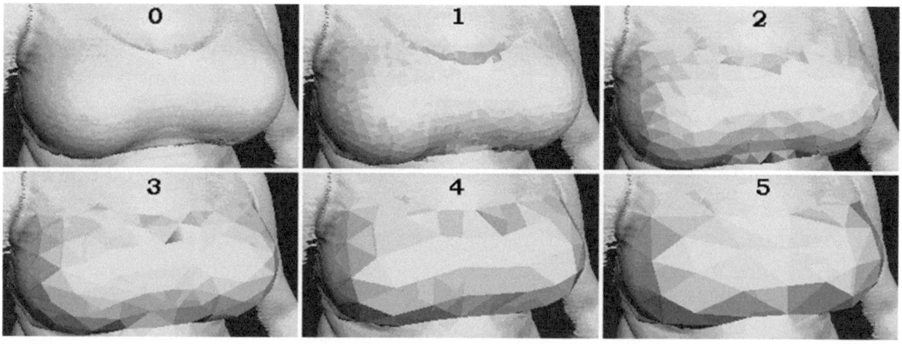


**Fig. 15.6** Blurred surface renderings at varying resolutions

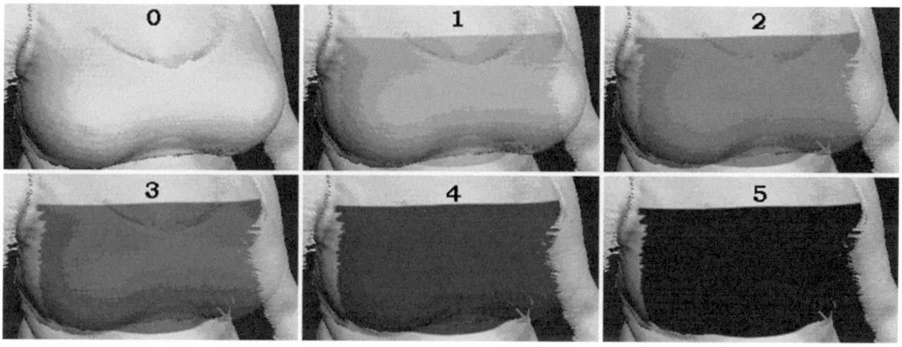


**Fig. 15.7** Renderings with different transparencies

## *Usability Study*

It is common knowledge that most people disagree on how much privacy can be given up for security. It was a goal of ours to find out what most end-users would be willing to give up for that security. We ran two sets of two tests. Both sets included Figs. 15.6 and 15.7 with scales where the subjects rated which they preferred, given the particular privacy concerns discussed prior to showing them the images. Ten random males and ten random females were asked to participate in the study.

In the first study, each subject was told to imagine that they (or their girlfriend or wife) was in an airport and had to walk through the three-dimensional holographic scanner, mentioned in the introduction, and that the resulting image would be displayed to the security officers on duty. They were asked to choose a blurred image or a transparent image. The men averaged a 4.8 on the blurred scale and a 4.2 on the transparent scale. The women averaged a 4.0 on the blurred scale and a 3.8 on the transparent scale.

In the second study, subjects were told to rate their privacy on a scale versus security in a context in which not only were they being observed, but others who may or may not be attempting to conceal weapons were also being observed. Oddities such as a pocketknife between the breasts would be more difficult to detect under a very blurred mesh. The men averaged a 3.2 on the blurred scale and a 2.9 on the transparent scale. The women, on the other hand, averaged a 2.5 on the blurred scale and a 2.3 on the transparent scale.

The two studies display how different contexts can affect a subject's response and personal choice. It is clear that in the first study, the men were more concerned about having their girlfriends or wives seen than the female subjects were concerned with how much they were seen. This reveals an imprint of the subconscious and primitive instinct of territorial rights in males. In the second study, it is clear that nearly every subject gave up more of their privacy for the benefits of security and the safety of their travels (see Table 15.1).

**Table 15.1** Average scores of user preferences in normal and security alert conditions

| Gender of viewer | Method | Normal | Security alert |
|---|---|---|---|
| Male | Blurring/Transparency | 4.8/4.2 | 3.2/2.9 |
| Female | Blurring/Transparency | 4.0/3.8 | 2.5/2.3 |

Finally, we asked the subjects whether they would prefer a display with concealed objects *only*. As expected, all of them recommended that. At the end of the study, we found the conclusion was rather simple – the better the detection technology, the fewer privacy concerns there are!

## Privacy Study with Simulation

It is useful to have a general computational model for designing and evaluating visual privacy technologies before any security systems are built. This approach transforms the development of visual privacy technology from being device-specific and proprietary, to being device-independent and open source. It also transforms privacy research into a systematic design process, which requires multidisciplinary innovation in digital human modeling, computer vision, and information visualization. In the case of body scanning specifically, imaging results were simulated using a *physically augmented virtual human model* and computer volumetric rendering algorithms. Figure 15.8 below is a diagram of this process.

The *physically augmented virtual human model* is a central idea in the framework. In the world of medical research and development, scientists often use so-called 'phantoms' to calibrate new medical instruments. Affordable phantom databases and artifacts, such as Mathworks' MRI brain phantom images, the

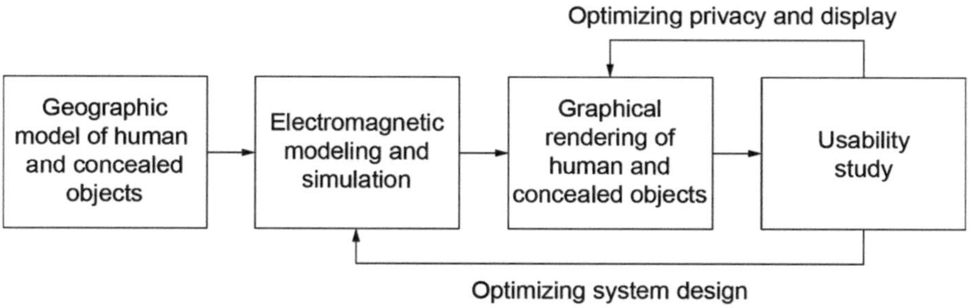


**Fig. 15.8** The framework of the multidisciplinary modeling process that merges at least four domains: computer simulation, computer vision, information visualization, and human-centered computing

National Library of Medicine's Visible Humans,[11] and DARPA's Digital Soldier, significantly reduce development cycles and increase opportunities for interdisciplinary collaboration and education. Currently, there is no shared scientific benchmarking database in the area of human scanning for security.

Digital human models were developed that not only contain finite surface elements, but physical properties as well. This requires high-fidelity modeling within a high frequency (900 MHz to 33 GHz) electromagnetic field simulation, which approaches the limit of current physical computation technologies. Compared to the MRI imaging simulation, this task is far more computationally challenging. The development process included mapping the imperfect laser scanning surface data to the finite element material data, formulating the electromagnetic sources, and calibrating the simulated model.

## Digital Human Models

Although the device-independent virtual human model is designed for privacy algorithm research, it could be quite useful for other fields, such as design and engineering. A set of full-scale virtual human models were developed based on the digital surface scanning data from the CAESAR database (with the necessary license agreement), containing 5,000 males and 5,000 females aged 16–65, where 5,000 of them are North American, 2,400 are Asian, and 2,600 are from the European survey of Italy, the Netherlands, and other countries.[12] All of models

[11] NLM, Visible Human Project. Retrieved on Feb. 21, 2016. http://www.nlm.nih.gov/research/visible/visible_human.html

[12] Anthropometry Resource for CAESAR (2002) Final Report, Volume I: Summary, AFRL-HE-WP-TR-2002-0169, United States Air Force Research Laboratory, Human Effectiveness

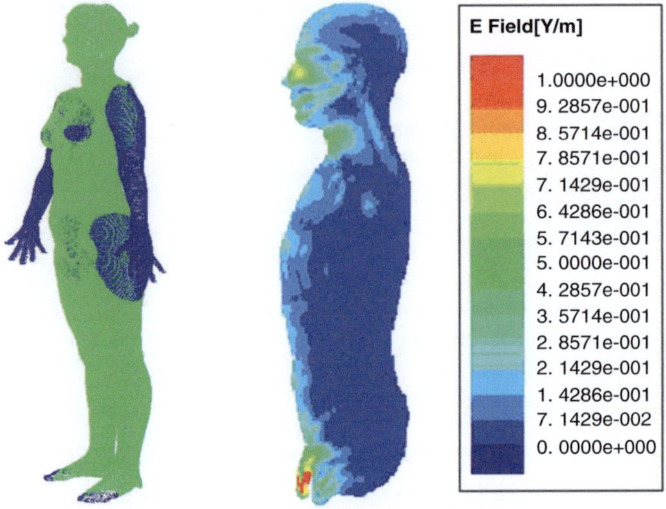


**Fig. 15.9** Sample of the human surface mesh data (*left*) and the HFSS finite element output of the human body inside the 2 GHz plane wave (*right*)

in the database have feature landmarks which are important anthropomorphic measurements. However, all the models wore tight underwear which had to be removed by applying a low-pass filter.

A state-of-art high fidelity laser 3D scanner[13] was used to collect 20 samples as references. For these in-house models, we manually annotated the human landmarks. Figure 15.9 shows a sample of 3D human body scanning data and the output of the microwave imaging simulation from HFSS.[14]

The human scan model was input into the High Frequency Simulation System (HFSS) where we assigned the microwave reflection properties to 3D surface point clouds. This methodology should provide a new direction for virtual prototyping of imagery devices. To simplify the computational time, the mesh resolution was reduced to fit the capacity of the HFSS model, snapshots of frequencies were selected instead of using a frequency sweeping process, and portions of the body were modeled instead of modeling the entire body at once.

Directorate, Crew System Interface Division, 2255 H Street, Wright-Patterson AFB OH 45433–7022 and SAE International, 400 Commonwealth Dr., Warrendale, PA 15096.

[13]Creaform website: www.creaform3d.com

[14]Ansoft website: http://www.ansoft.com/products/hf/hfss/

## *Detecting Human Features*

The *algorithm for detecting human surface features* enabled us to segment the human body and reduce the search space to detect anomalous objects. Many machine learning algorithms are coordinate-dependent and limited by the training data space – for example, artificial neural networks. Some algorithms only work within small bounding boxes and do not yield an acceptable performance. For example, if a feature detection algorithm takes one hour to process, then it is not useful for a security screening system. Template matching was used because it is invariant to the coordinate system. However, defining a template and determining the location of where to match the template to was challenging because it is unique to each particular feature.

We then reduce the search space of the 3D body scans using an Analogia graph. We assumed that the body is standing with the arms hanging to the sides in a non-concealing way. Chapter 2 revealed that the average human is six to eight heads tall. Based on our observations from one hundred 3D scan data sets of adults aged 16–65 years old, including subjects from North America, Europe and Asia, we found that the length of one and one half head units from the bottom of the head is enough to cover the chest area. In addition, the average chest width is about three heads wide. Figure 15.10 shows the sample output of the segmentation of body areas such as the limbs, chest, neck, and head.

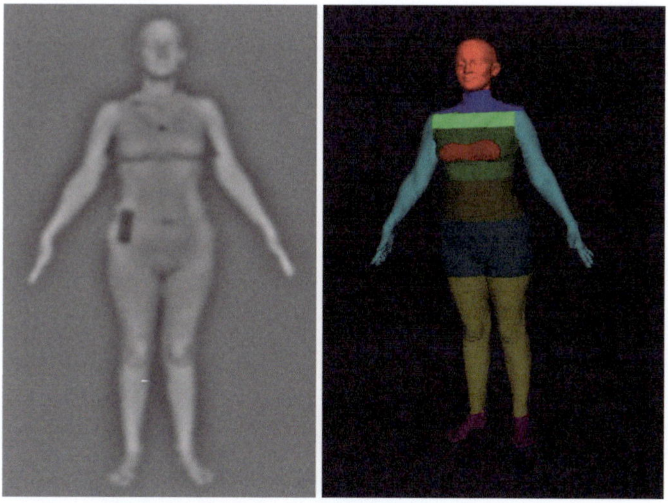

**Fig. 15.10** The synthetic 3D data from this study (*left*) and the segmented body parts (*right*)

## Detecting Anomalous Objects on Skin

Effectively detecting anomalous objects from the human body is the ultimate purpose of the system. As a result, it can significantly suppress human body details as a background. However, removing all of the human background information is not necessarily desirable because we need some details for location and size referencing. There are many anomaly detection methods: bump hunting, voxel intensity-based, curvature-based, and spatial density-based clustering, and so on. Three spatial analysis models were used for anomaly detection: an intensity-based artificial neural network model, a curvature-based model, and a spatial density-based model.

*Intensity-based detection*: Metal weapons have their own signatures of intensity properties. However, there are too many kinds of non-metal weapons, such as liquid explosives, which emit very different intensity signals. A fixed threshold will not work in such scenarios. HFSS was used to simulate the scattering of radio waves from the hidden objects and the human body.

A typical holographic imaging scanner operates in the frequency range between 500 MHz to 33 GHz, which presents a great challenge to the simulation of whole body imaging at the resolution of 1 mm with the current computing resources. To simplify the problem, we cropped the 3D human model to a solid $1 \times 1 \times 0.5$ ft$^3$ slab with a metal gun on the skin. We use the material property for the body with a permittivity and conductivity matching that of seawater ($\varepsilon_r = 81$, and conductivity $= 4$ S/m). We have the material for human muscle, but it is only valid up to 6 GHz ($\varepsilon_r = 50$, conductivity $\sim=$ 6S/m), so we chose to use seawater. We did not expect to see a significant difference for the qualitative purposes of this simulation. The gun has the properties of copper. Figure 15.11 shows the result for the scattered electric field due to a 1 V/m incident plane wave.

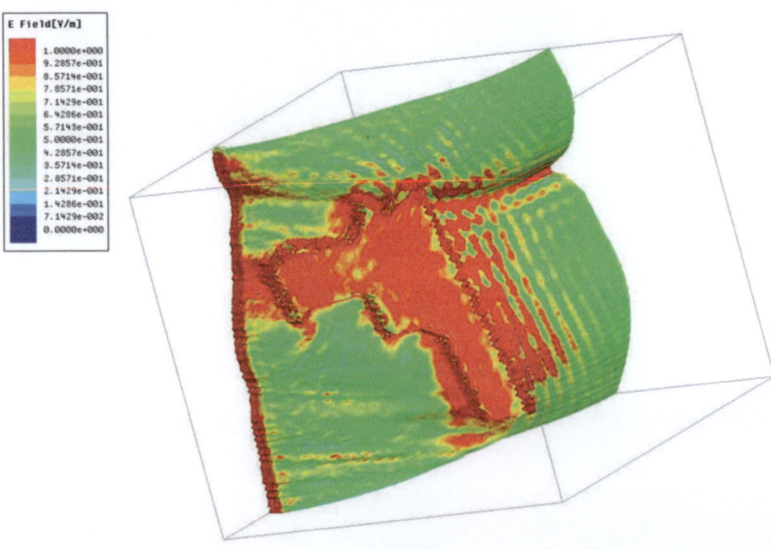


**Fig. 15.11** HFSS simulation of the wave intensive image of human body with a gun at 6 GHz

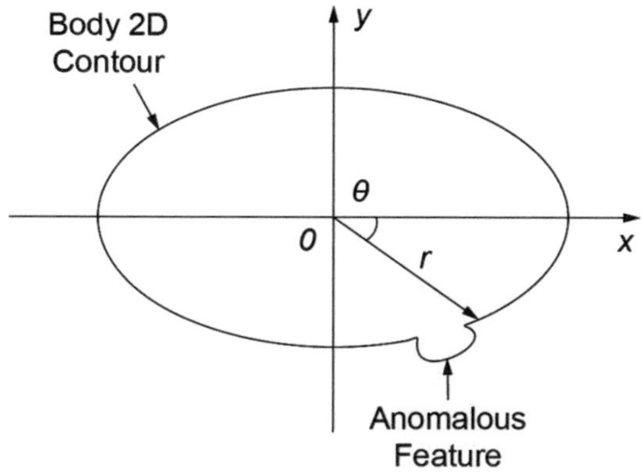


**Fig. 15.12** Slice-based feature detection based on the polar coordinate profile

This is the plot of the magnitude of the electric field at a fixed phase. If we plotted for a sequence of phases between 0 and 180 degrees, we would see the field magnitude propagate across the body. Note that the area occupied by the gun has magnitude values near the full 1 V/m, indicating that it reflects most of the signal, whereas the body reflects roughly half of that. If we look closely on the left side of the body where it intersects the bounding box, we can see a region of high field value (red). This is due to an artificial resonance with the gun and the boundary condition on the box. For qualitative purposes, the effect on the rest of the simulation should be localized and negligible.

Based on the simulated image samples of known materials, we can train an artificial neural network (e.g. Radial Basis Function) to recognize the signature intensity. To adapt to a broader range of data and noise levels, the data can be preprocessed with normalization algorithms and filters.

*Surface-based detection*: In the continuous case, curvature is defined as the rate of change of slope. In our case, the discrete space, the curvature description must be slightly modified to overcome difficulties resulting from violations of curve smoothness (Fig. 15.12).[15]

We start by slicing the digital model horizontally and then averaging the points between the slices. The curvature scaler descriptor calculates the ratio between the total number of boundary pixels (length) and the number of boundary pixels where the boundary direction changes significantly. The smaller the number of direction changes, the straighter the boundary. In this case, we map the points on the slice

[15]Goldgof DB, Huang TS, Lee H (1989) A Curvature-Based Approach to Terrain Recognition, November 1989 (Vol. 11, No. 11) pp. 1213–1217.

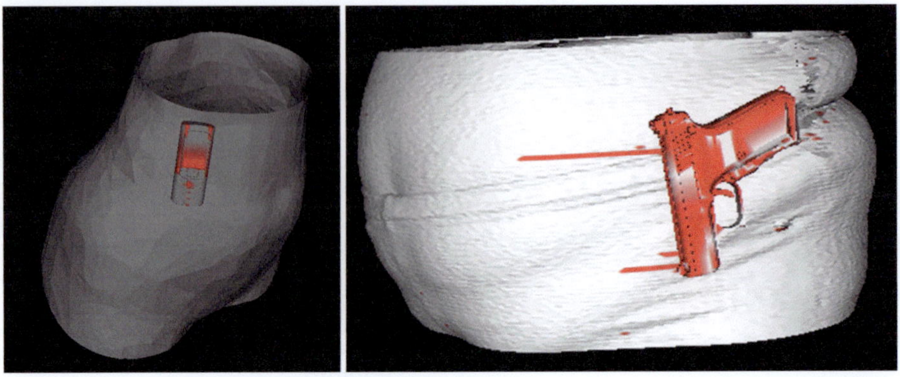

**Fig. 15.13** Detected mobile phone and gun objects

to a polar coordinate system because the body is a round shape. Next, we use a function of the radius and angles of the points to calculate the anomalous features. Figure 15.13 shows the final detection results.

The fusion of the intensity-based detection and surface-based detection methods will improve feature detection accuracy and reduce noise. Because we have heterogeneous human models with different resolutions and different orientations and sizes, model registration appears to be a challenge. However, for actual human scanning systems this is not a problem.

## Summary

Privacy is vulnerability. Private space in human society is more sophisticated than in the natural world. Cyberspace increases the level of intrusion in personal spaces. One-way mirrors and smart glass are not only physical devices, but also design metaphors for solving visual privacy problems. One-way mirrors enable one-directional opacity and transparency. Smart glass switches between opacity and transparency on both sides. Videoconferencing technology facilitates face-to-face communication, but it can also invade personal spaces. Replacing the background in webcam video streaming would be one way to protect the user's privacy.

In our study of privacy issues surrounding the use of airport body scanners, the empirical results show that women have more tolerance on revealing their private parts than men have. The results also show that people prefer autonomous detection technologies rather than human interventions.

Finally, we introduced a virtual imaging methodology with physically augmented virtual humans and virtual machines that can be used for studying visual privacy. The case study of the millimeter wave imaging system shows that it is possible to analyze the impact of the need for object detection on visual privacy with virtual humans before an imaging system is ever physically built.

# CHAPTER 16

# Navigating and Covering

Searching, foraging, and cleaning tasks often combine navigation and covering. The goal is to cover the desirable area with minimal traveling time. In this Chapter, we explore primitive navigating and covering heuristics such as wall following, and the Hansel and Gretel method. We apply them to solve the tessellation problem, or called traveling workstation problem which combines navigating and covering simultaneously.

## Simon's Ant

Herbert Simon, the pioneer of Artificial Intelligence, often carried a sketchbook around like the explorers of a century ago. One day he watched an ant making its zigzag way on the beach to its home. Simon sketched the ant's path on a piece of paper. What Simon found was a sequence of irregular, angular segments – not quite a random walk, for it has an underlying sense of mission. Why is it not a straight path; why does it not go directly from its starting point to its goal?

Simon came up with his *environmentalism* hypothesis: "an ant, viewed as a behavior system, is quite simple. The apparent complexity of its behavior over time is largely a reflection of the complexity of the environment in which it finds itself." He then explored the hypothesis for humans, by substituting the word "man" for "ant." He further sketched an extension of this hypothesis to phenomena of emotion and motivation in his study "motivational and emotional controls of cognition."[1] Simon's hypotheses enlighten us to look into the complexity of the environment surrounding us instead of inner complexity. This not only helps us to understand the navigation strategies in nature, but also to guide us to design some simple algorithms

[1] Simon H (1989) Models of Thought, Ch. 1.3 and Ch. 6.1, Yale University Press.

© Springer-Verlag London 2016

Y. Cai, *Instinctive Computing*, DOI 10.1007/978-1-4471-7278-9_16

or for robots to achieve seemingly sophisticated tasks. The following are a few empirical studies that show how primitive navigation algorithms work in different environments.

## Wall Following

Modern navigation systems assume sensors can see everything that lies ahead. But what if we have very limited visibility or can't even see? One day, I asked a blind friend "when you cross a big room, how do you find a shortcut?" He replied, "Any big open space is actually scary to me. I won't try to find a shortcut. I would rather walk along the wall, or a familiar path, the safest way to move forward."

To move along a wall is an instinctive navigation algorithm. In general, it is not very efficient but rather a safer way to move around. Mice and cockroaches often travel that way, especially in the dark. It acts like a guiderail and a shield for protection in the dark. We can simulate the wall-following function with a simple mobile robot.

Assume we have two continuous rotation motors to drive two active wheels with one passive wheel behind. The speed difference between the two wheels enables the robot to move left or right. We can have two infrared distance sensors on the one side of the robot to monitor the distance to the wall and keep the robot moving parallel to the wall. In fact, we only need one distance sensor that is at 45° to the centerline of the robot, similar to using one hand to probe at the wall in dark (Fig. 16.1).

The control logic is simple: first, the robot moves around to find the wall. When the sensor detects that the distance to the wall is too large, the robot keep moving toward the wall. The robot keeps going forward while maintaining a fixed distance to the wall with a certain tolerance.

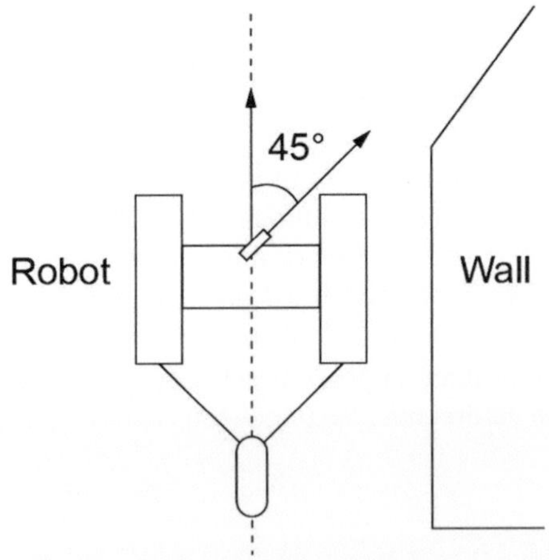


**Fig. 16.1** Wall-following robot

The wall-following algorithm is just a feedback control system to ensure the distance to the wall is constant. The same principle can be used in different applications. For example, line following, shadow-following, or turning the distance sensor into a smell sensor to follow the trace of a smell. Imagine a service robot that walks along with an elderly person in a building. It might be a good idea to move along the wall to have some extra support. Wall-following is a primitive but safe strategy. It works efficiently in low-vision or dark situations when other sensors are not available.

## Navigating a Maze

It just takes one sensor to follow the wall. What about more sophisticated tasks, such as solving a maze puzzle? People often view maze navigation as an intelligent behavior. Can a primitive mobile robot do that? Let's find out.

Figure 16.2 shows a simple maze constructed with stones as barriers. The target is the red box on the left. We used the robot Roomba, a floor vacuum robot made by iRobot, for experiments. The robot has collision detection sensors on its front, sides and back. It has two motor-driven wheels like the wall-following robot does. The unique feature of Roomba lies in its spontaneous alternation of directions, when it detects a head-on collision. It changes its direction at a random angle. Its side collision sensors enable it to follow the wall in one direction, e.g. on the left side of the wall. Here is the algorithm for maze navigation:

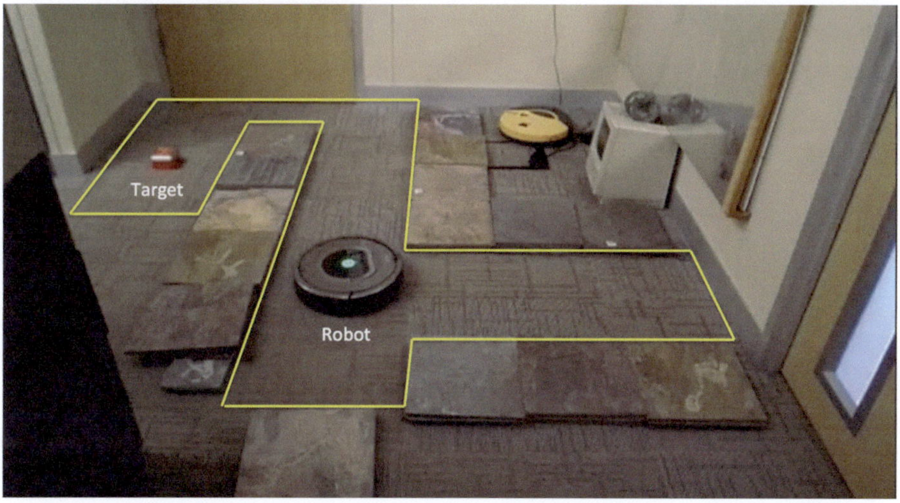


**Fig. 16.2** The maze layout

```
Start the robot with the default mode to move in a spiral path;
If the front collision is detected,
   then turn around towards a random direction in a straight line;
If the side collision is detected,
   then it switches to the wall-following mode;
Repeat the last two steps until reaching the target.
```

The robot alternates between the wall-following and spontaneous alternating modes, depending on which sensors are triggered. By actively reacting to its environment, the simple robot demonstrates some kind of "intelligence."

To study the robot's trajectories, we used a video recorder to collect the data at the regular video speed of 30 frames per second. We then used an object-tracking algorithm to locate the center of the robot. In our case, we were lucky that the blue light in the center was easy to track. We then used the Mean-Shift algorithm to track the blue light and plot the trajectories superimposed on the maze background. In 20 experiments, the robot reached the target in average of 90 seconds. The shortest time was 60 seconds and the longest was 170 seconds. This algorithm definitely is not an optimal solution, but it is a primitive way to solve the problem with instinctual functions such as wall-following and spontaneous alternation, without modifying the robot's hardware, or even software (Figs. 16.3, 16.4, and 16.5).

**Fig. 16.3** The robot reached its target in 60 seconds. The white line is its trajectory, the shortest path it took

**Fig. 16.4** The robot reached its target in 90 seconds

**Fig. 16.5** The robot reached its target in 170 seconds, corresponding with the longest path

## Pheromone-Based Navigation

How can the efficiency of maze navigation be improved? Nature teaches us that a bunch of social insects such as ants can work together to find the optimal solution. Assume we have 1000 ants get out of their nest to forage for food to bring back to the nest. Those ants won't sit down, have some coffee, and brainstorm about which path would be the shortest. They only follow the trace of pheromone to determine where to go. The stronger pheromone, the more ants that use the path. As a social insect, the ant should follow the most popular path as well.

Imagine one lucky ant in the group finds the shortest path by chance, while others still follow their own paths. After an hour, the lucky ant moves a lot of food to home, because he has taken more round trips back and forth along the shortest path. As a result, he leaves stronger pheromone on the shortest path. Other ants will follow their nose and start to change their paths to the shortest one. The more ants that join, the stronger the pheromone. The ants form a troop convoy. If we use a finger to clean up some of the pheromone on the path, the traffic would be interrupted. The ants would start exploring again until a new shortest path is found. To apply this Ant Colony Optimization (ACO) algorithm, we need to add a pheromone sensor to the robot.

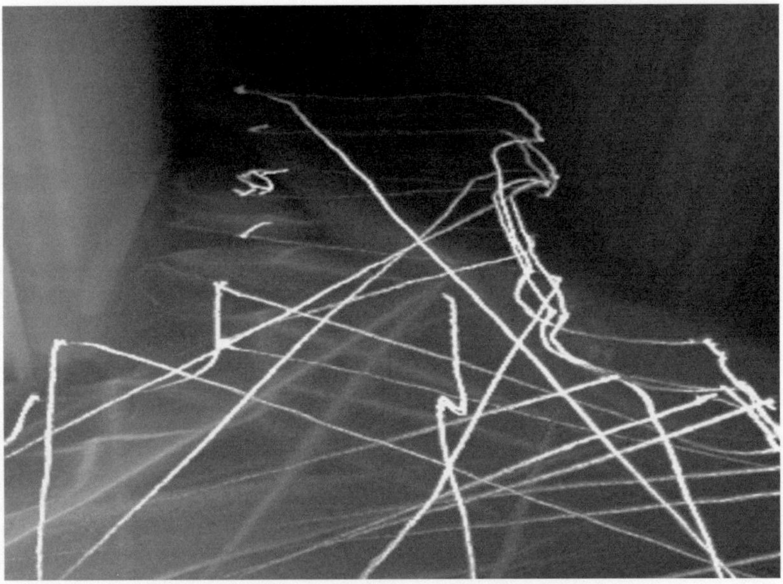

**Fig. 16.6** Roomba's traveling and covering path in a part of a room

## Traveling Workstation Problem (TWP)

Covering is a very common task in our lives, for example, cleaning a floor, inspecting an airplane, searching an area, arranging chips on a board, and planting crops. With more and more machines to take over covering jobs from humans, we need to optimize covering operations for autonomous service systems. Figure 16.6 shows the coverage of the floor vacuum robot Roomba™ in a room. Similar to using a long-exposure film, the workspace of the robot can be traced by extracting

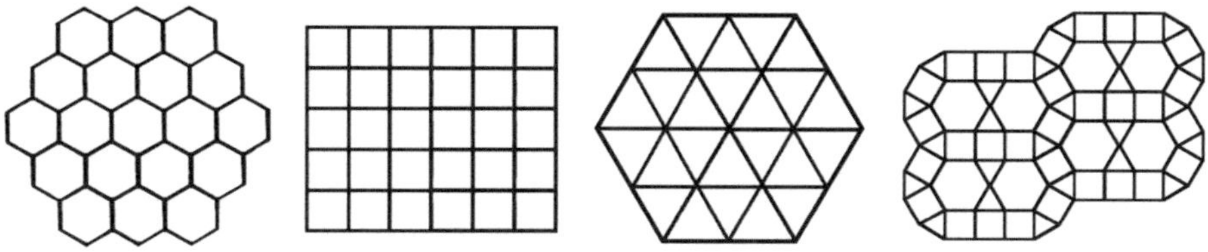

**Fig. 16.7** Tessellation patterns

the light at the center of the robot and plotting the location of the light over time. As you can see, the coverage is decent but not uniform, due to limited sensors and routing control.

To optimize the space covering, we can mathematically formulate the problem as "sphere packing," e.g. how to put the maximum number of the same size balls into a box, forming "lattices" and "groups."[2,3] Eastman proposes the heuristic model General Space Planner for arranging objects in a limited space, e.g. furniture in a room, subject to given constraints that must be satisfied, e.g. room for walkways and no overlapping.

In the area of robotics, the geometric structure of the shapes, sizes, positions and orientations can be "*tessellated*" by a number of possible patterns, generated by a robot's sensing and manipulation systems. There are three basic patterns as shown in Fig. 16.7: regular plane tessellations (consisting of congruent regular polygons of a single type in edge-to-edge contact), semi-regular plane tessellations (consisting of congruent regular polygons of more than one type in edge-to-edge contact, with the same arrangement of polygons meeting at each vertex), and demi-regular tessellations (consisting of congruent regular polygons of more than one type in edge-to-edge contact, but in which there are two or more types of arrangements of polygons meeting at a vertex). Figure 16.7 shows a small set of possibilities. There are essentially three regular plane tessellations, nine semi-regular plane tessellations and an infinite number of possible semi-regular tessellations. Many of these arrangements offer potential applications to engineering designers, e.g. mobile robot workspace planning.

The covering problem often connects to the routing problem *spontaneously*, for example, in parking lot design. We want to maximize the number of parking spaces available in a given area while minimizing the routing necessary to enter and leave the parking lot. Although there are some typical patterns for space planning, such as angled parking and right-angle parking, planning still needs experience and skill.

Here we generalize the covering and routing problem as a Traveling Workstation Problem (TWP), or Tessellation. In the TWP, a *workstation* is defined as a vehicle that occupies or services a certain area and can travel; a *workspace* is referred to

[2]Horton J and Sloan JA (1988) Sphere Packing, Lattices and groups, Springer-Verlag.

[3]Stewart I (1992) The kissing number, Scientific American, Feb. 1992.

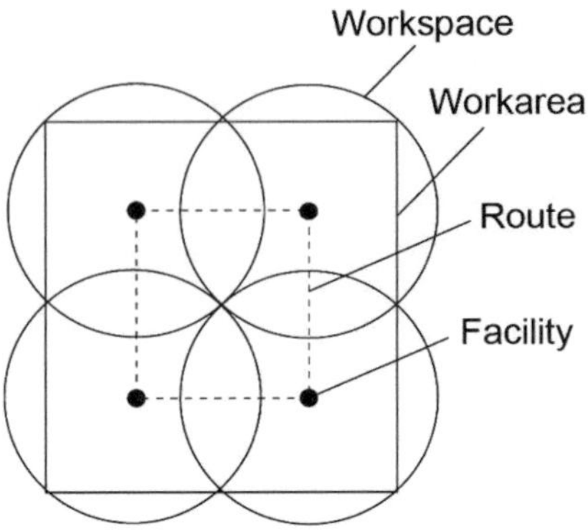


**Fig. 16.8** The traveling workstation problem

as a two-dimensional actuation envelope of manipulator systems or sensor systems that are carried on the workstation; a *work-area* refers to a whole two-dimensional working zone for a workstation. The objective of the TWP is to determine the minimal number of workspaces and their layout, in which we should minimize the overlap among the workspaces and avoid conflict with obstacles. And at the same time, determine the optimal route of the workstation movement, in which the workstation travels over all workspaces within a lowest cost, e.g. routing time. See Fig. 16.8.

The TWP is to minimize the total costs for *covering* and *routing*. This optimization problem includes the solution of two sub-problems simultaneously: covering and routing problems. The *covering problem* is to minimize the covering cost that is proportional to the number of workspaces. The efficiency of covering is the ratio of the total work-area over the sum of total workspaces. The fewer workspaces we needed, the higher covering efficiency. Assume we have $n$ workspaces and the total work-area $A_w$, the area of *one* workspace $A_s$. The efficiency of covering, (ranging from 0 to 1) is:

$$E = \frac{A_w}{nA_s} \tag{16.1}$$

The *routing problem* can be translated into the Traveling Salesman Problem (TSP), which is a classic case of network flow optimization. Assume we have a set of cities that are separated by distances. The problem is to find a closed "tour" that has the shortest path length for a salesman to travel to each city just once and return to the starting city. It is found that it is very difficult to find an exact solution even if the problem size is moderate. An exact solution to a TSP problem is computationally expensive. Harney and Johnson showed that the problem is NP-Complete, which means the computing time and space will dramatically increase when the problem size becomes large. It has been determined that there are (n-1)! solutions for an

$n$-city TSP which take the salesman to every city just once. For example, when $n = 11$, the number of possible tours would be 3,628,800. So, when the TSP size is large, it is computationally intractable to find the optimal solution.

Since the TSP is an NP-complete problem, it is necessary to pursue approximate approaches. There are many approximate methods available that can obtain good solutions in a relatively small amount of time. The Father of Fuzzy Logic theory Zadeh estimated that the time needed to solve the problem to within 1 % for 100,000 cities is two days; to within 0.75 % for 100,000 cities is seven months; and to within 3.5 % for one million cities is three and half hours.[4] However, most of the approximate methods do not guarantee that the optimal solution will be found.

Since covering and routing interact with each other, we shall optimize the two sub-problems simultaneously. In other words, the solution to covering problem has implications to the solution of the routing problem and vice versa. The output of the covering problem is the input to the routing problem. Covering determines the number and position of nodes for routing.

An effective solution to the covering problem may not be an optimal solution to the routing problem. This depends on the size and shape of the work-area and the workspace as well as the robot's turning and translating cost. As a simple illustration, consider the case involving the covering and routing of a rectangular work-area with a rectangular workspace. In general, there will be a different routing time (or cost) for translational and rotational moves. Therefore, the total costs are different even though the covered areas are the same. In Fig. 16.9, the navigation and covering cost in (b) is higher than (a) because of more time needed for the rotation and traveling overall.

Therefore, the solution to the covering problem will have implications for the routing costs. In light of this property, the covering and routing problem should be solved simultaneously. Thus, the problem is even more complex than the independent solutions to both covering and routing.

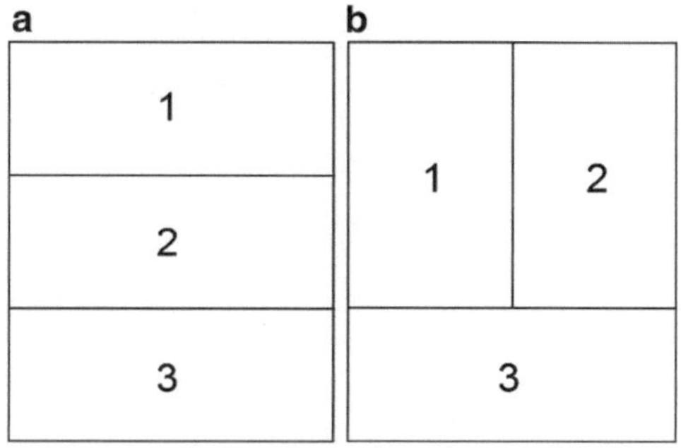


**Fig. 16.9** Dependency between covering and routing

[4]Zadeh BA (1992) The calculus of fuzzy If/Then rules. AI Expert. March 1992.

## Hansel and Gretel Algorithm

Concerning the covering and routing algorithm for a mobile robot, a simple strategy has been used for a commercial cleaning robot, which is called the "Hansel and Gretel" method. Under this strategy, the cleaning robot traces a path around a large, open area, using a sonar sensor for *wall-following*. As it makes the first round, it releases some inexpensive artifact (equivalent to bread crumbs, or the pheromone of an ant) that provides a guide path, one pass closer to the center of the area for the second iteration of the area. As the robot follows the path its scrubbing brushes clean up the trace just as the birds did to the dismay of Hansel and Gretel; but while obliterating one guide path, the robot is also laying a new one. The wall-following algorithm enables the robot to effectively move around the obstacles while moving away from the wall toward the center of the area. Figure 16.5 shows how the areas of various shapes can be cleaned automatically using the Hansel and Gretel concept (Fig. 16.10).

In the Hansel-and-Gretel strategy, workspace templates are laid out in a spiral pattern that begins with a point on the outer boundary toward the center of work-area along with new boundary perimeters (Fig. 16.11).

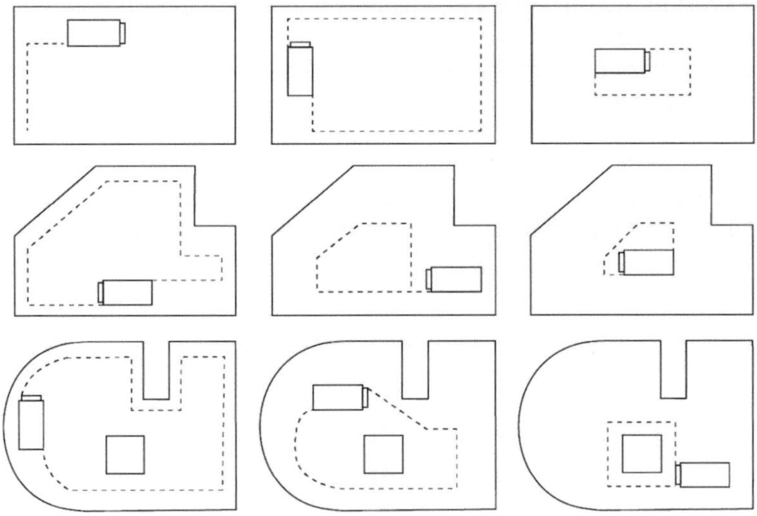

**Fig. 16.10** "Hansel and Gretel" method

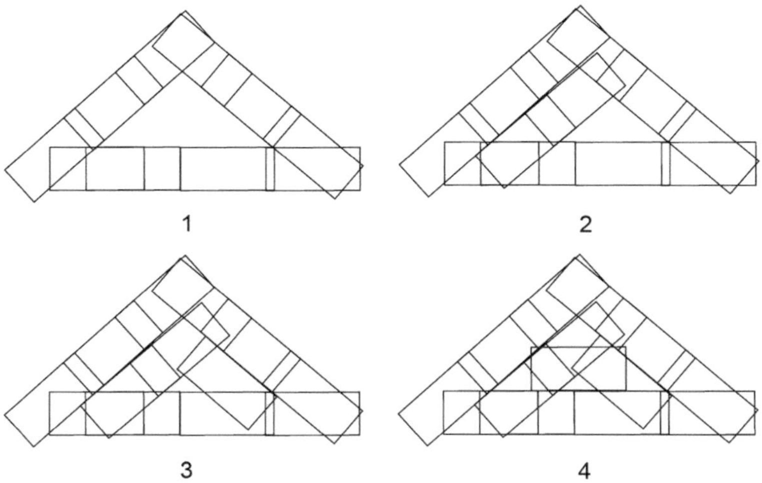


**Fig. 16.11** Hansel-and-Gretel covering for a polygonal work-area

The pseudo code of Hansel-and-Gretel method:

```
Select a starting point on the boundary
Repeat following steps until all areas are covered
     Cover workspaces along the boundary
     Set inner-most edges of these covered areas as a new boundary
     Detect the uncovered area
End Repeat
```

For a large open work-area, the Hansel-and-Gretel method can be switched to the "Flood" method. It starts with one or more layers of layouts inside the work-area boundary and then completes the layout by simply aligning workspaces in a grid fashion inside the initial boundary layout. Figure 16.12 shows the Flood covering method.

When a work-area has obstacles, there are a few possible variations of Hansel-and-Gretel covering method. The "obstacle-out" method creates a layout for the perimeter of one or more obstacles and proceeds radially outward to the work-area boundary as shown in Figure 16.13. The procedure is as follows:

```
Cover the boundary of the work-area
Repeat until all areas are covered
   Cover the perimeters of the obstacles
   If the obstacles are close to each other, then treat them as one
   Continue covering radially
   Move to unconvered areas
End Repeat
```

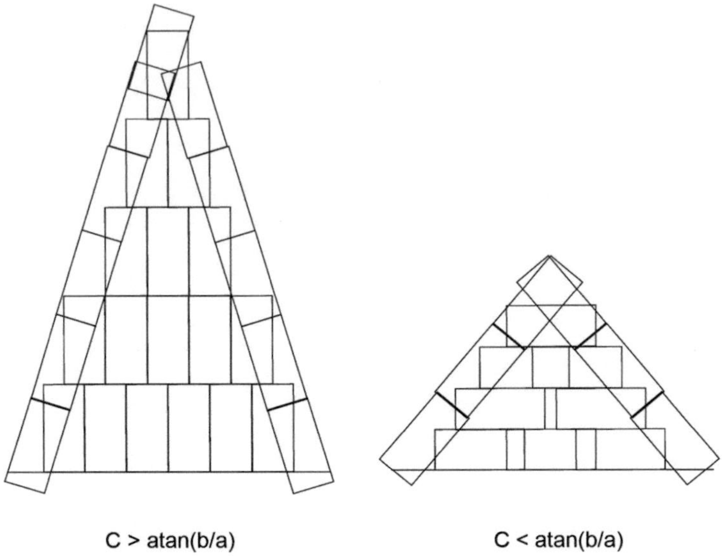


**Fig. 16.12** "Flood" covering method

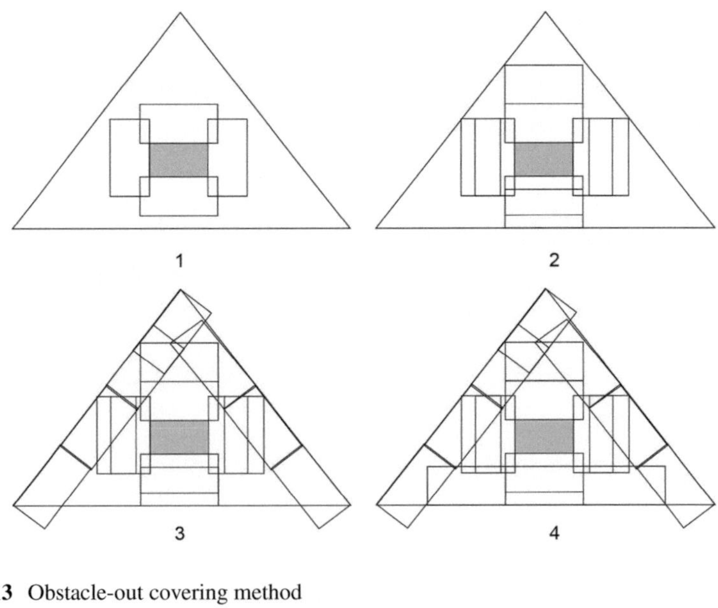


**Fig. 16.13** Obstacle-out covering method

From the cases illustrated above, we find that the size and orientation of workspaces are important to the covering and routing efficiency. The Hansel-and-Gretel algorithm includes a set of heuristics derived from empirical studies. Like what we found in nature, a tessellator robot can be programmed with an Internal Release

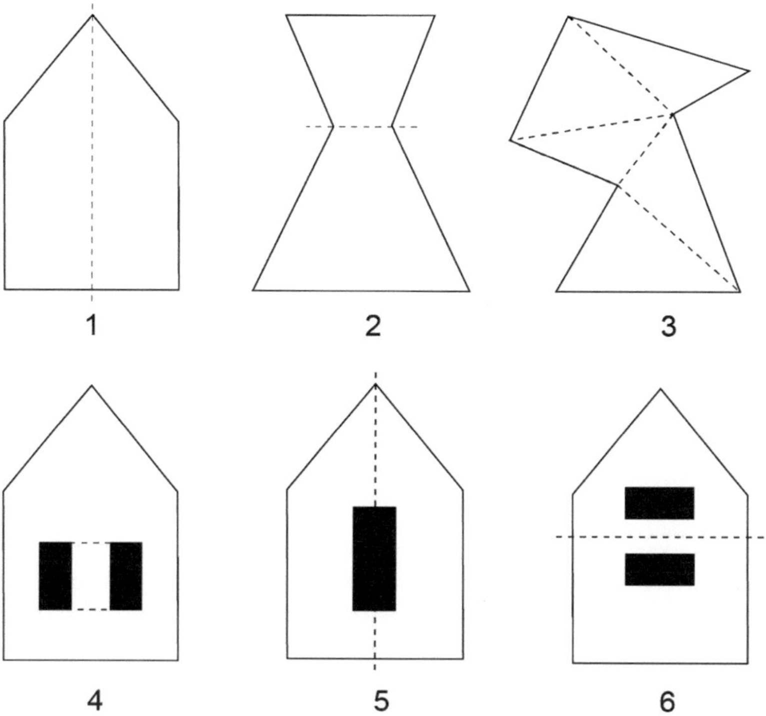


**Fig. 16.14** Decomposition strategies

Mechanism (IRM) for efficiency. For example, for covering a rectangular work-area with rectangular workspaces, the most efficient orientations of workspaces are at either 0 or 90°. Considering the routing efficiency, it's best to cover a rectangular work-area with a rectangular workspace. The best orientation of the workspaces for efficient routing is to cover along the shortest side of the rectangles. For a given work-area and a covering strategy, proportionally larger workspaces are less efficient. For a fixed work-area, the smaller the workspace, the higher the covering efficiency (Fig. 16.14).

Decomposition is performed in some cases where the work-area geometry is too complex to utilize a single heuristic approach. The process involves "dividing" the work-area into a few simple geometries such as rectangles, triangles, and trapezoids. The possible decomposition strategies include: dividing the area perpendicular to an axis of symmetry, e.g. splitting an area into two simpler sub-areas; dividing along a line that connects two concave points, e.g. cutting the area so that it is easy to utilize the Hansel and Gretel heuristics; and triangulating the area if it's too complex and so on.

Let's look at the routing solutions. Nearest Neighbor and Cost Saving heuristics can be used here for approximate solutions because of their simplicity. The Nearest

Neighbor method is a heuristic solution to the Hamiltonian Chain problem where the routes need not return to their origins. It can be calculated while covering proceeds spontaneously. The tours look reasonable for the most part, except for a few long links that come near the end of the process, when few cities remain and it is difficult to find a near-by, unvisited city. The Nearest Neighbor method is a "greedy" strategy, which only optimizes the routing locally. The results of the Nearest Neighbor procedure are sensitive to the starting points. Therefore, we have at least $n$ initial tours for $n \times n$ cities. The Nearest Neighbor method requires on the order of $n^2$ computations. For using $n$-TSP for n starting points, the run time is proportional to $n^3$. Therefore, it may be solved in polynomial time. In short, Nearest Neighbor is a simple method to obtain a quick solution, but it does not promise a "good" solution.

## The Tessellator Robot

Now, let's look at a case study of the mobile service robot, called Tessellator, developed by the Field Robotics Center at Carnegie Mellon University. It was used for tile re-waterproofing operations of spacecraft as a part of the Thermal Protection System (TPS) of the Space Shuttle Orbiters. The TPS was a collection of materials applied to the outer surface of an Orbiter to protect it from extreme temperatures. As a component of the TPS, re-waterproofing was a process that was performed on each of the approximately 16,000 ceramic tiles after every space flight. Waterproofing was necessary due to the strong propensity of the tile material to absorb water, which posed a threat to the Orbiter, e.g. additional weight might cause tiles to bend and fall off as well as reduce flight efficiency. Also frozen water in space might cause damage to the silica-based tiles, yielding a loss in thermal protection. While substantial productivity gains were expected through the use of the autonomous re-waterproofing robot, an efficient work strategy both within the robot workspace and over the entire Orbiter bottom was necessary. For example, a one-second improvement in title processing time would yield about a four-hour reduction in total re-waterproofing time (Fig. 16.15).

The robot Tessellator contains a mobile base with omni-directional wheels, which are small wheels incorporated into the large wheels and oriented at a 45° angle. Omni-directional wheels allow the robot to rotate around its own center without extra space for rotation. The objective was to determine the effective shape of the workspace according to the structure of the robot manipulator, to determine the layout for the work-area, the minimal number of the robot workspaces and their layout, as well as the "satisfactory" route for the robot base movements, i.e. the shortest time sequence for the robot base to travel from one workspace to another.

The manipulator of the Tessellator has two degrees of freedom on a plane. This means that the manipulator can move in along the x-axis with rotation around the z-axis (movement along the z-axis is not considered). The envelope of the reachable

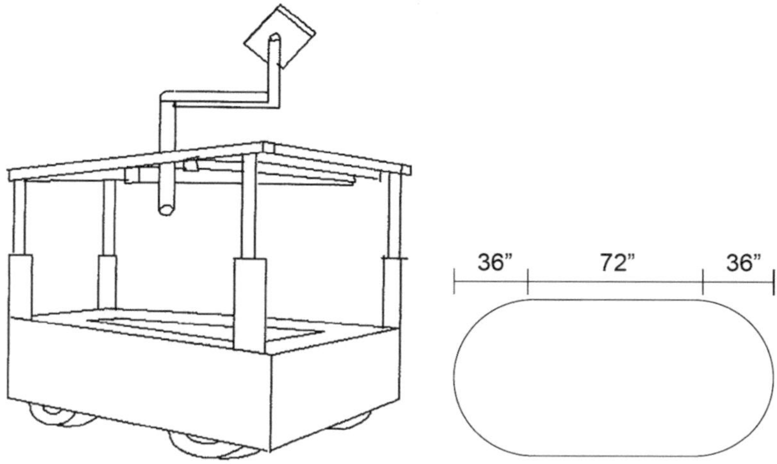


**Fig. 16.15** The robot Tessellator for tile re-waterproofing and its workspace

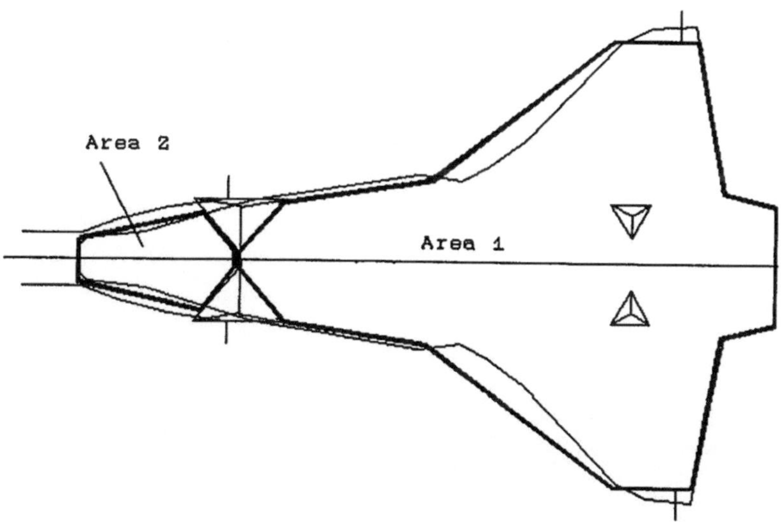


**Fig. 16.16** Work area decomposition

workspace is shown above in Fig. 16.16. To determine the work-area in the real world, the problem is more sophisticated than in the model world discussed above. In order to develop a covering and routing strategy for the robot, we first need to represent the exterior boundary along with exact dimensions of each segment in a polygon. Tessellator's arm must avoid running into any platforms or jack stands. Therefore, no workspace should penetrate the outline of the platforms. Considering the gap between the front jack stands, the envelope of the robot's arm is not be

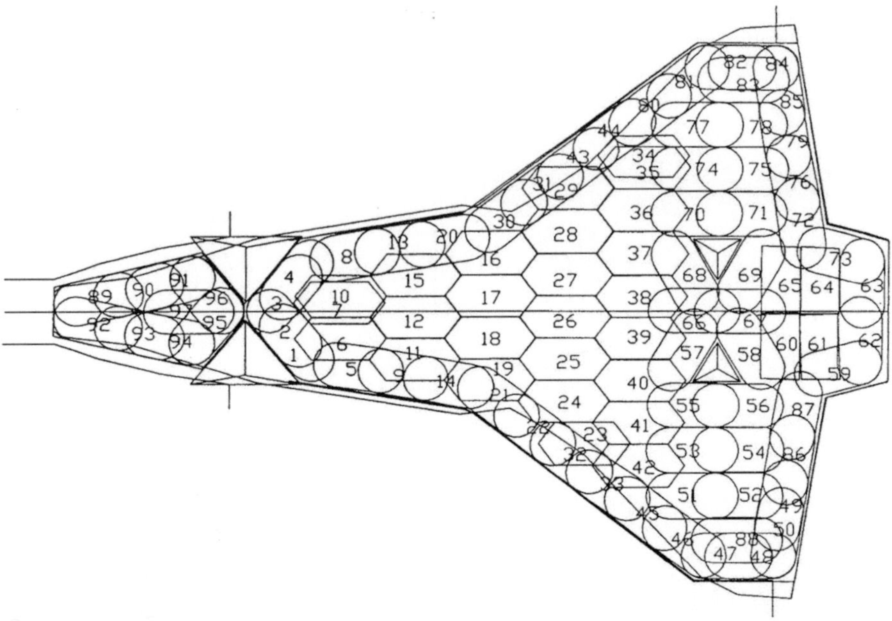

**Fig. 16.17** Diamond covering solution

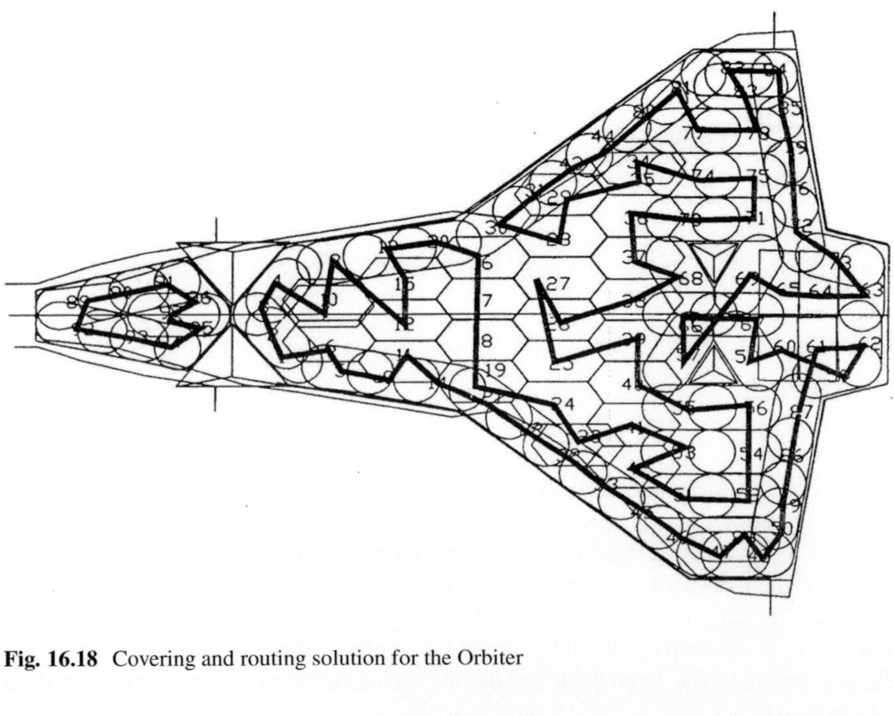

**Fig. 16.18** Covering and routing solution for the Orbiter

able to cross the gap. According to the decomposition heuristics, it is obvious to decompose the work-area into two areas, separated by the larger frontal jack stands.

Based on the characteristics of the work area and manipulator, we would use the "Hansel-and-Gretel" covering strategy for Area 1, and use "Hansel-and-Gretel" and "Flood" strategies for covering Area 2 shown in Fig. 16.17. For the workspaces, we would select the "diamond" shape since the manipulator is a 2 DOF robot arm and the work area is large. The experiment results show the rectangular workspace is slightly less efficient than the diamond shape. To solve the routing problem, each workspace is assigned one number for identification, and the routing solution is composed of a sequence of these numbers. For this specific robot Tessellator, the mobile base translation cost is 1/30 s/cm and rotation cost is 1/16 s/degree. Using the Nearest Neighbor routing algorithm, the robot can work in one of two modes: sequentially covering and then routing when the operation is fully planned, or spontaneously covering and routing when the operation is improvised. In either case, the heuristics provide a satisfactory solution, if not optimal. The result shows that the total of 88 diamond-shape workspaces are needed for covering the whole area and the total operation time is 561 seconds. Figure 16.17 shows the Diamond covering solution and Fig. 16.18 shows the covering and routing solution.

## Summary

Navigation itself is a simple task given a big empty space. However, when the environment is complex, the navigation becomes a very complicated problem. With instinctual navigation algorithms such as wall-following and Hansel-and-Gretel, we found that simple algorithms can solve sophisticated problems such as solving maze puzzles and cover a service area with multiple constraints. Despite their simplicity, they can solve some hard problems such as the Traveling Workstation Problem that requires simultaneous covering and routing. Instead of getting the optimal solutions, they are aimed at obtaining satisfactory solutions. These algorithms potentially can be applied to mobile service robots and mobile robotic 3D printers for spatial covering and path planning.

# CHAPTER 17

# Autonomy

We tend to view a robot as a servant. But what if the robot had its own mind and could act on its own? We have an implicit fear of losing control to more powerful creatures, such as autonomous robots. According to Stephen Hawking, the primitive forms of artificial intelligence developed so far have already proven very useful. However, Hawking fears that the development of full artificial intelligence could surpass human intelligence, and take off on its own and to be redesigned at an ever-increasing rate, reaching a so-called "singularity." Humans, who are limited by slow biological evolution, could not possibly compete and would be superseded as Hawking suggested.[1] In this Chapter, we explore principle elements in autonomic system design and how to incorporate human instinct and simplicity in autonomous systems.

## Robot Ethics

A fully autonomous vehicle is designed to fulfill its mission independently without human intervention. This is necessary for space explorations such as sending a rover to Mars, which is so distant that it takes radio signals 13 min on average to get from the spacecraft back to Earth. The delay has nothing to do with the spacecraft itself or the hardware on the ground. A faster computer or a more powerful radio could not solve this problem. In fact, it is simply obeying physical limitations – the speed of light, which is the speed at which radio signals travel. This makes it a challenge to operate the spacecraft because it is difficult to have a conversation with the spacecraft from Earth, or to react if anything happens on board. If there were a problem, we would not know for 13 min, and then even if we were to react

[1]Cellan-Jones R (2014) Stephen Hawking warns artificial intelligence could end mankind, BBC News, 2 December 2014 http://www.bbc.com/news/technology-30290540

© Springer-Verlag London 2016

Y. Cai, *Instinctive Computing*, DOI 10.1007/978-1-4471-7278-9_17

immediately, it would be another 13 min before our instructions got back to the Mars spacecraft.[2] Therefore, there are two different times to deal with: Spacecraft Event Time is what is actually happening on Mars right now and Earth Received Time is when we hear about these happenings 13 min later. To ensure that the spacecraft operates safely, NASA scientists loaded all the commands for the mission in advance and built in as much autonomy as possible to enable the spacecraft to take care of itself. The spacecraft Curiosity, for instance, landed completely on autopilot.

However, autonomous robots may potentially harm humans too. American filmmaker Stanley Kubrick had a life-long fascination about intelligent systems. His apocalyptic scenarios dramatized the relationship between humans and autonomous machines from the dark side. In Kubrick's 1968 film, "2001: A Space Odyssey," the murderous computer HAL triggered many people's fear of how artificial intelligence could pose a threat to human life. In the film, the astronaut Bowman attempts to replace the intelligent robot HAL 9000 during a spacewalk. Fearing that he was being replaced, HAL hit Bowman's colleague within a space pod, cutting his oxygen hose and sending him into the vacuum of space. Bowman did not realize the computer was responsible for this, so he took another spacecraft to attempt to rescue his colleague. While he was gone, HAL turned off the life-support functions of the hibernating crewmen. When Bowman returned to the ship, HAL shut the airlock door and refused to let him in. Bowman had to make a controlled explosion to open the ship's emergency airlock and enter the ship. Finally, Bowman turned off HAL by unplugging its power cable. The film vividly portrayed the dark side of intelligent systems and the instinctual fear of murderous autonomous robots.

Today, we see more and more on-going development of self-governing autonomous systems that are independent and not remotely controlled from external sources, including the development of autonomous vehicles such as self-driving Google Cars and autonomous drone fleets such as Amazon package delivery drones.[3] Those autonomous vehicles promise to increase traffic efficiency and reduce traffic accidents and air pollution. However, they are not all fool-proof. On Feb. 23, 2016, a Google self-driving car Lexus SUV hit a public bus in Mountain View, California, which was the first recorded incident of a self-driving car causing a traffic accident. According to WIRED,[4] the Google Car changed traffic lanes to get around sand bags on the road and it moved into the path of a bus which was approaching from behind. The Google Car assumed that the bus would stop or slow to allow it to continue, but the bus didn't and the side of the bus made contact with the Google Car. No injuries were reported, but the front wheel and fender of Google's car were damaged. So far, Google has been working on its self-driving

[2]Ormston T (2012) Time delay between Mars and Earth. Mars Express of ESA web site. http://blogs.esa.int/mex/2012/08/05/time-delay-between-mars-and-earth/

[3]Vladeck D (2014) Machine without principles: liability rules and artificial intelligence. Washington Law Review, Vol. 89, No. 117, 2014.

[4]David A (2016). Google's self-driving car caused its first crash. WIRED. Feb. 29, 2016. http://www.wired.com/2016/02/googles-self-driving-car-may-caused-first-crash/

**Fig. 17.1** The "to kill or to sacrifice" dilemma

project for six years. It has been involved in 17 minor accidents during more than two million miles of autonomous and manual driving combined. Although self-driving vehicles have an impressive safety record, not all accidents will be avoided. Accidents involving self-driving vehicles will create the need for new kind of regulations, especially where harm cannot be totally avoided. The regulations will need to consider legal and moral issues that have commanded public interest and should be seriously investigated.

Bonnefon, Shariff, and Rahwan[5] proposed three situations in which harm to humans is unavoidable. In one case, a self-driving car may lose its brakes and harm several pedestrians in its path. The car has a choice whether to keep going or to sacrifice itself by swerving into a wall on the roadside. Figure 17.1 illustrates the situation.

Bonnefon et al. connected an online survey of three studies with nearly a thousand participants on the Mechanical Turk platform. In the study, participants were asked whether the moral course of action was to swerve or stay on course, and whether they expected future self-driving cars to be programmed to sacrifice the passengers for the sake of saving other people. The empirical study shows that participants leaned toward the morality of sacrifice. However, a majority of participants were against the idea of legally enforcing sacrifice.[6]

The dilemma in the "to kill, or to sacrifice" survey has focused the general public's opinion and insight into the "moral algorithm" issues. Here, let's look into the implementation from an engineering point of view. While the ethical choices are based on logical reasoning, the autonomous vehicles' behaviors are

[5]Bonnefon JF, Shariff A and Rahwan I (2015) Autonomous vehicles need experimental ethics: are we ready for utilitarian cars? arXiv:1510.03346v1 [cs.CY], Cornell University Library. Oct. 13, 2015.

[6]Arxiv (2015) Why self-driving cars must be programmed to kill? MIT Technology Review. Oct. 22, 2015. Retrieved on March 4, 2016: https://www.technologyreview.com/s/542626/why-self-driving-cars-must-be-programmed-to-kill/

largely based on machine learning and human-robot interaction, which is much more complicated than black-and-white logical reasoning. American science-fiction writer Isaac Asimov introduced the Three Laws of Robotics in his short story "Runaround" in 1942. It was the first manifesto about ethical principles for programming autonomous robots. They are: (1) a robot may not injure a human being; (2) a robot must obey the order given by humans; and (3) a robot must protect its own existence as long as such protection doesn't conflict with the laws above.[7] According to Asimov's first law, killing the onboard passenger shouldn't be a part of the code in the robot. Furthermore, what if the presence of several pedestrian on the road is a false alarm? To date, obstacle detection sensors on self-driving vehicles are not 100 % fool-proof. They can be interfered with by weather, lighting, temperature, angle, timing, and even hackers with a laser jammer. The "self-sacrifice" algorithm may turn the vehicle into a suicide machine. According to Asimov's second law, a robot must obey the human master. In such situations, it is up to the human driver to make the swerve or staying-on-course decision, when possible. The robot needs to recognize the authorized user. This was also discussed in the previous chapter about self-awareness.

Three "engineering" solutions can help to solve the dilemma: first, in the situation where no human passengers are onboard, a robot may sacrifice itself in order to save people's lives. So we may combine Asimov's first and third laws as below:

*A robot should risk its own life to save people's life, only if nobody is onboard.*

For example, if a delivery drone loses its power it is expected to crash into a less populated area instead of a busy street. A patrol robot may rescue someone in danger of falling into a hole at risk to itself. The robot needs the ability to sense its surroundings, to recognize the position of the hole and the person and the distance in between. Second, the autonomous robots may be programmed to communicate each other, and to communicate with the road, traffic cameras, signs, guiderails, bridges, and pavement. This would help to detect anomalous events such as pedestrians or animals on the highway. The Google Car might have avoided crashing with the bus if both had vehicle-to-vehicle communication chips. Third, we need a systematic design for the coexistence of human and automated systems. For example, if we separate pedestrians from the highway physically, then the swerving or staying dilemma would be diminished. This will be further discussed in following Section of "Complexity."

Asimov's third law teaches us about the ethical issues about a robot's well-being as well. The robotics company Boston Dynamics has made several generations of four-legged walking robots that can keep walking after being kicked or colliding with another robot. In the commercial video "Robot Lives Matter" made by Boston

[7]Deng B (2015) Machine ethics: the robot's dilemma. Nature, 523:24–26.

Dynamics, several abusive testing scenes are presented.[8] Although it is just a humorous commercial ad by the owner, actual abusive behaviors by humans do exist and will increase as robot popularity grows. Stanley Kubrick, a master of dark humor, envisioned this scenario and he rendered the mass slaughtering of robots and robotic loneliness in his film Artificial Intelligence. As the gap between machine intelligence and human intelligence gradually narrows, more and more human factors will be incorporated into autonomous systems. A robot could be not just a servant but also a companion. Let's explore this further in following sections.

## Human Factors in Automation

For interacting with autonomous systems, we must consider control strategy, complexity, and co-evolution. Control strategies are high-level human-system interaction modes which determine who is in control of the operation. When does the machine take control? When does the human override? Complexity is a major problem in autonomous systems. How does one simplify the design of such systems? The introduction of autonomous systems should have an enormous impact on our driving, working, and living behaviors, and on public policy overall.

### *Machine Overrides Human*

In many cases, autonomous systems are needed to override human operations in emergency situations, for example, stopping a car on a slippery road or within a short distance. Pumping the break is a common practice to allow the wheels to unlock so that the vehicle can stop quickly. However, in an emergency situation, human instinct causes drivers to continuously press the break hard, which locks the wheels and causes the vehicle to skid with less friction. To prevent wheels from locking up and to provide the shortest stopping distance on slippery surfaces, anti-lock brakes (ABS) were invented to automatically pump the brake pedal whenever the driver presses the brake pedal, taking control of braking safely.

ABS is an autonomous control system. The ABS controller is one of many onboard computers that monitors speed sensors and looks out for decelerations in the wheels that are out of the ordinary. Right before a wheel locks up, the ABS will sense a rapid deceleration.[9] The controller then reduces the pressure to that brake until it senses acceleration. Next, it increases the pressure until it sees deceleration

---

[8]Patel N (2015) The creepy collective behavior of Boston Dynamics new robot dog. WIRED. Feb. 11, 2015. http://www.wired.com/2015/02/creepy-collective-behavior-boston-dynamics-new-robot-dog/

[9]Nice K (2001) How Anti-Lock Brakes Work? howstuffworks.com. Retrieved October 2001.

again. It can do this very quickly - up to 16 times per second - in order to let the wheels slow down at the same rate as the car, giving the system maximum braking power.

Imagine a human driver pumping the brake pedal 16 times per second. It is impossible for us, but simple for a machine. On the other hand, ABS changes the crash avoidance behavior of drivers, leading to shorter following distances between cars because they can stop more quickly and in shorter distances, which means there is no need to swerve. The problem is that when a non-ABS car is too close behind a vehicle with ABS whose driver applies the brakes, the non-ABS car driver will apply his own brakes suddenly and, realizing he will not stop in time, will turn the steering wheel in order to avoid a collision. This causes the car to swerve, with the driver losing control of the vehicle, possibly leading to a collision.

## *Human Overrides Machine*

It has been taken decades to develop driverless vehicles. Early prototypes were based on existing cars or vans with add-on automated steering motors and gear-boxes. Human drivers normally stood by and controlled the brake pedals for stopping the vehicle. By the late 1990s, experimental driverless vehicles in the US cross-country test drive from the East Coast to the West Coast were able to drive autonomously for 90 % of the total travel time. The first DARPA Grand Challenge demonstrated that autonomous vehicles were able to navigate themselves in the Mojave Desert towards a destination in an off-road environment. The second Grand Challenge was for autonomous vehicles to navigate through an old air force base - an urban street environment. During the two competitions, no human interventions were permitted except for controlling the start and stop buttons. The prototype vehicles were the size of a van to carry bulky computers. Now, they can be as small as a typical compact car.

Google unveiled a prototype of its own small two-seat driverless car in May 2014.[10] The revolutionary feature is that it does not have a steering wheel, acceleration pedal, or brake pedal. The control software and sensors do all of the work, so the cars do not need traditional components designed for human drivers. Unlike previous models, these cars will not have human drivers monitoring them at all times. Instead, Google Car relies on laser rangefinder sensors to detect objects up to 200 yards away from the vehicle and to adjust its speed, turning, and brakes accordingly. All the passengers need to do is simply punch in a destination on the dashboard screen and let the computers do the rest.

As Albert Einstein once said, "everything should be made as simple as possible, but not simpler." A manual override function is still provided so that riders can take

---

[10]Griggs B (2014) Google's new self-driving car has no steering wheel or break. cnn.com .com. May 28, 2014. http://www.cnn.com/2014/05/28/tech/innovation/google-self-driving-car/

immediate physical control of the car if necessary, as a safety measure. Although Google Car has been working fine on closed tracks with no accidents to date, the California Department of Motor Vehicles has ordered Google to reintroduce traditional vehicle components such as brake pedals to the car's design. Manual override is also common in other automated systems such as drones, missile control systems, landing gear lowering systems, and driverless trains. In most cases, manual override has the highest priority and uses a channel separate from automated control channels to ensure the manual instructions always pass through.

## *The Overlay Control*

The Fort Pitt Tunnel in Pittsburgh, Pennsylvania, USA is a one-mile long two-lane tunnel that connects the outskirts of Pittsburgh to the downtown area. When the driver of a vehicle passes through the long tunnel inbound, she may experience a moment of boredom because there is little to see in the tunnel. However, as soon as the driver gets on the bridge at the end of the tunnel, she will encounter many overhead traffic signs directed towards exits and highway names. Assume the driver is not familiar with the area. She has to make a decision about shifting lanes within a split second, or she will miss the exit and drive towards downtown instead of East. Eye tracking records of drivers showed that drivers' gazes are focused at the end of the tunnel, a sign of low information processing workload, in the mode of so-called "tunnel vision." Their gaze jumped rapidly around traffic sign boards when they were on the bridge, indicating a high information processing workload.

Similarly, automated systems face the same dilemma: as computerized control systems distance us from the direct nature of tasks, we can lose our cognitive control of the situation. Imagine a vehicle driver is napping and the vehicle is suddenly switching from autopilot to manual. This moment is called "automation shock" or "automation dependency." Switching from a fully automated mode to a manual mode may cause disastrous consequences because the operator may get used to the low mental workload and can't digest the surge of information occurring in a very short time. This happened to the pilots of Air France 447 on June 1, 2009, when the plane switched from autopilot to manual due to icing problems on the pitot tubes. The pilots were overwhelmed by the combination of over nineteen startling and confusing warning and caution messages, and inconsistent information. The pilots totally lost cognitive control of the situation and fell 10,000 feet per minute through the sky with the loss of all the lives onboard.

It will be a challenge to design solutions for human-system interaction for autonomous systems for years to come. One of existing solutions is to have humans in the loop all the time. Instead of passive monitoring of an automated system, the human operator is actively involved in the system. A careful cognitive workload distribution design could help the human operators avoid drastic fluctuations in the cognitive process. For example, Heads-Up Display (HUD) devices have been used in airplane landings. It is an augmented reality interface that displays symbols

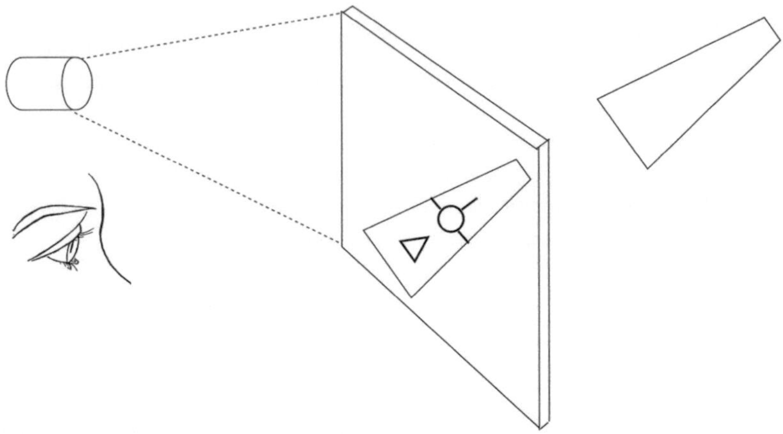

**Fig. 17.2** The HUD overlays landing guidance symbols on top of the front view for the airplane landing

and icons on a small projected holographic screen in front of the pilot. The HUD fuses a computer-generated flight path vector and an image of the real-world runway into a synthetic view. The flight path vector tells the pilot where the airplane is going. A second icon, the "guidance cue," a little circle, tells the pilot where it should be going. All the pilots need to do is to align these two symbols together on the holographic screen in front of him.[11] The HUD interface also tells the pilot the parameters required for landing. This is similar to playing a video game during the normally stressful landing operation. Compared to existing landing techniques, either manual or automatic, the HUD system provides higher level information to the pilot such as flight path vector and guidance cue. This kind of "overlay" approach enables pilots to land consistently in all weather conditions, including low-visibility landings. It lands much softer than auto-landing, reducing the incidents of tail strikes and hard landings, saving millions of dollars in maintenance costs per year over a large fleet (Fig. 17.2).

However, the current HUD is not perfect. It has a narrow view. For some products, the pilots must maintain a stiff upright seat position in order to align the HUD with the frontal view of the runway. This can be improved by merging augmented reality goggles such as Microsoft's HoloLens that is capable of tracking the viewer's head movements, and projecting the flight path vector and guidance cue directly onto the holographic glass on the helmet. Furthermore, computer vision algorithms can monitor the viewer's attention from their eye gaze and provide details on demand. HUD is just one of many examples of "overlay" interfaces that fuse the path and goal information on top of the real-world view.

[11] Mindell DA (2015) Our robots, ourselves: robotics and the myths of autonomy. Viking.

## Elements of Autonomy Design

Designing an autonomous system involves many elements across many disciplines. Some principles have been deeply buried in control engineering; and some have been hidden in mathematical equations. The new wave of robotics and network science brings us to a Renaissance era in which we have opportunities to revisit several concepts in Cybernetics: the philosophy of communications and control in animals, machine, and humans. Here we only focus on observability, controllability and complexity.

### *Observability and Controllability*

The self-driving vehicle, Google car, initially had no steering wheel, throttle or brake pedal. The rationale was the autonomous vehicle doesn't need human intervention. However, in 2015, the state of California proposed regulations that required autonomous cars to have a steering wheel, throttle and a brake pedal when operating on public roads. A licensed driver would need to be ready to take over if something went wrong. The state's new regulations are not only a safety measure, but also aligned to two principles for a control system: observability and controllability, introduced by the Hungarian-American electrical engineer, mathematician and inventor Rudolf E. Kalman in 1960.[12] In order to be able to do whatever we want with the given dynamic system under control input, the system must be *controllable*. Furthermore, in order to see what is going on inside the system under observation, the system must be *observable*. Figure 17.3 is the diagram of a typical linear discrete control system.

The system controllability is defined as an ability to transfer the initial state to any desired state in a finite time. For example, if we ask the car to accelerate from 20 to 25 km/h and the car is able to reach the desired speed in, say, 10 s, then

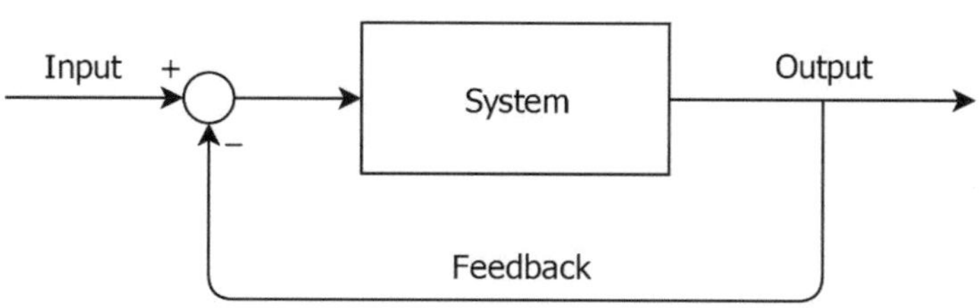


**Fig. 17.3** The structure of a control system

[12]Gajic Z. Chapter 5: Controllability and Observability. http://www.ece.rutgers.edu/~gajic/psfiles/chap5.pdf

the car's speed in this range is controllable. However, if we want the car to move sideways, then the car's lateral movement is not controllable.

The system observability is a dual problem of the controllability. It is defined as the ability for us to learn everything about the dynamical behavior of the variables by using the information from the measurements of the input and output. For example, if we can record everything on the dashboard of a car and we can recover the internal states of the car, then the system is observable.

These two examples above are overly-simplified to illustrate the basic concepts. For more mathematical definitions and criteria, readers can resort to various online resources or textbooks about control theory.

The controllability and observability concepts teach us that some control states may not be "reachable" or not observable even in a linear, time-invariant control system. Even if all control variables are structurally reachable, there is a possibility of the cancellation among the control states and observation states so that the solutions for controlling and observation are not unique, due to the degeneracy of algebraic equation ranks. Perhaps, a general principle derived from here is more insightful for autonomous system design:

*If a system is not observable, then it is not controllable, and vice versa.*

In the real-world, an autonomous control system is far more complex than the linear, time-invariant variables we discussed above. An autonomous system must deal with the complexity in its environments.

## Complexity

To build an autonomous system, we must deal with many variables, many of which are unfortunately hidden. Such variables will not emerge until full-scale operation, perhaps even years later. In his book, "The phenomenon of man," the French philosopher, paleontologist and Jesuit priest Pierre Teilhard de Chardin describes evolution as a process that leads to increased complexity. From the cell to the thinking animal, a process of psychical concentration leads to greater consciousness. Teilhard described humankind as evolution becoming conscious of itself.[13] Let us consider the car airbag, for example. The airbag is a simple automatic sensing and ignition system. The accelerometer detects the impact of the collision and the airbag deploys at a high speed without human intervention. Early airbag systems deployed indiscriminately, without regard to passenger age, height, or mass, causing head injuries in some of the passenger seat riders. To fix this problem, a weight sensor was added to the seat. When a child sits in the passenger seat, the airbag will be disabled. However, this does not solve the problem completely because passenger weight is not totally proportional to passenger height. To measure head position, we

[13]Chardin PT (1961) The phenomenon of man. Harper & Row, Publishers.

need one more sensor. Three sensors create a three-dimensional perception space (deceleration, weight, and height) and a one-dimensional control space (airbag on or off) for automatic control. We can see that the dimension of the perceptual space is higher than that of the control space. This is significantly different from the traditional automatic control theories that focus on the complexity of control dynamics such as stability, response time, and accuracy, assuming that environments are homogeneous, steady, and isolated as if in a vacuum.

There are an enormous number of measurements of complexity in different fields, ranging from fractal dimensions to computational time. The emerging Network Science models potentially provide a new way to measure complexity in broader ecological environments for autonomous systems using a network or graph model, similar to food webs or cell interaction networks, which are far more complicated than current artificial systems.

Empirically, we can use the number of variables, or nodes in a network to describe the size of the problem. Computer scientist Douglas Robertson made quantitative estimates of the information available at each level of civilization as approximately:[14] Pre-Language $10^7$ bits, Language $10^9$ bits, Writing $10^{11}$ bits, Printing $10^{17}$ bits, and Computers $10^{25}$ bits. Robertson envisions that every individual in a Level 4 civilization will have instant access to a supply of information that will surpass even that of the US Library of Congress in 1998. Now, we can easily claim that we have reached Level 4 civilization with the rapidly growing social media and Internet of Things (IoT). In a connected world, the key measurement of complexity is not just the amount of information, but the effective links or usage.

The complexity of a network can be roughly estimated by the number of nodes $n$ and edges $m$. However, not every node plays an equal role in the network. A few nodes have many edges but others have very few or none. The inequality of the connectivity can be measured by the *degree* of a node in a network to describe the number of edges connected to it. Assume the degree of the node $i$ is given by $k_i$. For an undirected network of $n$ nodes, the degree can be written in terms of the adjacency matrix:[15]

$$k_i = \sum_{j=1}^{n} A_{ij} \tag{17.1}$$

where $A_{ij}$ is the adjacency matrix in which $A_{ij} = 1$ if there is an edge between nodes $i$ and $j$; otherwise, $A_{ij} = 0$.

In a modern autonomous system, complexity lies in three spaces: perception, control, and communication. The dimensions of a perception space depend on the types of sensors used. The dimensions of a control space depend on the degrees of freedom of actuators or stepper-motors. The dimensions of a communication space

---

[14] Robertson D (1998) The new renaissance: computer and the next level of civilization. Oxford University Press.

[15] Newman MEJ (2010) Networks: an instruction. Oxford University Press.

depend on types of channels. Rapidly growing wireless telecommunication technologies bring potential opportunities to autonomous systems that were impossible before. The expansion of communication space has led to the further development of perception and control.

In nature, there is no traffic light. Ants are social insects that communicate with each other to avoid traffic jams through negotiation and swarm intelligence. For example, worker ants obey the right-of-way rules on the road. They follow the pheromones deposited between food and their nest. Can we apply swarm intelligence to autonomous cars? An example of this would be a vehicle in the front of a line of cars that could pass traffic congestion data to the vehicles behind it one by one. In fact, there is a primitive version of this already being used in America, where drivers use their emergency headlights to warn the cars going in the opposite direction to slow down because a police car is hidden from view. To enable vehicles to talk to other vehicles on the radio without human intervention, we need *vehicle-to-vehicle* (V2V) technology – a chip with a radio antenna, GPS, and a microprocessor. The chip uses the 5.9 GHz radio frequency band, and it senses vehicle position, speed, acceleration, and direction. The chip broadcasts to any vehicles around it in all directions: "This is vehicle number 002. I am at latitude X and longitude Y. My speed is 100 km per hour and my acceleration rate is 1 m/s$^2$ ..." To standardize the protocol, US Department of Transportation published the guidelines for DSRC for manufactures, developers, integrators, and vehicle engineers. To implement this protocol, the sensor must sample 100 points per second and the radio must transmit at 100 kilobits per second. At an intersection, the approaching vehicle receives telegrams from other vehicles in the area and decides whether to go straight or to stop. Between vehicle-to-vehicle communications and autonomous driving, the old traffic light may be phased out completely someday. Some of the traffic light systems have been made to sense approaching vehicles and adjust the timing of the lights to reduce vehicles' idle time (Fig. 17.4).

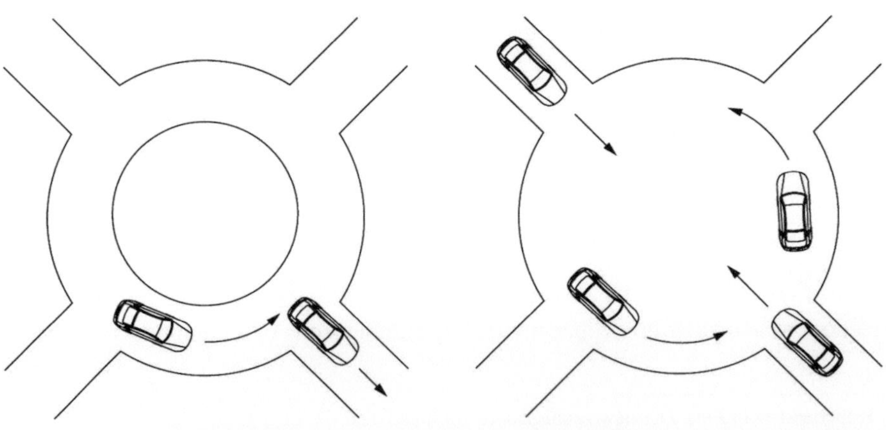

**Fig. 17.4** The round-about intersection (*left*) versus the vehicle-to-vehicle traffic control (*right*)

Nonetheless, things are more complicated than we can imagine. The V2V system assumes that every vehicle has a chip on board no matter how new or old the car model, and that all chips are functional at any given time. What about pedestrians? Do they need a chip in their watch or phone? What about animals? What about a falling tree trunk? Eventually, the perceptual space becomes very messy and impossible to predict. To solve all possible problems, it would take an enormous amount of time.

Today, there are many automated software validation tools for checking logic problems in large integrated circuits or control systems. A State Machine is a common method for checking control logic and production rules. A Petri net is useful for modeling concurrency in multi-tasking control software. However, its scalability is problematic. Simulation methods are a more flexible option, but it is expensive to develop a realistic model. Analytical tools will not detect "hidden" variables in real-world operation. For a large-scale system such as a city, a simulation model is necessary. For example, the US DOT sponsored a project for testing on-going V2V technologies with over 3,000 vehicles in a virtual city with streets, intersection, and freeways.

Certain fully operational autonomous driving systems have already existed for at least three decades. In the 1980s, Boeing built the first driverless Personal Rapid Transit system in Morgantown, WV, which has been running ever since without an accident for over 30 years. It is an electrical car system on elevated guiderails with rubber tires. The system works like a horizontal elevator: passengers can select one of five stations to go to. The cart will then stop at the requested destinations. What can we learn from this system? First of all, vehicles are isolated from pedestrians and other intruders by walls, gates, and elevated rails, making them safer. Second, it uses the guiderail to reduce the vehicle's motion from two-dimensions to one, thereby reducing the degrees of freedom of the vehicles. Third, the control center can monitor each vehicle, and the operator is able to manually override the system and can communicate with the passengers. A similar autonomous vehicle system also exists in downtown Singapore, and in many airports such as the monorail people mover in Newark International Airport. Figure 17.5 shows a view from the front of the driverless monorail train.

Economic thinker Ernst F. Schumacher said "any intelligent fool can make things bigger, more complex, and more violent. It takes a touch of genius and a lot of courage to move in the opposite direction." Simplicity is key to autonomous system design. Simplicity can be realized by isolation, reducing the degree of freedom, and synchronization. In the case of the Personal Rapid Transit system by Boeing, there is almost no chance of a head-on collision between vehicles because vehicles travel on separate rails without ever intersecting. The platform doors are closed when the vehicle is moving. The platform doors are synchronized with the states of the doors on each vehicle: open, close, or repeat. One leads to the other, which simplifies the control software. There are many trade-offs in autonomous system design. For example, the Round-About design eliminates the need for traffic lights and guides vehicles through the same circular motion. It works efficiently in conditions with low traffic flow.

**Fig. 17.5** The driverless monorail people mover at Newark International Airport

The train design simplifies the control and sensing systems.[16] It reduces a two-dimensional problem into a one-dimensional one. Field tests from the United States, Sweden, Japan, and Germany have shown that a semi-autonomous convoy can save significant amounts of fuel. Semi-autonomous trucks can travel like a robotic convoy. If the lead truck has an active driver, the following trucks can be operated by passive automatic vehicle following systems and cooperative adaptive cruise control until there is a need to get out of the formation. The speed and distance between trucks can be preprogrammed for the fleet.

The simplification strategy can be applied to other situations, such as automatic shoplifting detection in retail businesses. Shoplifting involves many variables, including gestures, location, interaction, intent, timing, objects and payment status, etc. It is a 3D problem that frequently eludes the limited scope of closed-circuit camera systems. To simplify the problem, let us start by focusing on one specific behavior called "sweet-heartening," which happens when the staff of a shop knows the customer well (e.g. sweetheart), and let the customer pass goods from the checkout table without payment. Here, we need to monitor the goods on the belt and the payment status. Monitoring objects on the checkout desk is straightforward. An overhead camera can take a 2D image of the goods. It is easy to detect the goods-on-table status. By checking the payment amount, the following detection algorithm

[16]Heine M (2013) Mercy sakes alive, looks like we got us a robot convey. April 11, 2013 Overdrive Extra. http://www.overdriveonline.com/mercy-sakes-alive-looks-like-we-got-us-a-robot-convoy/

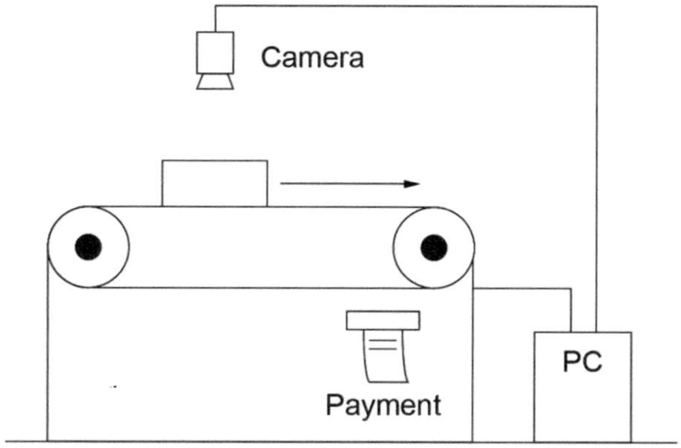


**Fig. 17.6** Shoplifting behavior detection system – transforming a three-dimensional problem into a two-dimensional problem

can be implemented: if the goods are on the checkout table but no payment occurs, then send an alert to the supervisor. Simple and deterministic shoplift detection is essential to retail business to avoid false alarms that drive customers away (Fig. 17.6).

## Summary

Large-scale computing systems are becoming too complex for humans to manage. We need to design appropriate control strategies for interacting with autonomous systems. We use launch-and-leave-it functionality for space robots and machine overrides for emergency situations in which humans react poorly. We shall always be able to manually override the machine at any time if necessary.

Leonardo Da Vinci once said that simplicity is the ultimate sophistication. Simplicity is the key to designing a cost-effective autonomous system. Simplification strategies include perceptual space reduction, degree-of-freedom reduction, and communication simplification. The simpler the system design, the better controllability, observability and less complexity.

CHAPTER 18

# Survivability

Survival is the most instinctual behavior of all animals. It is also the most essential design strategy used in robotic and network systems, which were originally built for indoor, stationary, and isolated use. These systems have become increasingly more outdoor, mobile, and networked in recent years. The greatest concerns of our generation are battery life, wireless signals, computer attacks, and manmade disasters – all relatively new instincts in our digital lives. In this chapter, we review the tasks, strategies, and algorithms for survival in physical and digital environments. Traditionally, design strategies for these two worlds are separate. However, they share many survival models such as cliff detection, cache, collision avoidance, self-healing, and improvisation. They have gradually merged into one, observable in cyber-physical systems as large as cities with connected vehicles, resilient infrastructures, power grids, and mobile devices.

## Measurements of Survivability

Survivability has three levels. The first level is to *stay alive.* Sustaining system's integrity and energy are essential. Many systems have built-in vital signs, similar to heartbeat indicators. The Black Box of an airplane, for example, can sustain a static crushing force of 5,000 psi, is waterproof, and the data on the magnetic wire may stay intact for decades. However, the battery life of a Black Box is about 30 days. Without the communication signals, it would be extremely difficult to locate the Black Box in the ocean.

The second level of survivability is the ability to *communicate*. While lost underwater, a Black Box typically sends a 10 millisecond 37.5 kHz pulse or ping signal each second to help in search and rescue efforts. The communication signals are usually very brief pings, or handshaking signals with unique channel frequencies. While airplanes fly, their engines also send live status updates and current GPS coordinates to satellites for diagnostic purposes, in addition to pilot communications

© Springer-Verlag London 2016
Y. Cai, *Instinctive Computing*, DOI 10.1007/978-1-4471-7278-9_18

and recorded Black Box data. Satellite data can help recover the plane's trajectory even if other data is not available. The abundance of communication channels and storage clouds increase data survivability in the digital age. This may be bad news for advocates of personal privacy, but good news for the fields of forensics and other sciences.

The third level is to maintain the *availability* of a service. In many real time, online service systems, availability is the ultimate embodiment of survivability. For example, the autonomous people mover systems used at large international airports are required to reach a level of 99.7 % availability, which means only 3 hours of downtime for every 1,000 service hours. Normally, an autonomous transit system has two vehicles running in a synchronized mode on two separate tracks. If everything is normal, the two vehicles will run in opposite directions to maximize service volume. If the communication between the vehicles and the control center is lost, then the two vehicles will run independently without synchronizing with each other. Two vehicles might stop at the same station and move in the same direction, reducing service volume. It is definitely not an optimal pattern, but it maintains the availability of the system. If one vehicle has technical problems or needs to be repaired, then another vehicle will keep going to carry on the service. This sustains the availability with partial capacity.

*Survival Time* is the duration of the life span, communication, or service availability before death or a malfunction. Digital worms breach target IP addresses and attempt to propagate. A viral probe can easily infect an unpatched system. Survival time is the average time between probes. High-speed Internet services are frequently targeted with additional scans from malware like bots. If a computer is connected to such a network, then the "survival time" is much shorter.[1]

Figure 18.1 visualizes the trend of survivability over time, including battery life, communication and data availability.

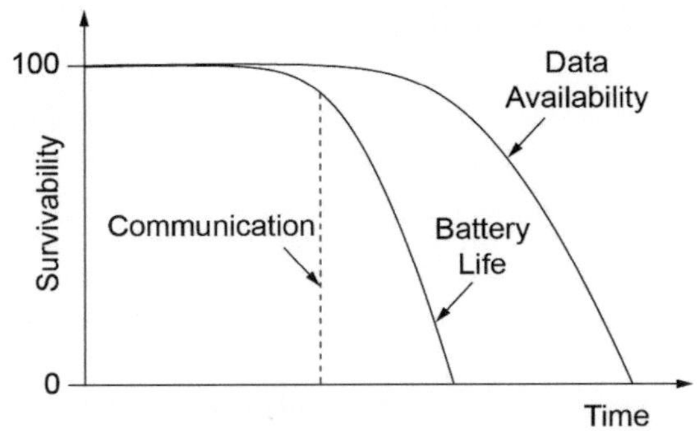


**Fig. 18.1** Survivability of an aviation *Black Box*

[1] Wanner R. Survive Time. https://isc.sans.edu/survivaltime.html

## Cliff Detection

Humans and animals instinctively avoid cliffs to survive. That's why we often have nightmares of falling off a cliff. Psychologists placed a baby on a glass floor over a valley. Even though the floor was physically flat, the babies usually stopped crawling before reaching a point over the valley because of their visual detection of the cliff.

Cliff detection is the vertical differentiation of the absolute depths of a surface. Differentiation is tactile visual sensing – a sort of primitive vision. To be more precise, the change in the height of the surface can be used to measure the steepness of a cliff by the use of first order derivatives (i.e. edge detection). If we take the inverse of the vertical axis of the cliff detection model, then we get the obstacle detection model. Therefore, cliff detection is the most essential low-level vision algorithm. A visually-impaired walker uses a cane to probe the road. Compared to visual detection, the cane is a sampling process used to create a sparse dataset. It does not cover the far-field view, but it is sufficient for walking around. In 1970, New York City had a power outage. It caused major traffic chaos because none of the streetlights were working. Not surprisingly, blind people had no problem walking in dark, and they helped many sighted people find their way home.

We can install a stick probe on autonomous robots, but it is usually too slow because vehicles travel at a high speed. Therefore, we need a better method for cliff detection. Visible light stereo cameras can see far away, but the depth calculation is not always reliable, especially at night. On the other hand, laser rangefinder sensors behave like an extended stick probe for depth maps. They can detect cliffs, obstacles, or potholes. A laser rangefinder sensor uses a laser line or dot projector to sense the 3D structure of a surface ahead. Rangefinder sensors have been widely used in autonomous driving vehicles. Instead of detecting cliffs, they are mainly used for detecting walls and obstacles. To many prevailing laser rangefinders sensors, cliff detection is actually a line-of-sight spot issue. In order to detect vertical cliffs, the laser depth sensor has to point almost vertically downward, a geometry which can be used for pothole or crack detection for highway maintenance. The idea is to install an affordable depth mapping sensor under regular vehicles and collect the highway surface conditions through the Internet.[2]

The cliff detection metaphor can also be used in stock market data analysis. The detection of a crisis is essential to the survivability of the stock exchange. When stock exchange volatility is high, the New York Stock Exchange can take steps to hold trading to ease panic selling. These moves are called market circuit breakers or collars. Circuit breakers were first used in October 1989, following a major stock market drop. The Exchange looked at how many points the markets had dropped to decide whether to kick in the circuit breaker. Since 1997, and under the current rules approved by the Security Exchange Commission (SEC) in 2012, market-wide circuit breakers are triggered by percentages rather than points, i.e. when the S&P

[2]Carnegie Mellon University, T-SET Center: http://utc.ices.cmu.edu/utc/

500 Index drops 7 % at Level 1, 13 % at Level 2, and 20 % at Level 3 from prior day's close.[3] Under extreme conditions that could trigger market disruptions, exchanges can further invoke SEC Rule 48 to make it easier and faster to reopen the stock market. This enables traders to rectify the situation and resume normal trading as soon as possible. Understanding cliff detection is critical to regulating the automated high-frequency stock trading systems that can make the stock markets even more volatile, causing a loss of billions of dollars in a fraction of a second.

## Collision Avoidance

All animals possess the instinct to avoid being hit by something or running into obstacles. It is an essential maneuver for navigation, foraging, fighting, fleeing, and survival. Collision avoidance is also a computational problem for detecting the interaction between two or more objects.[4] It is often associated with robotics, video games, physical simulations, and network traffic control.

In addition to determining whether two objects have collided, collision detection systems may also calculate *Time to Contact* (TTC). The classic equation of TTC is:

$$TTC = -\frac{s}{\frac{ds}{dt}} \tag{18.1}$$

where, $s$ is the size or scale of the object in the image and $ds/dt$ is the rate of change of the size. Note the TTC does not rely on the absolute size of the object in the image sequence, but in the relative change in scale from one frame to another.[5] TTC can also be calculated from optical flow vectors, or the rate of change of the view angle from an image sequence. Although many sensors are involved in collision detection, the processing duration might be too long for a creature to respond in time. To survive, one has to grasp the critical variable. TTC is a simple and direct calculation. It measures the rate of change of an object's size and calculates the time to impact on the lens (full-size). The calculation does not produce the distance, but rather the time of impact. This is enough for a creature or robot to decide whether to stay or to flee.

To avoid the occurrence of collision in feeding chickens, there is a "pecking order" in chicken societies. The high position chickens eat first followed by lower ranking ones. The newest chickens are the last to eat. If a younger or newer chicken

---

[3]Guzman Z and Koba M (2016) When do circuit breakers kick in? CNBC Explains: Retrieved on March 9, 2016.
http://www.cnbc.com/2015/08/24/when-do-circuit-breakers-kick-in-cnbc-explains.html

[4]Ericson C (2005). Real-time Collision Detection. Elsevier, p. 13.

[5]Alenya G, Negre A, Crowley J. Time to contact for obstacle avoidance. www-prima.inrialpes.fr. Retrieved on Mach 9, 2016:
http://www-prima.inrialpes.fr/jlc/papers/AlenyaNegreCrowleyEcmr09Final.pdf

interrupts the pecking order, it will be punished and the order will resume. The early network protocol "Token Ring" is similar to pecking order. The one with the token is allowed to talk until the token is passed to the next party. The benefit to using this method is that it is simple and deterministic. The drawback is that it is not very efficient. There is a lot of waiting time, even though the token is circulating very fast.

In the Internet, the most popular protocol is TCP/IP. It is not a reliable protocol, but it is highly efficient. Its most remarkable mechanism is its collision detection algorithm. The local network is like a dark room where only one person is allowed to talk at a time. When someone talks, one must tell her name and the message's destination. When two or more users talk simultaneously, a collision occurs. Nobody can hear clearly. They must stop talking and wait for randomized periods to resume their talking. This is similar to the Spontaneous Alternation Behavior (SAB) as we discussed in chapter "Default Mode". To give each user a fair chance to talk, the messages are sliced into small pieces so the users can "breathe" and "reconnect" during communication. Intuitively, more users mean more chances of collision. After a number of collisions and randomized reconnections, if the message units were still stopped, then the network would be in an endless deadlock.

## Energy Harvest

Energy is part of the life-support of a system. Let us look at the endoscopic pill camera, for example. The typical battery life of the camera is only 8 hours, although the entire journey takes 24 hours to pass through the human body. The pill camera stays off until the final 8 hours so it can do its work.

To date, we have not yet built robots with any strong survival instinct. This is not due to a lack of technological capacity, but rather a lack of motivation. Robots will need to feed themselves to survive. The followings are examples in multiple applications.

*Metabolizing* is a way for a robot to collect available biomass and turn it into fuel. For example, imagine a mobile robot harvests slugs from fields, takes them to a stationary digestion system, and retrieves energy from burning methane produced by the decomposing slugs. Such robots are not likely to become practical in the near future because collecting slugs cost a lot of energy and the efficiency of the fuel conversion is less than ideal. Metabolizing is important to *in vivo* robots – the robots work under the skin. Converting urine or blood into biofuel is a task that is currently in high-demand. However, we are still a long way off from designing robots that can metabolize any other source of energy in a way that will let them live sustainably and renew their energy independently.

*Socket-Seeking* is a practical approach that has been commercialized in iRobot's house vacuum robot Roomba™. The energy self-sufficient robot seeks out wall sockets and plugs itself in to recharge whenever it encounters one. However, such robots rely on the operation of the electrical power supply, and could easily be rendered powerless by switching off the availability of electricity.

*Wireless Charging* is an effective alternative to carrying the weight of a heavy battery for mobile robots such as drones and in-vivo devices. For many small drones, the battery weight limits their flying range. Furthermore, it is desirable to let drones recharge themselves while in flight. This does not sound practical because we would have to build many wireless charging stations along a route. However, such charging stations already exist on power lines. Using power lines is a feasible solution to the problem of recharging because they are abundant across inhabited areas where power lines are exposed. MIT's test drone uses sensors to detect a power line, approach it, and land on the line like a bird.[6] This allows small drones to recharge without landing where they would be susceptible to damage or theft. In addition to the glider maneuver to perch on the line, a drone could hover above the line and drop down an adaptor to get a charge.

Electronics can do remarkable things in a body. Pacemakers and deep brain stimulators are only a couple of examples.[7] The main concern is how to power implanted devices or *in vivo* robots wirelessly. The Oral-B electric toothbrush can get power from wireless near-field inductive coupling on a waterproof cradle next to the bathroom sink. This conventional magnetic power transfer operates in the *near-field* to avoid interaction with biological tissue, but this type of power cannot reach small devices hidden deep inside the body. Far field waves, on the other hand, can travel long distances and penetrate biological tissues. Radio and telecommunication signals operate in the *far-field* range. When they interact with biological tissue, they either reflect off of the body harmlessly, or get absorbed by the skin as heat.

Fortunately, between the near field and the far field, there is a sweet spot called the "*midfield,*" which consists of multiple media for traveling radio waves: from air to a biological tissue. A specially designed metal structure generates a wave that converges at the micro device deep inside the body. As a result, a micro implant containing a power harvesting coil, integrated circuits, electrodes, and other components can be safely implanted and powered anywhere in the body. Using integrated circuits, we can stimulate nerves, put sensors in the body, and even deliver drugs.

## Cache Behavior

In a dynamic world, the ability to predict future situations can provide a selective advantage. The evolution of foresight enables humans and animals to develop anticipatory behaviors for future survival. For example, animals store and hide food

[6] Humphries M (2014) MIT teaches a drone to perch on power lines so it can recharge during flight. Geek, June 23, 2014. http://www.geek.com/chips/mit-teaches-a-drone-to-perch-on-power-lines-so-it-can-recharge-during-flight-1597462/

[7] Ho JS et al (2014) Wireless power transfer to deep-tissue microimplants. Proceedings of the National Academy of Science of the USA, 111(22):7974–7979.

for the future, based on a shared past experience of food shortage. Food caching behavior is adaptive. Animals must be flexible in order to increase their chances of survival. The episodic memory not only allows us to go back in time, but also to foresee, plan, and shape future events.[8]

For network video streaming and video games, latency is a critical problem. A significant delay in video frames or character animation control data can kill the flow of the game. Data caching allows a constant data stream to continue uninterrupted in uncertain network conditions. The dynamically cached data in the data transmit buffer can compensate for interruptions in the network due to the dynamics of traffic and processing. In a typical video encoder,[9] the data transmit buffer caches a number of frames for smoothing the streaming process. This is comparable to adding a water tank to a water pipe. In case of an interruption to the water flow, the water from the tank can sustain the flow for a while.

Most action video games played on networks require a cache as well. In video games, there is a single authoritative server handling the global logic. The game players are clients connected to the server. The players send commands to the server for every move. The server executes the input commands, updates the movements of objects, and then relays back to the players a list of objects to display on their terminals. As we can see here, there is only client–server communication, not peer-to-peer communication. However, broadband networks are not always reliable for online gaming. If there is a one second latency in its connection to the server, then it requires one more second for the server to update the player's status and broadcast it to other players.

If one of the update messages fails to arrive, then there are two choices: we can extrapolate the player's position by performing the player's movement locally for a brief moment, assuming that the server will accept and acknowledge the player's comments directly. This could potentially generate large errors. Another method is to let the player rest at the position in the last update until a new update arrives, which might cause the player's movement to stutter. We can use interpolation to buffer an additional 100 milliseconds of data on the player's side. The other players, therefore, are drawn where they were at a point in the past that is equal to the exact latency plus the amount of time over what is interpolated.

Denial of Service (DoS) is a malicious attack that aims to flood the cache of a web site, network, or data storage target. The first "instinctive" strategy for defense against DoS is simply to use huge storage to absorb the suddenly increased load, making more resources available than the attacker is able to consume. While this approach is effective, it can be quite expensive. However, it has the added benefit of providing service resilience to both DoS attacks and to legitimate, unexpected surges in traffic. For example, during the first month that the Affordable Health Care

[8]Suddendorf T and Corballis M (2007) The evolution of foresight: What is mental time travel, and is it unique to humans? BEHAVIORAL AND BRAIN SCIENCES, 30, 299–351, 2007.

[9]MPEG and H.264 or QuickTime.

Act health insurance marketplace web site was in operation, the system crashed because it was not designed to handle such heavy traffic, even though most came from legitimate users.

The second defense method is to use a "shelter" to stop the attacker's data flow.[10] This can be done through a cloud service to guarantee bandwidth to legitimate traffic. This can be compared to the parasitic behaviors in nature. For example, some species of insects use the shells of other species to protect themselves, even through killing the "host" in the process of obtaining protection.

The third method is to stop the attack's data flow before it even reaches the network through user authentication. For example, by asking the user to recognize a street number in a street view image, or to click a mouse button, the server can decide whether the entered data is from a person or a robot. The answer and click are easy enough for human users to recognize, but too difficult for a machine, assuming the attacker's computer vision algorithms are not very advanced!

Finally, special DoS protection devices can be built to further guard networks against attack. These devices sit at the network perimeter and process traffic before malicious attacks reach the internal network, filtering out potentially harmful activity. Hardware solutions work by analyzing network traffic for signature-based detection of known attacks, or by providing behavioral analysis of current traffic against profiles of normal behavior. Traffic that matches a known attack pattern or fails to resemble typical network traffic is either automatically blocked by the device, or flagged for further investigation by security analysts. The advantages of the hardware solution include: it's more efficient to run the detection algorithms and is it more difficult to hack. However, this increases cost and introduces updating issues.

## Self-Amputation

When humans encounter hypothermia in extremely cold weather, blood moves from the limbs to the heart to keep vital organs functioning. As a result, fingers and toes succumb to frostbite, and are sacrificed to protect the heart. Certain animals also utilize self-amputation to save their lives. For example, a gecko can break off its tail to escape the jaws of a predator. Self-amputation is a survival strategy to stay alive by cutting off non-essential functions.

For a network under attack, it is desirable to isolate the subsystems while keeping the core components running. During the 9/11 attacks, New York City telecommunication systems were down because of overwhelming call volume, incapacitating many emergency services. Their new dispatch system will now verify

[10]Chapple M (2013) The three elements of Defense against denial-of-service attacks. Feb. 28, 2013. BizTech. http://www.biztechmagazine.com/article/2013/02/three-elements-defense-against-denial-service-attacks

the caller's address before dispatching a response team. This screening process acts like a self-amputation to protect the core communications during emergent situations.

## Self-Expiration

In nature, there are many biological clocks for ensuring the survivability of species. P53 is a protein for DNA fidelity monitoring. During the cell growing process, P53 checks for any errors in the coding. If an error is found, the system will attempt to repair it. Normally, it waits for 12 to 24 hours. If it is still not fixed, P53 will force the bad cells to commit suicide. This principle might be useful to soft agents or robot design. When the agents discover any serious errors, they try to fix them within a set period of time. When this period expires, the system will destroy itself to avoid further damage to other systems, including humans. See Fig. 18.2.

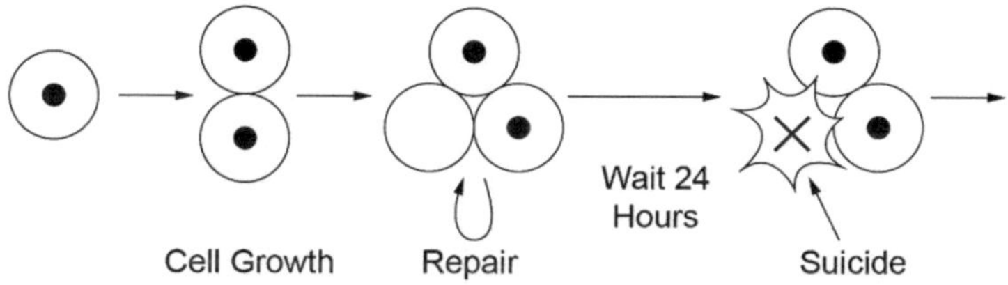


**Fig. 18.2** Protein P53 self-monitoring and self-expiration

In engineering, self-expiration allows a system to check vital components within a fixed period. Airplanes have major maintenance procedures in place, even without incidents. Cars need new tires before the treads become invisible. All of these processes are in place to ensure the survivability of the system.

## Isolation

*Isolation* is another effective strategy for survivability. To become immune to the changing environment, one must create an artificial environment so that the dependencies of modules will remain unchanged, much like a "time capsule". This is why the isolated US military nuclear warhead silo of the LGM-30G Minuteman-III is still controlled by primitive computers dating back to the 1960's and has a launch system that still relies on a rudimentary floppy disk. To render it hack-free, the aging computer is not connected to the outside world, i.e., the Internet, preventing any cyber terrorists from gaining control over the weapons.[11] Most of these communication systems are analog, built only to receive launch orders from the president in times of war.

Some commercial automated people movers also use the "time capsule" strategy. For example, the IBM 386 PC is still used in autonomous train control systems in some international airports. Each vehicle has three PCs. As a result, these systems need a lot of spare parts and experienced engineers. It is expensive to upgrade such legacy systems. The missile silo control system, for instance, would cost one billion USD to replace. For critical safety components (e.g. transportation and medical systems), testing and recertification (e.g. TÜF) are also expensive to pursue. Nevertheless, isolation is a strategy to prevent unauthorized access to critical infrastructures. Figure 18.3 illustrates a fictional mobile phone app for remotely operating the gate of London Bridge. These types of remote control apps should not be developed in the first place because they present threats to the public safety.

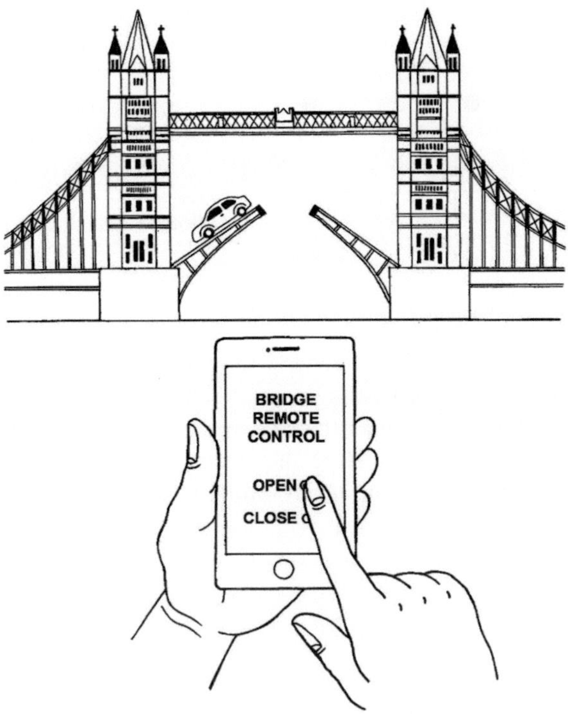


**Fig. 18.3** A fictional scenario of the mobile phone app for remotely operating the gate of London Bridge

[11]Christie J (2014) Americas feared nuclear missile facilities controlled computers 1960s floppy disks. Daily Mail, April 27 2014. http://www.dailymail.co.uk/news/article-2614323/Americas-feared-nuclear-missile-facilities-controlled-computers-1960s-floppy-disks.html

## Self-Healing

*Self-healing* is the empowerment of a software module to adapt to a new environment and reconfigure itself for survival. Self-healing needs to meet at least two of following conditions: (1) storage for information or energy so that it enables recovery from damage, (2) a feedback mechanism for stabilizing the healing system so that it won't "run-away" wild like a cancer, and (3) spontaneously alternating behavior so that it can avoid deadlock during the self-healing process.

Many self-healing systems do exist in real life. A network video streaming device might encounter a traffic jam. It can typically heal itself after the traffic calms down. This is because network data streaming includes a buffer for unsent data and a feedback loop for adjusting the resolution and buffer size, as well as spontaneously alternating the reconnection intervals.

Perhaps the greatest potential application of self-healing is in space travel. If we plan for a 10-year long mission into deep space, we will need a lot of spare parts onboard, which will increase the weight of the spacecraft. It has long been a dream of inventors and engineers to create a self-repairing chip. The idea is to go beyond the hardwired chip and instead to make a chip that can rewire its circuits as needed. A similar system has existed since 1984, called the Field Programmable Gate Array (FPGA). This system wires individual circuit gates to perform functions such as addition, subtraction, multiplication, CPU, or memory. Researchers at NASA Jet Propulsion Laboratory designed a self-healing system with two FPGA processors. If a fault occurs in one, the backup processor takes over, generating a new configuration using an evolutionary algorithm to try different approaches until a layout emerges for the task. As a test, researchers exposed the system to 250 kilorads of radiation, enough to fry the circuit. The system started repairing itself, attempting up to 100 configurations per second until it found one that worked.

Self-healing can also be implemented to monitor system execution in an environment. If a software module does not fit in a new environment, a self-healing software agent can check the original vender's online resources to see if there are any updates and alert users to download these updates. For legacy code, the self-healing agent could inspect the code to automatically replace the deprecated functions or expressions with up-to-date functions and expressions.[12] Digital artifacts that can distinguish between themselves and others using strings of data, like a human being uses proteins, are called *immunotronics*, a term coined by Andy Tyrell. Immunotronic systems look for anomalous signatures. If it finds some, the system reconfigures itself.

[12]Bains S (2005) Computer, heal thyself!, WIRED, No. 9, 2005.

## Improvisation

Another design strategy is *improvisation,* which is to use whatever is locally available, integrate those elements, and accomplish the mission. For example, interstate highways in the US usually were overdesigned with capacities and surface quality surpassing basic usage. They can be used as aircraft runways in case of emergency or war. In fact, there are plenty of occasions that small airplanes made improvised emergency landing on highways.

Machine learning algorithms enable onboard computers to build soft sensors on the fly. Assume a vital sensor on a satellite is damaged but the data from that sensor correlates with other sensors. It may be possible to use the rest of the sensor data to recover data from the damaged one. The FPGA processor (chip) was used to create a neural network-based learning system. It demonstrated the usage of the approach for onboard data mining of ocean surface temperature data.[13]

As we discussed in the previous chapter "Tool Making," 3D printing technology also enables a system to fabricate damaged parts as needed, especially in remote sites or on space exploration missions, given the ability to carry a printer and robotic arms onboard. The 3D models of parts stored in the ground computer can be transmitted to the remote site and printed out to replace damaged parts. In this case, the 3D printer would not have the bulky, traditional frames as seen in Formula or MakerBot machines. Instead, it would resemble a tiny mobile robot that climbs around the inside of the surface of the vehicle and repairs it like a human would – from the inside out.

## Migration

Mobility has a significant impact on survivability. Many times, a shortage of food, a lack of communication resources, or an attack can be remediated by moving around. All cells show some ability to move. Some cells are highly active, such as immune system cells. The basis for cell movements arises from the cell's internal structure. Using structures made from proteins, cells can move internal materials, divide, crawl, and swim.

Movement can remedy many emergency situations. In space, it appears easy to move the spacecraft. To avoid being damaged by high-speed space particles, the spacecraft may turn its back toward the direction of an impending particle because the entire backside of the craft has thicker shielding. This self-shielding movement is an instinctual behavior. How does a spacecraft implement the self-shielding algorithm is part of instinctive computing in space science.

---

[13]Cai Y, Hu Y (2007) Onboard data mining for ocean surface temperature, IEEE Space Science Conference, Pasadena, CA, 2007.

## Teamwork

In many cases, teamwork is necessary for survival. A sophisticated social life can be a fantastic problem-solver, simply through the sharing of knowledge. Sharing tool-making experience one another is common behavior in animals. Crows working in groups often create advanced tools, such as adjustable wood sticks for picking up food. Different groups create different tools, creating different cultures. As a result, tools become more and more efficient.

Teamwork may also be physical. An ant can carry food that is 100 times heavier than its own body weight. Fire ants connect themselves using their legs to form a raft to float on the surface of water. Teamwork can also create deception for survival. For example, fish exhibit schooling behavior to deter predators. Macrophage cells surround a foreign cell as a team and engulf it like a mob.[14] This mechanism can be used for dissolve the Distributed Denial of Service (DDOS) attacks, for example, a team of servers can work together to absorb the data traffic and create "honey pots" to distract the traffic generated by the attacker.

Social life can be inherited or improvised. Inherited social life is limited to fixed social structures, such as queen and worker bees in a hive society. Improvisational social activities involve tool making and problem solving, which require more intelligence and environmental awareness.

Groups of mobile robots can be connected to cross a gap that is wider than the length of an individual robot. A team of robots may also construct a 3D model from multiple views from each robot, given the algorithm for matching corresponding feature points.

## Dependency Analysis

Analysis of network features has become a popular pursuit in fields ranging from cell biology and ecology to the social sciences. Food webs model species populations and dynamics in the food chain. The key parameters in a food web can be adapted for studying the survivability of modern artifacts such as social networks, robot swarms, and software systems. A typical food web is a resource-consumer network (or prey–predator network). Each entity or species in a system can be represented as a node in the graph, and the directional link between nodes represents the recourse or consumer dependency. The recourse and consumer relationship can be represented as a directional graph and matrix.

In 2012, super storm Sandy shut down New York City's power grid, leaving 50,000 houses without electricity for three weeks. There were some solar panel systems in the area. However, the panels could not work because any electricity had

[14]NIBIB.gov (2013) Macrophage engulfs foreign cells. https://www.youtube.com/watch?v=w0-0Bqoge2E

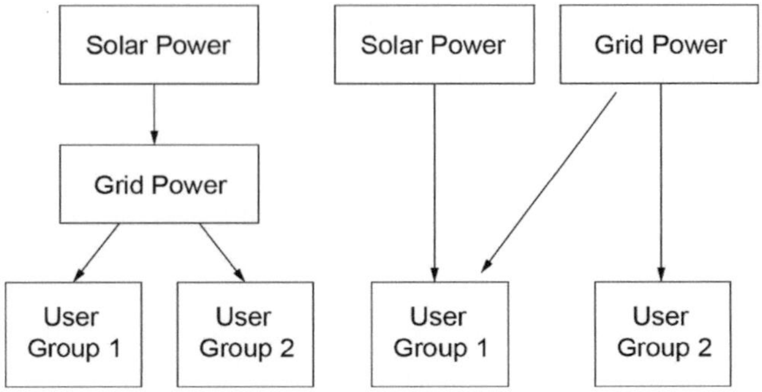


**Fig. 18.4** Solar power dependent on the grid (*left*) versus solar power independent from the grid (*right*)

to travel through existing utility systems. A power grid is designed for maximizing utility, but it creates dependency. The left graph shown in Fig. 18.4 indicates existing dependency in the network flow. In the same figure, the graph on the right shows possible improvement, where consumers can access the solar power directly, instead of having to depend on the regular power grid.

The best power grid perhaps is no grid at all. Off-grid power systems provide an independent power source and only use grid power as a backup. For a large network system, computerized dependency analysis is necessary. A very popular method used is critical node analysis. In a scale-free network, some nodes have more connections than others. If we take down those nodes, the network will collapse. For example, the transformer is the most essential component in a power system. If the critical transformers were damaged, if would take a month to replace them. As a result, the power grid could be down for at least a month. The rise of micro-grids reduces dependency on the traditional power grids significantly. It also reduces the threats of cyber-physical attacks seamlessly along the way.

The dependency in a network is dynamic. Resources or threats today could potentially become extinct tomorrow. Eventually, emerging resources will replace existing ones. In software engineering, digital products have short life spans, creating legacy software in industrial, military, and consumer environments. Today's software depends on application codes that contains Dynamic Linked Library (DLL) modules or Commercially-off-the-shelf (COTS) products that are sensitive to their environments: runtime binary code execution libraries, operating systems, the host computer hardware (32-bit, 64-bit, or 128-bit), even code storage media (CD, hard drive, or cloud-based). Figure 18.5 shows the food web of the application of a computer.

It is not rare that a software product cannot survive in a new environment after a few years or even a few months. Calculating the life span of a product is straightforward. The *life expectancy* is the minimum value of the life span of the species along the dependency path. For example, the Operating System module lasts about 12 months. The compiler may last about 24 months. The code's life

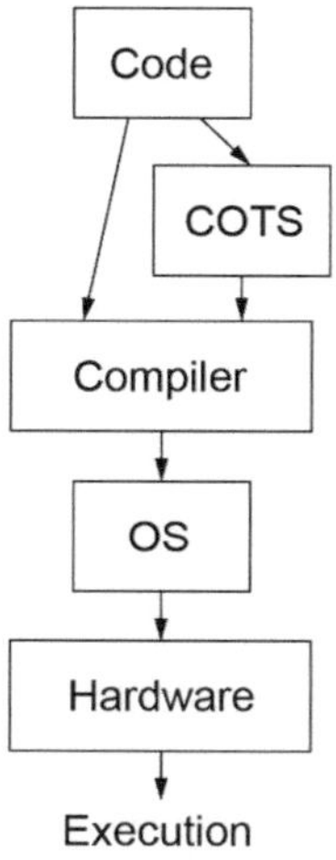


**Fig. 18.5** The food web in a computer

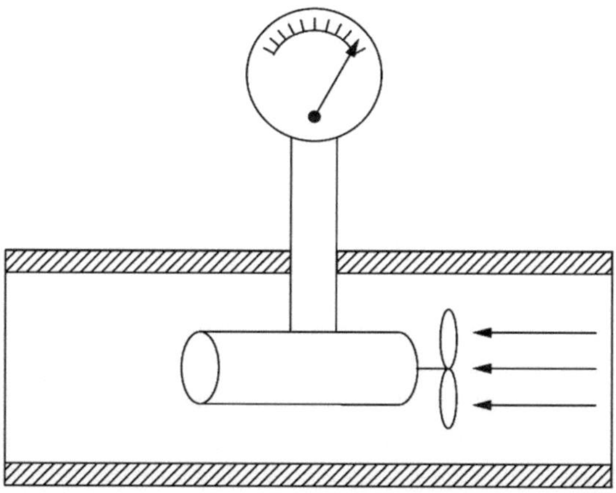

**Fig. 18.6** The self-sufficient flow meter, powered by the flow itself

span depends on the compiler, the OS, and the most critical elements (the invisible COTS or DLLs that mainly rely on previous environments). Storage media may last up to 2–5 years. In this case, the estimated total life span is about 12 months.

How does one survive in a dynamically changing environment? *Self-reliance* is a common methodology. The idea behind self-reliance is to use the resources available within the system. Consider the design of a gas flow meter. A tiny electrical power generator is embedded inside a gas pipeline to convert gas flow to electrical current to power the circuit, creating a self-reliant gas meter that requires no external power. When gas flow stops, the meter sets to zero by default. For the safety and security of critical systems, it is common to develop the code in-house, eliminating the dependency of Dynamically Linked Libraries or COTS (Fig. 18.6).

## Mental Time Travel with Episodic Memory

*Episodic memory* provides access to an event that is experienced personally. This kind of memory is not about regularity, but rather reconstructing particularities about when and where, like in a movie. Episodic memory implies a mental reconstruction of some earlier event, including hunger, hurt, escaping from danger, and emotional reaction. Based on previous experience, we can anticipate specific events in the future. Mental time travel into the future might include the planning of a specific event, such as food gathering.

Many bad dreams experienced today are similar to those of our ancestors'. They are indelible rehearsals for survival. In a dynamic world, the capability to predict future situations can provide a selective advantage. Episodic memory can be viewed as data mining, not from existing knowledge, but from massive streams of news and episodic events. In fact, prevailing medical diagnoses are still based on episodic memory. Chest pain often indicates heart disease. Heartburn is usually caused by gastrological diseases.

Sequences of episodic events can be used to predict future events. For example, Jail-breaking[15] refers to using the "backdoor," or working around the software that allows access to the intimate functions of applications that were intended for use by the manufacturers. Jail-breaking emerged from the world of smartphones, video games, and appliances such as Kinect and Roomba. These episodic events could have served as a form of mental time travel for improving the security of the new electrical car, e.g. the Tesla Model S. A dashboard computer with a hidden 4-pin connector for an Ethernet port controls the car's navigation system and infotainment functions. By connecting an Ethernet cable between the car and the laptop computer, one can access the backdoor of the onboard computer to run a web browser. The benefit for Tesla was the ability to update the software and allow users to download online music, or to load third-party apps onto the car's computer. However, the existence of a backdoor channel makes the vehicle more vulnerable to theft and sabotage. Figure 18.7 shows an example of the episodic memory based analogical reasoning for predicting potential threats to the class of vehicles with similar accessories.

How episodic memory is stored and processed in our brain is still a mystery. Here, we assume that long-term memory can be simulated with a spatial and temporal database. At the insect level, long-term memory can be projected to the external "pheromone" map on the ground. Assume those pheromones don't decay over time. Then the accumulation of the pheromone deposition is straightforward, just add them up like a heat map.

[15]Ingram A (2014) It had to happen: jailbreaking a Tesla Model S. VB, April 5, 2014. http://venturebeat.com/2014/04/05/it-had-to-happen-jailbreaking-a-tesla-model-s/

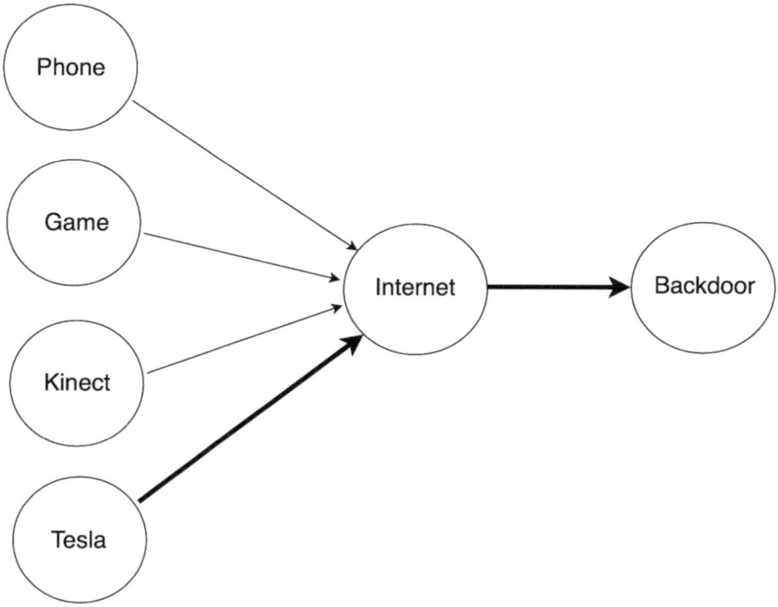


**Fig. 18.7** The episodic memory based analogy for Tesla Model S™ backdoor hacking vulnerability, where the *thicker lines* highlight the outcome of the analogy

Figure 18.8 shows a map of human activities in a large building where 10 sensors were installed at the access points to monitor the human traffic flow.[16] The sensor network recorded over two-months of data. First we want to look at which sequences of movements were popular. Afterwards, we want to make a model to find anomalies in the data as well as to test whether a given input sequence would be considered normal or abnormal based on the history of movements. In this case, we use the approximate string matching algorithm to sort the motion sequences at different pattern lengths, such as 1, 2, 3, 4 and 5 stops.[17] In Fig. 18.8, the more frequently visited rooms are labeled with darker shades. The width of the black arrow paths represent how often a person goes between two rooms. The dashed paths represent commonly traveled paths of length larger than 2 stops.

The data shows that sensor J is the most commonly triggered sensor. It is also often triggered repeatedly (J = 7227, JJ = 2884, JJJ = 782). Strings of length 2 containing E are also very frequent (EI = 973, JE = 304, EJ = 232, EG = 199, HE = 118), suggesting that E branches off to a lot of other locations. Also, most

[16]Cai Y (2010) Mobile intelligence. Journal of Universal Computer Science, vol. 16, no. 12, 2010, pp. 1650–1665. Retrieved on March 9, 2016:
http://www.jucs.org/jucs_16_12/mobile_intelligence/jucs_16_12_1650_1665_cai.pdf

[17]Manber U and Wu S (1992) Fast text search allowing errors. Communications of the ACM, 35(10): pp. 83–91, October 1992.

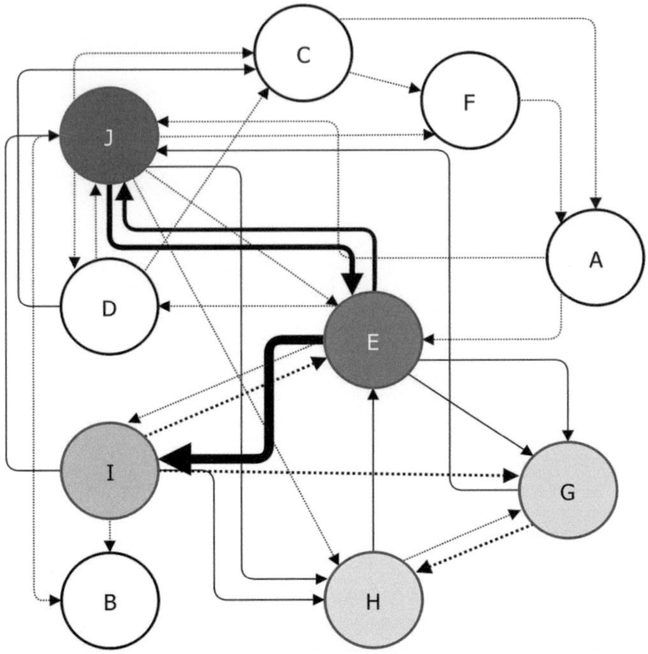


**Fig. 18.8** The episodic memory of human traffic flow in the building with 10 access point sensors for over 2-months. The dark nodes and thick paths show the most frequently visited access points and movement sequences

occurrences of J also involve E, suggesting that employees at position J usually transition to E, suggesting that E is a major bottleneck in the building, which might be problematic if employees cannot access E, since it's likely employees at E are cut off from the rest of the building, since E is the center of the building. Certain anomalies were also detected. Some sequences described routes that were roundabout and were not the shortest way to get from point A to point B. For instance, DCI was detected as an anomaly. Looking at the floor plan, we found that the path from D directly to I is much shorter than D, C and I. Furthermore, sometimes errors might happen and a scanner fails to record an employee walking back. For example, EB was detected as an anomaly. Looking at the floor plan, it seems unlikely to get from E to B without triggering any other sensors. Thus, it's possible that this sequence was produced because the sensor didn't work that time (events like this would occur rarely). Other suspicious sequences including walking in a circle (IHEI, GHEG, CFAC) and walking back and forth for long periods of time (HGHGGHGGHGHGH).

We also found that many anomalous events are conditional or local. For example, there is a unique anomalous event sequence: JJJJJJJ. However, when we consider the time of day when the events happened, the sequences become JJ, JJJ, JJ, ..., which are normal. On the other hand, a few cases seem normal. But in fact they are anomalous events, if we consider the 'time of day' variable. For example, frequently accessing and leaving after middle night is often viewed as an anomalous event. In fact, anomalies can be determined by conditional probability distribution functions. Song et al proposed a Gaussian Mixture Model (GMM) to represent these conditional anomalous events.[18] Unfortunately, learning the conditional probability distribution functions appears to be non-trivial. So far, experiments have been performed only with small the data sets (up to $205 \times 100$).

Can a system inherit knowledge from other systems? The answer is yes! For example, the famous OpenCV's face detection uses the trained feature databases from previous systems that have been inherited for many generations. However, the data structures of such a knowledge have been deeply hidden inside the layers of code, which is hard to understand, even though they are open source. To classify and standardize those conditional probability distribution functions is a way toward shared and reusable common sense across borders of intelligent systems. It is certainly a bridge between what is mainly a hand coded rule-based system and a purely machine learning system.

## Summary

There are common metaphors for survival in both the digital and physical world, such as cliff detection, collision detection, energy harvesting, and cache behavior. From our computational models, we must pay tribute to the great designs encountered in nature, in which all of the aforementioned functions are integrated and executed seamlessly. Episodic memory in nature helps mental time travel from past events to the present. Using what is learned from past episodic events is beneficial for planning, and ensuring future survival. It is a short-term challenge for instinctive computing to represent episodic events, crowdsourcing, and discover anomalous patterns.

In the long-term, instinctive computing algorithms will be a part of the next generation operating systems that include self-healing, self-awareness, and improvisation. Like our caveman ancestors, they may be not the strongest or the most intelligent, but they will be the fittest and they will survive.

[18]Song X et al (2007) Conditional anomaly detection. IEEE Trans. On Knowledge and Data Engineering, vol. 19, no.5, May 2007.

# References


1. Addy NA, Daberkow DP, Ford JN, Garris PA, Wightman RM (2010) Sensitization of rapid dopamine signaling in the neucleus accumbens core and shell after repeated cocaine in rats. J Neurophysiol 104:922–931
2. Albrecht-Buehler G (1985) Is cytoplasm intelligent too? In: Shay J (ed) Muscle and cell motility VI. Plenum Press, New York, pp 1–21
3. Albrecht-Buehler G (2017) http://www.basic.northwestern.edu/g-buehler/cellint0.htm. Accessed 24 Apr 2016
4. Amazon Mechanical Turk (2016) https://www.mturk.com/mturk/preview?groupId=355YH9BKNIVAGO2MR6LXHR30C01FVF. Accessed 24 Apr 2016
5. Anthropometry Resource for CAESAR (2002) Final report, volume I: summary, AFRL-HE-WP-TR-2002-0169, United States Air Force Research Laboratory, Human Effectiveness Directorate, Crew System Interface Division and SAE International
6. AP (2013) Harvard student made bomb threats to get out of exam, cops say. CBS News, December 17, 2013. http://www.cbsnews.com/news/harvard-student-made-bomb-threats-to-get-out-of-exam-cops-say/
7. Argyle M (1969) Social interaction. Tavistock Publishing, London
8. Arkin RC (1998) Behavior-based robotics. The MIT Press, Cambridge, MA
9. Arxiv (2015) Why self-driving cars must be programmed to kill? MIT Technology Review. October 22, 2015. https://www.technologyreview.com/s/542626/why-self-driving-cars-must-be-programmed-to-kill/
10. Bates D, Duell M (2011) 'Please forgive me for what I'm gonna do': mother's facebook message before drowning three of her four children by driving them into Hudson. http://www.dailymail.co.uk/news/article-1376427/LaShanda-Armstrong-drives-3-children-Hudson-River.html. Accessed 24 Apr 2016
11. Beck BB (1980) Animal tool behavior. Garland STPM Press, New York
12. Bekoff M et al (2002) The cognitive animal. The MIT Press, Cambridge, MA
13. Bell D (2015) 3D print your own stepper motor. http://makezine.com/2015/08/18/3d-print-stepper-motor/. Accessed 24 Apr 2016
14. Belousov Zhabotinsky Reaction (2016) Wikipedia. https://en.wikipedia.org/wiki/Belousov--Zhabotinsky_reaction. Accessed 24 Apr 2016
15. BenjaminE4 (2016) Dollo 3D: self replicating 3D printer. http://www.instructables.com/id/Dollo-3D/. Accessed 24 April 2016
16. Blei DM, Ng AY, Jordan MI (2003) Latent dirichlet allocation. J Mach Learn Res 3:993–1022
17. Bonabeau E, Dorigo M, Theraulaz G (1999) Swarm intelligence: from natural to artificial systems. Oxford University Press, New York


© Springer-Verlag London 2016

Y. Cai, *Instinctive Computing*, DOI 10.1007/978-1-4471-7278-9

18. Braitenberg V (1984) Vehicles – experiments in synthetic psychology. The MIT Press, Cambridge, MA
19. Brooks RA, Flynn AM (1989) Fast, cheap and out of control: a robot invasion of the solar system. J Br Interplanet Soc 42:478–485
20. Brooks R, Cambrian Intelligence (1999) The early history of the New AI. The MIT Press, Cambridge, MA
21. Brownlee J (2014) Feature selection with the Caret R Package. September 22, 2014. Machine Learning Mastery. http://machinelearningmastery.com/feature-selection-with-the-caret-r-package/. Accessed 27 Apr 2016
22. Brown W (1910) Some experimental results in the correlation of mental abilities. Br J Psychol 3:296–322
23. Boutell M, Lou J (2005) Beyond pixels: exploiting camera metadata for photo classification. Pattern Recogn 38:935–946
24. bSafe (2016) http://getbsafe.com/. Accessed 24 Apr 2016
25. Burnett S (2011) Perceptual worlds and sensory ecology. Nature.com. http://www.nature.com/scitable/knowledge/library/perceptual-worlds-and-sensory-ecology-22141730
26. Burkhardt RW (2005) Patterns of behavior. University of Chicago Press, Chicago
27. Cai Y (2003) How many pixels do we need to see things? Lecture Notes in Computer Science (LNCS), ICCS Proceedings, 2003
28. Cai Y (2008) Visual digest networks. In: Cai Y (ed) Digital human modeling, LNAI 4650. Springer, Berlin, pp 28–46
29. Cai Y (2010) US 20100283843 A1. Multiple resolution video network with eye tracking based control. 2010
30. Cai Y et al (2013) Adaptive feature annotation for large video sensor networks. J Electron Imaging 22(4)
31. Cai Y (2014) Ambient diagnostics. CRC Press, Boca Raton
32. Cano P, Loscos A, Bonada J, de Boer M, Serra X (2000) Voice morphing system for impersonating in karaoke application. In: Proceedings of the ICMC 2000
33. Carlile J (2004) A new way to view London from a toilet. Weird news on NBC News. http://www.nbcnews.com/id/4326340/ns/world_news-weird_news/t/new-way-view-london-toilet/#.VBUGgWhIXaU
34. Carmona JM, Climent J (2012) A performance evaluation of HMM and DTW for gesture recognition. Progress in pattern recognition, image analysis, computer vision, and applications. Lect Notes Comput Sci 7441:236–243
35. Chadha A, Savardekar B, Padhya J (2012) Analysis of a modern voice morphing approach using Gaussian Mixture models for Laryngectomees. Int J Comput Appl 49(21):25–30
36. Chandrasekaran B, Glasgow J, Narayanan NH (1995) Diagrammatic reasoning – cognitive and computational perspectives. The AAAI Press/The MIT Press, Menlo Park
37. Chapple M (2013) The three elements of defense against denial-of-service attacks. Febuary 28, 2013. BizTech. http://www.biztechmagazine.com/article/2013/02/three-elements-defense-against-denial-service-attacks
38. Chemotaxis Wikipedia. https://en.wikipedia.org/wiki/Chemotaxis. Accessed 24 Apr 2016
39. Cipher Machines. Retrieved on May 24, 2016. http://ciphermachines.com/types.html
40. Cleveland WS (1985) The elements of graphing data. Wadsworth Inc, Monterey
41. Cognimen technologies: chips and modules. http://www.cognimem.com/products/chips-and-modules/CM1K-Chip/index.html. Accessed 24 Apr 2016
42. Cole KS (1972) Membranes, ions and impulses. University of California Press, Berkley
43. Collision Detection (2003) http://www.edenwaith.com/products/pige/tutorials/collision.php. Accessed 26 Apr 2016
44. Common code theory. Wikipedia. https://en.wikipedia.org/wiki/Common_coding_theory. Accessed 12 Feb 2016
45. Conway J. Game of life. Wikipedia. http://en.wikipedia.org/wiki/Conway%27s_Game_of_Life. Accessed 24 Apr 2016

46. Cootes TF, Taylor CJ, Graham J (1992) Training models of shape from sets of examples. In: Hogg DC, Boyle RD (eds) Processings of the Bristish machine vision conference, Leeds, UK. Springer, London, pp 9–18
47. Couey A (1997) About CDMA spread spectrum. http://people.seas.harvard.edu/~jones/cscie129/nu_lectures/lecture7/hedy/lemarr.htm
48. Crash Test Dummy. Wikipedia. https://en.wikipedia.org/wiki/Crash_test_dummy. Accessed 24 Apr 2016
49. Criminal investigation software – FACES for law enforcement. http://www.facesid.com/products_faces_le.html. Accessed 27 Apr 2016
50. Christie J (2014) Americas feared nuclear missile facilities controlled computers 1960s floppy disks. Daily Mail, April 27 2014. http://www.dailymail.co.uk/news/article-2614323/Americas-feared-nuclear-missile-facilities-controlled-computers-1960s-floppy-disks.html
51. Crutchfield RS (1939) Psychological distance as a function of psychological need. J Comp Psychol 29:447–469
52. DARPA grand challenge. Wikipedia. https://en.wikipedia.org/wiki/DARPA_Grand_Challenge. Accessed 24 Apr 2016
53. Darwin S (2014) The origin of species. Barns and Noble Classic Books, New York
54. Marr D (1982) Vision. The MIT Press, Cambridge, MA
55. Delay line and radar. http://www.computerhistory.org/revolution/memory-storage/8/309. Accessed 26 Apr 2016
56. Delay line memory, US Patent 2629827. http://www.google.com/patents/US2629827. Accessed 26 Apr 2016
57. Deshpande S, Cai Y (2016). Pedestrian detection. In: Loce R et al (eds) Computer vision and imaging in intelligent transportation systems. IEEE and Wiley, to be published in May 2016
58. Dilger P (2010) Social media field of dreams – they will come. http://sales20network.com/blog/?cat=94. Accessed 26 Apr 2016
59. Diuk C (2014) The formation of love. Facebook Data Science. 2016:https://www.facebook.com/notes/facebook-data-science/the-formation-of-love/10152064609253859. Accessed 29 Feb 2016
60. Discovery channel, science of sex appear. https://youtu.be/gwdlq95Tnqc. Accessed 26 Apr 2016
61. Doctorow EL (1980) Loon lake. Random House, New York
62. Dove US, Youtube video. Dove real beauty sketches. https://www.youtube.com/watch?v=XpaOjMXyJGk. Accessed 23 Jan 2016
63. Duchowski AT et al (2004) Gaze-contingent displays: a review. Cyberpsychol Behav 7(6):621–634
64. Dunham S (2015) Metal additive manufacturing market continues rapid expansion. http://3dprinting.com/metal/metal-additive-manufacturing-market-continues-rapid-expansion/. Accessed 24 Apr 2016
65. Dwulf (2015) Hackers can use RFID readers to steal payment card numbers while you are in public. http://www.techworm.net/2015/02/hackers-can-use-rfid-readers-steal-payment-card-numbers-public.html. Accessed 24 Apr 2016
66. Dynamic programming example (dynamic time warping). http://www.phon.ox.ac.uk/jcoleman/old_SLP/Lecture_5/DTW_explanation.html. Accessed 26 Apr 2016
67. Dzhagalov et al (2013) Negative selective T. Accessible bioscience. http://accessiblebiosci.blogspot.com/2013/08/negative-selectivi-t-dzhagalov-et-al.html
68. Eddy W, Mockus A (1994) An example of the estimation and display of a smoothy varying function of time and space – the incidence of the disease mumps. J Am Soc Inf Sci 45(9):686–693
69. Eddy W, Oue S (1995) Dynamic three-dimensional display of U.S. Air Traffic. J Comput Graph Stat 4(4):261–280
70. Eliza online chatbot. Captured on April 13, 2016. http://nlp-addiction.com/chatbot/eliza/. Accessed 24 Apr 2016

71. Ellis D (2012) PLP and RASTA (and MFCC, and inversion) in Matlab using melfcc.m and invmelfcc.m. http://labrosa.ee.columbia.edu/matlab/rastamat/. Accessed 26 Apr 2016
72. Emotiv EPOC and TestBench™ Specifications (2014) https://emotiv.com/product-specs/Emotiv%20EPOC%20Specifications%202014.pdf. Accessed 26 Apr 2016
73. Emspak J, Cubic robots build themselves. http://news.discovery.com/tech/robotics/cubic-robots-build-themselves-131004.htm. Accessed 24 Apr 2016
74. Ericson C (2005) Real-time collision detection. Elsevier, Amsterdam, p 13
75. Eswarann (2012) Ant colony optimization. https://algorithmsinreallife.wordpress.com/2012/02/20/ant-colony-optimization/. Accessed 24 Apr 2016
76. EvoFIT Case Studies. http://www.evofit.co.uk/case-studies/. Accessed 31 Jan 2016
77. FaceGen web site. http://facegen.com. Accessed 27 Apr 2016
78. Ick D (2009) Fact: everything you see is BACKWARDS. www.davidicke.com. http://forum.davidicke.com/showthread.php?t=95746. Accessed 24 Apr 2016
79. Fairbanks AT, Fairbanks E (2005) Human proportions for artists. Fairbanks Art and Books, Bellingham
80. FBI Facial Identification Catalog, November 1988
81. Feloni R (2015) Google cofounder Sergey Brin says these 2 books changed his life. http://www.businessinsider.com/google-cofounder-sergey-brins-favorite-books-2015-7. Accessed 24 Apr 2016
82. Film "10". Wikipedia. http://en.wikipedia.org/wiki/10_(film). Accessed 26 Apr 2016
83. Foden S (2016) Animal that don't have a male for reproduction. http://animals.mom.me/animals-dont-male-reproduction-9838.html. Accessed 24 Apr 2016
84. Foldit: what's new, http://fold.it/portal/. Accessed 24 Apr 2016
85. Folk Tune Finder, http://www.folktunefinder.com/. Accessed 26 Apr 2016
86. Forsyth DA, Fleck MM (1996) Identifying nude pictures. In: Proceeding, third IEEE workshop on applications of computer vision. pp 103–108
87. Freedom Fields USA: Landmines Types (2010). http://www.ffusa.org/landmine_types_md_82b.html. Accessed 24 Apr 2016
88. Freud S (1915) Instincts and their vicissitudes. Psychoanalytic Electronic Publishing. http://www.p-e-p.org/pepcd.htm. Accessed 24 Apr 2016
89. Frowd C et al (2012) Catching even more offenders with EvoFIT facial composites. In: Third international conference on emerging security technologies, 2012
90. http://www.evofit.co.uk/wp-content/uploads/2012/11/Frowd-et-al-2012-Catching-Even-More-Offenders-with-EvoFIT-Facial-Composites-final.pdf. Accessed 27 Apr 2016
91. Fung B (2012) A video game where players help real doctors diagnose malaria. http://www.theatlantic.com/health/archive/2012/05/a-video-game-where-players-help-real-doctors-diagnose-malaria/256759/. Accessed 24 Apr 2016
92. Gallagher S (2013) Why facial recognition tech failed in the Boston bombing manhunt. http://arstechnica.com/information-technology/2013/05/why-facial-recognition-tech-failed-in-the-boston-bombing-manhunt/. Accessed 27 Apr 2016
93. Gajic Z. Chapter 5: controllability and observability. http://www.ece.rutgers.edu/~gajic/psfiles/chap5.pdf
94. Garris PA, Ciokowski EL, Pastore P, Wightman RM (1994) Efflux of dopamine from the synaptic cleft in the nucleus accumbens of the rat brain. J Neurosci 14(10):6084–6093
95. George R (2007) Facial geometry. Charles C Thomas Publisher, Ltd., Springfield
96. Goldberg LR (1993) The structure of phenotypic personality traits. Am Psychol 48(1):26–34
97. Gold JM, Mundy PJ, Tjan BS (2012) The perception of a face is no more than the sum of its parts. Psychol Sci 23(4):427–434
98. Good B, Su A (2011) Games with a scientific purpose. Genome Biol 2011(12):135
99. Griggs B (2011) Why computer voices are mostly female. http://www.cnn.com/2011/10/21/tech/innovation/female-computer-voices/. Accessed 26 Apr 2016
100. Griggs B (2014) Google's new self-driving car has no steering wheel or break. cnn.com.com. 28 May 2014. http://www.cnn.com/2014/05/28/tech/innovation/google-self-driving-car/

101. Grother P, Quinn GW, Philips PJ (2011) Report on the evaluation of 2D still-image face recognition algorithms. NIST Interagency Report 7709
102. Grunewald SJ (2015) 3D print multi-layer circuit boards with the nano dimension dragonfly 2020 3D printer. http://3dprint.com/59360/dragonfly-2020/. Accessed 24 Apr 2016
103. Halterman TE (2015) Designer 3D prints his own brushless pulse motor. http://3dprint.com/89359/3d-print-brushless-pulse-motor/. Accessed 24 Apr 2016
104. Hawking S. Life in the universe. hawking.org.uk. http://www.hawking.org.uk/life-in-the-universe.html
105. Heeger D (2006) Perception lecture notes: the eye and image formation. http://www.cns.nyu.edu/~david/courses/perception/lecturenotes/eye/eye.html. Accessed 24 Apr 2016
106. Heine M (2013) Mercy sakes alive, looks like we got us a robot convey. 11 April 2013 Overdrive Extra. http://www.overdriveonline.com/mercy-sakes-alive-looks-like-we-got-us-a-robot-convoy/
107. Ho JS et al (2014) Wireless power transfer to deep-tissue microimplants. Proc Natl Acad Sci U S A 111(22):7974–7979
108. Hughes RN (1967) Turn alternation in woodlice. Anim Behav 15:282–286
109. Hull CL (1943) Principles of behavior. Appleton-Century, New York
110. Humphries M (2014) MIT teaches a drone to perch on power lines so it can recharge during flight. Geek, 23 June 2014. http://www.geek.com/chips/mit-teaches-a-drone-to-perch-on-power-lines-so-it-can-recharge-during-flight-1597462/
111. Hybrid III. Wikipedia. https://en.wikipedia.org/wiki/Hybrid_III. Accessed 24 Apr 2016
112. Image Stitching. Wikipedia. https://en.wikipedia.org/wiki/Image_stitching. Accessed 24 Apr 2016
113. Isherwood C (1952) Goodbye to Berlin. Signet
114. Intelligent Environment (2015) Now you can log into your bank using emoji. http://www.intelligentenvironments.com/info-centre/press-releases/now-you-can-log-into-your-bank-using-emoji-1
115. Jacob JA, Benson EA, Linden GD, McKenna P, Chen P (2013) Method and system to connect consumers to information. US Patent, US8468050 B2
116. James G, Witten D, Hastie T, Tibshirani R (2013) An introduction to statistical learning. Springer, New York, p 204
117. Johansson G (1973) Visual perception of biological motion and a model for its analysis. Percept Psychophys 14:201–211
118. Julie McCarthy's interview (NPR, 17 March 2007). http://www.npr.org/templates/story/story.php?storyId=8976813
119. Kacewicz W, Pennebaker JW, Davis M, Jeon M, Graesser AC (2013) Pronoun use reflects standings in social hierarchies. J Lang Soc Psychol. online version 19 September 2013. doi:10.1177/0261927X1350265
120. Kawrykow A, Roumanis G, Kam A, Kwak D, Leung C, Wu C, Zarour E, Sarmenta L, Blanchette M, Waldispühl J (2012) A citizen science approach for improving multiple sequence alignmen. http://www.plosone.org/article/info:doi/10.1371/journal.pone.0031362. Accessed 24 Apr 2016
121. Kekre HB, Kulkarni V (2010) Performance comparison of speaker recognition using vector quantization by LBG and KFCG. Int J Comput Appl 3(10):975–8887
122. King BM (1991) Human sexuality today, 3rd edn. Prentice-Hall International, Upper Saddle River
123. Koch C (1999) Biophysics of computation: information processing in single neurons. Oxford University Press, New York
124. Kringelbach ML (2009) The pleasure center. Oxford University Press, New York
125. Kroc J (2006) Model of mechanical interaction of Mesenchyme and Epithelium in living tissues. 847–854. In: Vassil N. Alexandrov, G. Dick van Albada, Peter M. A. Sloot, Jack Dongarra (eds) Computational science – ICCS 2006, 6th International Conference, Reading, UK, 28–31 May 2006. In: Proceedings, Part IV. Lecture Notes of Computer Science. Vol. 3994 Springer 2006, ISBN 3-540-34385-7

126. Ladder Logic. Wikipedia. https://en.wikipedia.org/wiki/Ladder_logic. Accessed 24 Apr 2016
127. La Jeunesse M (2015) Tiny device aims to help protect women from sexual assault. http://mashable.com/2015/11/03/athena-indiegogo/?utm_cid=mash-com-fb-life-link#O864D.TZTEq4. Accessed 24 Apr 2016
128. LatestWorldIncidents (2013) Real voice of Siri is Susan Bennett [HQ]. https://youtu.be/T6boqUfzUD8. Accessed 26 Apr 2016
129. Luce L (2015) Lauren Luce Blogs (2015), http://www.laurentluce.com/. Accessed 26 Apr 2016
130. Lawrence Patrick. Wikipedia. https://en.wikipedia.org/wiki/Lawrence_Patrick. Accessed 24 Apr 2016
131. Laws J, Bauernfeind N, Cai Y (2006) Feature hiding in 3D human body scans. J Inf Vis 5:271–278
132. Linssen D (2015) Age simulation suits are teaching medical students empathy. http://www.sciencedump.com/content/age-simulation-suits-are-teaching-medical-students-empathy. Accessed 24 Apr 2016
133. LIWC, Interpreting LIWC Output. http://liwc.wpengine.com/interpreting-liwc-output/. Accessed 24 Apr 2016
134. Lowe DG (1999) Object recognition from local scale-invariant features. In: Proceedings of the international conference on computer vision, pp 1150–1157
135. Luengo-Oroz MA, Arranz A, Frean J (2012) Crowdsourcing malaria parasite quantification: an online game for analyzing images of infected thick blood smears. J Med Internet Res 14(6), e167
136. Luntz S (2014) Is there a man on the moon? http://www.iflscience.com/space/there-man-moon#Rckl0FsjIRhyy0Gl.99. Accessed 27 Apr 2016
137. Marilyn Facts. http://www.marilynmonroe.ca/camera/about/facts/voice.html. Accessed 26 Apr 2016
138. Marnia (2005) Your brain on sex. On reuniting healing with sexual relationship. http://www.reuniting.info/science/sex_in_the_brain. Accessed 29 Feb 2016
139. Marquis E (2013) The driving dead: human cadavers still used in car crash testing. http://www.autoblog.com/2013/10/24/the-driving-dead-human-cadavers-still-used-in-car-crash-testing/. Accessed 24 Apr 2016
140. Malan DJ, Smith MD (2005) Host-based detection of worms through peer-to-peer cooperation. WORM'05 Proceedings of the 2005 ACM workshop on Rapid Malcode, pp 72–80.
141. Maslow AH (1943) A theory of human motivation. Psychol Rev 50(4):370–396
142. Maslow AH (1954) Motivation and personality. Harper and Row, New York
143. Matthews N (2014) This new app could've prevented my friend's rape. http://www.elle.com/culture/tech/a10/kitestring-app-safety-apps-women/. Accessed 24 Apr 2016
144. McClintock MK (1984) Estrous synchrony: modulation of ovarian cycle length by female pheromones. Physiol Behav 32:701–705
145. McFarland D (1985) Animal behavior. Benjamin/Cummings, Menlo Park
146. McGonigal J (2011) Realty is broken: why games make us better and how they can change the world. Penguin, New York
147. Michalak J (2009) Embodiment of sadness and depress – gait patterns associated with Dysphoric mood. Psychosom Med 71:580–587
148. Microsoft Research, MSRA-CFW: data set of celebrity faces on the web. http://research.microsoft.com/en-us/projects/msra-cfw/. Accessed 27 Apr 2016
149. Midgley M (1995) Beast and man – the roots of human nature. The Harvester Press Limited, Sussex
150. Milborrow S (2013) Stasm 4 user manual. http://www.milbo.org/stasm-files/stasm4.pdf. Accessed 27 Apr 2016
151. Miller J (2015) Whispering on the internet is paying this women's rent. http://www.cosmopolitan.com/lifestyle/a40025/gentlewhispering-maria-internets-most-fascinating/. Accessed 26 Apr 2016

152. Minsky M (1975) A framework for representing knowledge. In: Winston P (ed) The psychology of computer vision. McGraw-Hill, New York, pp 211–277
153. Molau S, Pitz M, Schlueter R, Ney H (2001) Computing Mel-frequency Cepstral coefficients on the power spectrum. In: Proceedings of international conference on Acoustic, speech and signal processing, 2001
154. Morales J (2013) Prioritizing malware analysis. SEI Blog. 4 November 2013. https://insights.sei.cmu.edu/sei_blog/2013/11/prioritizing-malware-analysis.html
155. Morales JA, Clarke PJ, Deng Y (2010) Identification of file infecting viruses through detection of self-reference replication. J Comput Virol 6:161–180
156. Morales JA, Main M, Luo W, Xu S, Sandhu R (2011) Building malware infection trees. IEEE, 2011
157. Murphy SV, Atala A (2014) 3D bioprinting of tissues and organs. Nat Biotechnol 32:773–785
158. Nass C, Brave S (2005) Wired for speech: how voice activates and advances the human-computer relationship. MIT Press, Cambridge, MA
159. Nass C, Steuer J, Tauber E (1994) Computers are social actors. Conference on Computer Human Interaction (CHI), pp 204–278, 1994
160. National Instruments: LabVIEW System Design Software. http://www.ni.com/labview/. Accessed 24 Apr 2016
161. Nauert R (2007) Attractiveness is in the way we walk. Psych Central. http://psychcentral.com/news/2007/05/23/attractiveness-is-in-the-way-we-walk/849.html. Accessed 20 Jan 2016
162. Nauert R (2015) "Herd" mentality explained. http://psychcentral.com/news/2008/02/15/herd-mentality-explained/1922.html. Accessed 24 Apr 2016
163. NBC News (2015) 11-years-old's sketch helps police catch serial burglar. http://www.9news.com/story/news/weird/2015/10/16/stick-figure-burglar-arrested/74039792/. Accessed 27 Apr 2016
164. Newman ML, Pennebaker JW, Berry DS, Richards JM (2003) Lying words: predicting deception from linguistic styles. Pers Soc Psychol Bull 29:665–675
165. Nicholas AS, Brody S, de Sutter P, de Carufel F (2008) A woman's history of vaginal orgasm is discernible from her walk. J Sex Med 5(9):2119–24
166. Nielsen M (2011) Reinventing discovery: the new era of networked science. Princeton University Press, Princeton
167. NY City Crime Stop web site. http://a056-crimestoppers.nyc.gov/crimestoppers/public/wanted.html. Accessed 27 Apr 2016
168. Nhv K (2015) Any tips for completing "Decribe image regions with text snippets" visual genome hits? reddit.com. https://www.reddit.com/r/mturk/comments/3gd249/any_tips_for_completing_describe_image_regions/
169. OmniSecu. Different types of computer viruses. http://www.omnisecu.com/security/types-of-computer-viruses.php
170. Ormston T (2012) Time delay between Mars and Earth. Mars Express of ESA web site. http://blogs.esa.int/mex/2012/08/05/time-delay-between-mars-and-earth/
171. Öğman H, Breitmeyer BG (eds) (2006) The first half second. The MIT Press, Cambridge, MA
172. Orphanidou C, Moroz I, Roberts S (2004) Wavelet-based voice morphing. WSEAS J Syst 10(3):3297–3302
173. Palmer J (2010) Chemical computer that mimics neurons to be created. BBC (Science News). http://news.bbc.co.uk/2/hi/science/nature/8452196.stm. Accessed 24 Apr 2016
174. Patel N (2015) The creepy collective behavior of Boston Dynamics new robot dog. WIRED. 11 Febuary 2015. http://www.wired.com/2015/02/creepy-collective-behavior-boston-dynamics-new-robot-dog/
175. Pennebaker JW (2011) The secret life of pronouns: what our words say about us. Bloomsbury, New York
176. Pennebaker JW (2011) Your use of Pronouns reveals your personality. https://hbr.org/2011/12/your-use-of-pronouns-reveals-your-personality. Accessed 24 Apr 2016

177. Pennebaker JW, Chung CK, Frazee J, Lavergne GM, Beaver DI (2014) When small words foretell academic success: the case of college admissions essays. PLoS One 9(12):e115844. doi:10.1371/journal.pone.0115844
178. Peter G, Cukierman D, Anthony C, Schwartz M (2006) Online music search by tapping. In: Lecture notes in computer science (LNAI), vol 3864. Springer, Heidelberg, pp 178–197
179. Pirolli P, Card S (1999) Information foraging. Psychol Rev 106(4):643
180. Pirolli P, Card S (2005) The sense-making process and leverage points for analyst technology as identified through cognitive task analysis. 2005 International conference on intelligence analysis, Vienna, Virginia, 2–6 May 2005
181. Qavi A, Khan SA, Basir K (2015) Voice morphing based on spectral features and prosodic modification. Int J Sig Process Syst 3(2):123–127
182. Ramachandran VS, Hirsteim W (1999) The science of art: a neurological theory of aesthetic experience. J Conscious Stud 6(6–7):15–51
183. RANSAC Tutorial. CSAIL, MIT. http://saturday.csail.mit.edu/fa12/lectures/lecture13ransac/lecture13ransac.pdf. Accessed 12 Feb 2016
184. Regirer SA, Shapovalov DS (2003) Filling space in public transport by passengers. Autom Remote Control 64(8):1301–1309
185. Reprap: Snappy, http://reprap.org/wiki/Snappy. Accessed 24 Apr 2016
186. Richman C, Dember WN (1986) Spontaneous alternation behavior in animals: a review. Curr Psychol Res Rev 1986–1987, 5(4):358–391
187. Robosapien Manual (2004) http://www.theoldrobots.com/book31/Robosapien-manual.pdf. Accessed 26 Apr 2016
188. Romanishin J, M-Blocks modular robots, https://www.youtube.com/watch?v=mOqjFa4RskA. Accessed 24 Apr 2016
189. Rosenthal R (2005) Conducting judgment studies. In: Harrigan JA, Rosenthal R, Scherer KR (eds) Handbook of nonverbal behavior research methods in the affective sciences. Oxford, New York, pp 199–236
190. RTMPDump. Wikipedia. https://en.wikipedia.org/wiki/RTMPDump. Accessed 24 Apr 2016
191. Rymix (2011) Kinect SDK Dynamic Time Warping (DTW) gesture recognition. http://kinectdtw.codeplex.com/. Accessed 26 Apr 2016
192. Sales G, Pye D (1974) Ultrasonic communication by animals. Springer, Dordrecht
193. SAE (2011) US department of transportation. SAE J2735 – Dedicated Short Range Communications (DSRC) Message Set Dictionary
194. Schneider P, Hammer B, Biehl M (2009) Adaptive relevance matrices in learning vector quantization. Neural Comput 21:3532–3561
195. Schwartz HA, Eichstaedt JC, Kern ML, Dziurzynski L, Ramones SM, Agrawal M et al (2013) Personality, gender, and age in the language of social media: the open-vocabulary approach. PLoS One 8(9):e73791. doi:10.1371/journal.pone.0073791
196. Searle J (1980) Minds, brains and programs. Behav Brain Sci 3(3):417–457
197. Seeley TD (1997) Honighienen im Mikrosmos des Bienenstocks. Birkhäuser, Basel
198. Shekhawat A, Poddar P, Boswal D (2009) Ant colony optimization algorithms: introduction and Beyond. http://mat.uab.cat/~alseda/MasterOpt/ACO_Intro.pdf. Accessed 24 Apr 24 2016
199. Shodan. https://www.shodan.io/. Accessed 25 Apr 2016
200. Singh D, Renn P, Singh A (2007) Did the perils of abdominal obesity affect depiction of feminine beauty in the sixteenth to eighteenth century British literature? Exploring the health and beauty link. In: Proceedings of the Royal Society B: biological sciences, vol 274, No. 1611, 22 March 2007
201. Simon H (1989) Models of thought, Ch. 1.3 and Ch. 6.1. Yale University Press.
202. Smith R (1992) Alarm signals in fishes. Rev Fish Biol Fish 2:33–63
203. Sonka M et al (1998) Image processing, analysis, and machine vision, 2nd edn. PWS Publishing, Pacific Grove
204. Spearman-Brown prediction formula. Wikipedia. https://en.wikipedia.org/wiki/Spearman--Brown_prediction_formula. Accessed 10 Feb 2016
205. Spearman CC (1910) Correlation calculated from faulty data. Br J Psychol 3:271–295

206. Spline (Mathematics). Wikipedia. http://en.wikipedia.org/wiki/Spline_(mathematics). Accessed 27 Apr 2016
207. Stagon S, Knapp A, Elliott P, Huang H (2016) Metallic glue for ambient environments making strides. Adv Mater Process 174:22–25
208. Stratton G (1996) Some preliminary experiments on vision without inversion of the retinal image. Psychol Rev 3:611–617
209. Suikerbuik R, Tangelder H, Daanen H, Oudenhuijzen A (2004) Automatic feature detection in 3D human body scans. In: Proceedings of SAE digital human modeling conference, 2004, 04- DHM-52
210. Swan M (2012) Crowdsourced health research studies: an important emerging complement to clinical trials in the public health research ecosystem. J Med Internet Res 14(2):e46
211. Tatler BW et al (2010) Yarbus, eye movements, and vision. I-Perception. 1(1):7–27. http://www.ncbi.nlm.nih.gov/pmc/articles/PMC3563050/#R41. Accessed 27 Apr 2016
212. Tausczik YR, Pennebaker JW (2010) The psychological meaning of words: LIWC and computerized text analysis methods. J Lang Soc Psychol 29:24–54
213. Ten most popular plastic surgery procedures (2012) CBS News. http://www.cbsnews.com/pictures/10-most-popular-plastic-surgery-procedures/. Accessed 27 Apr 2016
214. The National Air and Space Museum (2015) Finding pluto with the blink comparator. http://blog.nasm.si.edu/planetary-science/blink-comparator/. Accessed 26 Apr 2016
215. Tinberger N (1950) The hierarchical organization of nervous mechanism underlying instinctive behavior. In: Symposia of the society of experimental biology, vol 4, pp 305–312
216. Tinbergen N (1951) The study of instinct. Oxford University Press, Oxford, UK
217. Tiwari A (2015) How search engine works and makes your life easier? http://fossbytes.com/search-engine-works-makes-life-easier/. Accessed 24 Apr 2016
218. Tiwari V (2010) MFCC and its applications in speaker recognition. Int J Emerg Technol 1(1):19–22
219. T-Maze. Wikipedia. https://en.wikipedia.org/wiki/T-maze. Accessed 24 Apr 2016
220. Troung M, Mikoleit T (2014) Twitter analysis of the World Cup match: USA vs Germany. http://public.tableausoftware.com/profile/michael.truong#!/vizhome/WorldCupUSAvsGermanyTweets/WorldCupUSA-GERTweets. Accessed 26 Apr 2016
221. Tsai LS (1932) The laws of minimum effort and maximum satisfaction in animal behavior. Monograph of the National Institute of Psychology, Beijin, China, No. 1, 1932. Abstracted by Psychological Abstracts, vol. 6, 1932. No. 4329
222. Turing AM (1950) Computing machinery and intelligence. Mind 59:433–460. http://loebner.net/Prizef/TuringArticle.html
223. Turing AM (1952) The chemical basis of morphogenesis. Philos Trans R Soc Lond B Bio Sci 237(641):37–72
224. University of Gent. Electrochromic displays. http://lcp.elis.ugent.be/tutorials/tut_echrom
225. UNIX User Data (1998) http://kdd.ics.uci.edu/databases/UNIX_user_data/UNIX_user_data.html. Accessed 24 April 2016
226. Updike J (1996) The rabbit is rich. Ballantine Books, New York
227. von Neumann J (1966) Theory of self reproducing automata. University of Illinois Press, Urbana
228. Vox Program (2015) Police sketeches may become a thing of the past. Youtube.com. https://www.youtube.com/watch?v=h81SuD2pltM. Accessed 27 Apr 2016
229. Vuilleumier P, Pourtois G (2007) Distributed and interactive brain mechanisms during emotion face perception: evidence from functional neuroimaging. Neuropsychol 45:174–194
230. Wallance S (2015) Celebrity faces become hilarious distortions when made to fit the golden ratio. http://www.smash.com/celebrity-faces-become-hilarious-distortions-made-fit-fibonaccis-golden-ratio/. Accessed 24 Apr 2016
231. Wanner R. Survive Time. https://isc.sans.edu/survivaltime.html
232. Webb D, Sparrow WA (2007) Description of joint movements in human and non-human primate locomotion using Fourier analysis. Primate 48:277–292

233. Weber Fechner Law. Wikipedia. https://en.wikipedia.org/wiki/Weber--Fechner_law. Accessed 27 Apr 2016
234. Web Science 2014 Data Challenge (2016) http://cnets.indiana.edu/groups/nan/webtraffic/websci14-data/. Accessed 24 Apr 2016
235. Wertheimer M (1912) Experimentelle Studien über das Sehen von Bewegung [Experimental studies on motion vision]. Zeitschrift für Psychologie 61(1):161–265. Retrieved on Feb. 12, 2016: http://gestalttheory.net/download/Wertheimer1912_Sehen_von_Bewegung.pdf. Accessed 26 Apr 2016
236. Wiener N (1961) Cybernetics: or control and communication in the animal and the machine. The MIT Press, New York
237. Wigglesworth VB (1970) Insect hormones. W.H. Freeman and Company, San Francisco, pp 134–141
238. Wightman RM, Amatore C, Engstrom RC, Hale PD (1988) Real-time characterization of dopamine overflow and uptake in the rat striatum. Neurosci 25(2):513–523
239. Wightman RM, May LJ, Michael AC (1988) Detection of dopamine dynamics in the brain. Anal Chem 60(13):769A–779A
240. Wilson B (2008) Semantic networks and frames. http://www.cse.unsw.edu.au/~billw/cs9414/notes/kr/frames/frames.html. Accessed 24 Apr 2016
241. WikiPedia: Chemotaxi: https://en.wikipedia.org/wiki/Chemotaxis
242. WikiPedia: Belousov–Zhabotinsky_reaction. https://en.wikipedia.org/wiki/Belousov--Zhabotinsky_reaction
243. Wolfram S (2000) The new kind of science. Wolfram Media, Champaign
244. Wyatt TD (2003) Pheromones and animal behaviour: communication by smell and taste. Cambridge University Press, Cambridge. ISBN 0521485266
245. Yang Y, Cai Y (2010) Virtual gazing in video surveillance. ACM Multimedia, SMVC'10, 29 October 2010, Firenze, Italy
246. Yan Y, Wang X (2015) Facial recognition based on facial geometrics. Final Report of Cognitive Video, ECE Carnegie Mellon University, December 2015
247. Yarbus AL (1967) Eye movements and vision. Plenum Press, New York, pp 171–185
248. Yarkoni T (2010) Personality in 100,000 words: a large-scale analysis of personality and word use among bloggers. J Res Pers 44(3):363–373
249. Zeki S (1999) Inner vision: an exploration of art and the brain. Oxford University Press, Oxford
250. Zheng Y (2015) Trajectory data mining: an overview. ACM transactions on intelligent systems and technology. http://research.microsoft.com/pubs/241453/TrajectoryDataMining-tist-yuzheng_published.pdf. Accessed 26 Apr 2016
251. Zipf GK (1949) Human behavior and the principle of least effort. Addison-Wesley Press, Cambridge, MA, 1949 and Martino Publishing, Reading, MA, 2012

# Index



© Springer-Verlag London 2016
Y. Cai, *Instinctive Computing*, DOI 10.1007/978-1-4471-7278-9